Intelligent
Hybrid Systems

Intelligent Hybrid Systems

Edited by

Suran Goonatilake and Sukhdev Khebbal

University College London, UK

JOHN WILEY & SONS
Chichester · New York · Brisbane · Toronto · Singapore

Baffins Lane, Chichester
West Sussex PO19 1UD, England

National Chichester (01243) 779777
International (+44) 1243 779777

Reprinted April 1995

Other Wiley Editorial Offices

John Wiley & Sons, Inc., 605 Third Avenue,
New York, NY 10158-0012, USA

Jacaranda Wiley Ltd, 33 Park Road, Milton,
Queensland 4064, Australia

John Wiley & Sons (Canada) Ltd, 22 Worcester Road,
Rexdale, Ontario M9W 1L1, Canada

John Wiley & Sons (SEA) Pte Ltd, 37 Jalan Pemimpin #05-04,
Block B, Union Industrial Building, Singapore 2057

British Library Cataloguing in Publication Data

A catalogue record for this book is available from the British Library

ISBN 0 471 94242 1

Produced from camera-ready copy supplied by the editors
Printed and bound in Great Britain by
Redwood Books, Trowbridge, Wiltshire.

Contents

Foreword

"If the oak tree flowers before the ash, prepare for a splash: if the ash tree flowers before the oak, prepare for a soak"

(Source: English folklore)

This rule is an example drawn from hundreds of apparently simple and often contradictory rules for predicting future observations of the notoriously fickle English weather. On closer examination the rule is less simple than it first appears. The variables in the rule are 'fuzzy'; there is no definition of either what 'rain' is or what is meant by flower. Even the observation frequency is left to the user's discretion. People, of course, regularly use sets of rules to approximate to and infer from observations, the principles which govern the evolution of complex systems like the weather. Although this example is taken from folklore weather forecasting, it could just as easily have been drawn from the financial markets, where combinations of simple rules based on the analysis of time-series are routinely used to support trading decisions, or from medicine, where classifications of symptoms must be made in the light of incomplete readings and often conflicting information. It is in solving these types of difficult real-world problems that a new generation of intelligent computing methods, Intelligent Hybrid Systems, are now bearing fruit.

The grander claims for rule-based AI (Artificial Intelligence) in the 80's seemed to fail in part because specialist practitioners either failed to understand or failed to flag the limitations of their single approach. The contributors to this volume have supported a different view: there is no one definitive approach which will explain cognition or solve complex problems, and there is no 'philosopher's stone'; instead, there are a number of tools and models that can be applied under different circumstances. The proof of this philosophy is to be found in the following chapters that detail successes ranging from aircraft design optimisation to stock portfolio selection.

Increasingly, complex industrial and business problems are driving the research and development of new intelligent computing methods. At Reuters we use an intelligent system for checking data-errors in the flood of data we collect from Stock Exchanges around the world. Some of our financial market customers have recently experimented with neural network and genetic algorithm forecasting tools which draw data from our financial data-feeds. In other companies, credit evaluation and credit card fraud detection are now routinely subject to the analysis of intelligent techniques to help operators make better decisions. With the convergence of media and technology one expects further uses of hybrid systems to analyse, distil and help make decisions on the so-called information superhighway.

This timely book reviews the current state of intelligent hybrid systems. It offers both a classification scheme to analyse and describe hybrid models and presents clear arguments for their need. It is above all a book about how to move from theory to practical implementations. I believe it will inspire academics and technologists alike to take a dip in the waters of this emerging field.

John Taysom

Vice President
Reuters New Media Inc.
Palo Alto

Acknowledgements

This book owes much to many people. Firstly, we would like to take this opportunity to thank all of our contributors for their time and effort in preparing their papers. A very special thanks goes to Bob Ijaz for his enthusiasm and tireless effort in formatting the book. Many thanks to the staff at John Wiley for making the book possible, especially to Gaynor Redvers-Mutton for her patience and guidance throughout the project. We are also grateful to Charles, Sinclair and Co. for the use of their printing facilities.

We would also like to thank our colleague Konrad Feldman for proof reading all the chapters and for making constructive technical comments. We also like to acknowledge the Department of Computer Science at University College London and our colleagues in the Intelligent Systems Group, for providing a stimulating environment for our explorations into hybrid systems.

Finally, we would like to thank Jim Hendler, Lawrence Davis, Tariq Samad, Henry Tirri and Andreas Scherer for encouraging the very initial idea of a book on intelligent hybrid systems.

Contributors

Venkat Ajjanagadde,
Wilhelm-Schickard-Institut,
University of Tuebingen,
Sand 13 W-7400 Tuebingen,
Germany.

Wendy Foslien,
Honeywell SSDC,
3660 Technology Drive,
Minneapolis, MN 55418,
U.S.A.

Frederic Gruau,
Centre d'etude nucleaire de Grenoble
DRFMC SP2M PSC,
BP85X 38041
France.

Vasant Honavar,
Assistant Professor of Computer Science and Neuroscience,
Computer Science Dept.
226 Atanasoff Hall,
Iowa State University,
Ames, IA 50011,
U.S.A.

Charles Karr,
U.S. Bureau of Mines,
Tuscaloosa Research Center,
P.O. Box L,
University of Alabama Campus,
Tuscaloosa, AL 35486-9777,
U.S.A.

Randy Kerber,
Lockheed AI Center,
3251 Hanover Street,
Palo Alto,
CA 94394
U.S.A.

Casimer C. Klimasauskas,
Neuralware,
Penn Center West IV-227,
Pittsburgh,
PA 15276.
U.S.A.

Brian Livezey,
Lockheed AI Center,
3251 Hanover Street,
Palo Alto,
CA 94394
U.S.A.

David Montana,
Bolt Beranek and Newman
70 Fawcett Street
Cambridge,
MA 02138,
U.S.A.

Khai Minh Pham,
inferOne,
Immeuble Les Maradas,
1 Bd de l'Oise,
95030 Cergy Pontoise Cedex,
France.

David J. Powell,
Corporate Research and Development,
General Electric Company,
Schenectady, NY,
U.S.A.

Tariq Samad,
Honeywell SSDC,
3660 Technology Drive,
Minneapolis, MN 55418,
U.S.A.

Andreas Scherer,
Praktische Informatik I,
FernUniversitaet,
Postfach 940,
W - 5800 Hagen,
Germany.

Gunter Schlageter,
Praktische Informatik I,
FernUniversitaet,
Postfach 940,
W - 5800 Hagen,
Germany.

Danny Shamhong,
Department of Computer Science
University College London
Gower Street
London WC1E 6BT,
U.K.

Lokendra Shastri
International Computer Science Institute
1947 Center Street,
Suite 600 Berkeley,
CA 94704-1105,
U.S.A.

Evangelos Simoudis,
Lockheed AI Center,
3251 Hanover Street,
Palo Alto,
CA 94394
U.S.A.

Michael M. Skolnick
Assistant Professor
Department of Computer Science
Rensselaer Polytechnic Institute
Troy, NY., 12180
U.S.A.

Henry Tirri
Department of Computer Science
P.O. Box 26
SF-00014 University of Helsinki
FINLAND.

Siu Shing Tong,
Corporate Research and Development,
General Electric Company,
Schenectady, NY.
U.S.A.

Leonard Uhr
Computer Sciences Dept,
University of Wisconsin,
Madison,
Wisconsin,
U.S.A.

About The Editors

Dr. Suran Goonatilake is at the Department of Computer Science at University College London. He is a co-founder of SearchSpace Ltd., a London-based company that specialises in applying intelligent systems in finance and business. He holds a PhD in Computer Science from University College London and a bachelor's degree in Computing and Artificial Intelligence from the University of Sussex. His PhD research was on the use of genetic-fuzzy hybrid systems for supporting financial trading. His current research interests include the use of genetic algorithms for decision rule induction and their application in insurance risk assessment, direct marketing and portfolio management. Suran is also the co-editor of the book "Intelligent Systems for Finance and Business".

Dr. Sukhdev Khebbal is a co-founder of SearchSpace Ltd. and a project leader in the Department of Computer Science, University College London. His current research interests include the theoretical aspects of intelligent hybrid systems, object-oriented programming and intelligent agents . His PhD research was on the design of object-oriented environments for integrating neural networks and expert systems. He is the principal designer of the HANSA architecture (Heterogeneous Application geNerator Standard Architecture) which is the core framework of the largest European Community funded project in hybrid systems. HANSA allows software developers to rapidly generate decision support systems using industry standard tools and intelligent systems. To complement this research background, Sukhdev has ten years of practical experience in the application of computing techniques to business problems.

The authors can be contacted at:

S.Goonatilake@cs.ucl.ac.uk
S.Khebbal@cs.ucl.ac.uk

1

Intelligent Hybrid Systems:
Issues, Classifications and Future Directions

Suran Goonatilake and Sukhdev Khebbal

1.1 Introduction

Humans are hybrid information processing machines. Our actions are governed by a combination of genetic information and information acquired through learning. Information in our genes hold successful survival methods that have been tried and tested over millions of years of evolution. Human learning consists of a variety of complex processes that use information acquired from interactions with the environment. It is the combination of these different types of information processing methods that has enabled humans to succeed in complex, rapidly changing environments.

This type of *hybrid* information processing is now being replicated in a new generation of *adaptive machines.* The applications range from aircraft control systems that diagnose and repair themselves to systems that can successfully trade in foreign exchange markets. At the heart of these adaptive machines are *intelligent* computing systems, some of which are inspired by the mechanics of nature.

Neural networks, for example, are inspired by the functionality of nerve cells in the brain. Like humans, neural networks can *learn* to recognise patterns by repeated exposure to many different examples. They are good at recognising complex patterns, whether they be hand-written characters, profitable loans or good financial trading decisions. Genetic algorithms are also naturally inspired and based on the biological principle of "survival of the fittest". The main idea behind a genetic algorithm is the

Intelligent Hybrid Systems, Edited by S. Goonatilake and S. Khebbal. ©1995 John Wiley & Sons Ltd.

evolution of a problem's solution over many generations, with each generation having a better solution than its predecessor.

While these intelligent techniques have produced encouraging results in particular tasks, certain complex problems cannot be solved by a single intelligent technique alone. Each intelligent technique has particular computational properties (e.g. ability to learn, explanation of decisions) that make them suited for particular problems and not for others. For example, while neural networks are good at recognising patterns, they are not good at explaining how they reach their decisions. Fuzzy logic systems, which can reason with imprecise information, also have particular strengths and limitations. They are good at explaining their decisions but they cannot automatically acquire the rules they use to make those decisions. These limitations have been a central driving force behind the creation of *intelligent hybrid systems* where two or more techniques are combined in a manner that overcomes the limitations of individual techniques.

For example, there is now considerable interest in applying fuzzy logic for tasks such as loan evaluation. Typically, an expert in loan evaluation has to specify all the rules needed for the fuzzy logic system to make a decision on whether to grant a loan or not depending on the application details. However, this is a time consuming and error-prone process that should ideally be automated. Neural networks with their learning capabilities can be used to automatically learn these fuzzy decision rules, thus creating a hybrid system which overcomes the limitations of fuzzy systems.

Hybrid systems are also important when considering the varied nature of application domains. Many complex domains have many different component problems, each of which may require different types of processing. For example, the Kobe Steel Plant in Japan controls blast furnaces for making iron and steel by using a hybrid of different intelligent techniques to solve sub-tasks of the problem [Otsu92]. Neural networks are used to predict heat levels from sensor data gathered from the furnace. These predictions are then used by an expert system to infer the correct control adjustments so as to maintain the furnace at its optimal operating level. This complex control task was previously performed by a highly trained operator who had many years of experience.

Intelligent hybrid systems represent not only the combination of different intelligent techniques but also the integration of intelligent techniques with conventional computing systems such as spreadsheets and databases. For intelligent systems to add value to organisational decisions they must be able to extract and use information from a wide variety of sources. In addition, the decisions or results produced by the intelligent systems should be disseminated to existing applications or other systems for further processing. For these reasons it is vital that there are methods and protocols for integrating intelligent systems with other conventional computing systems. One such method is object-oriented programming [Wien88] which is a software engineering methodology that can provide the "glue" to join together different processing techniques. This forms a natural model for intelligent hybrid systems because individual techniques can be defined as objects which interact by sending a common set of messages.

This book presents a collection of work in this emerging area of intelligent hybrid systems, written by leading-edge researchers from around the world. It spans the combination of many different techniques including neural networks, genetic algorithms, expert systems, fuzzy logic systems and rule induction. Application examples are drawn from domains including industrial control, financial modelling, and cognitive modelling. The book's aim is to inform researchers and application developers, in both industry and academia, about the potential of intelligent hybrid systems for solving real-world problems.

The next section examines the strengths and weaknesses of different intelligent techniques which determine the need for hybrid systems. Section 1.3 introduces a classification scheme for hybrid systems and section 1.4 details a typical hybrid systems development cycle. Section 1.5 concludes by discussing potential areas of future development.

1.2 Properties of Intelligent Systems

In this section we compare and contrast different intelligent systems on four key information processing capabilities. These are knowledge acquisition, brittleness, higher and lower level reasoning and explanation.

1.2.1 Knowledge Acquisition

Knowledge acquisition is a crucial stage in the development of intelligent systems. As a process, it involves eliciting, interpreting and representing the knowledge from a given domain [Kidd87]. Knowledge acquisition for expert systems (from domain experts) is time consuming, expensive and potentially unreliable [Feig79]. Experts find it difficult to articulate particular types of "intuitive" knowledge [Part86] and are sometimes unwilling to participate in lengthy knowledge elicitation exercises. Other problems include the possible existence of gaps in an expert's knowledge and the correctness of an expert's knowledge [Jack86].

Furthermore, expert systems do not have mechanisms to deal with any changes in their decision making environment — they cannot adapt and learn from changes in their operating environment. Thus the *maintenance of knowledge* in expert systems is time consuming and expensive.

Due to these problems, techniques such as neural networks and genetic algorithms, which can learn from domain data, have certain advantages. In expert systems, the decision boundaries — the bounds used to make particular decisions — are specified by a domain expert, while in neural networks and genetic algorithms these decision boundaries are *learned* [Weis91]. Changes in the operating environment cause the decision boundaries to be shifted or changed. Systems that learn can detect and adapt to these changes.

Rule induction systems are another example of systems that have been applied to automate the knowledge acquisition process. They have been used to learn rules and decision trees from "raw" domain data and have been applied successfully in industrial and commercial domains [Quin86, Mich80].

1.2.2 Brittleness

Although there are notable successes of expert systems, many of these systems operate in very narrow domains under limited operational conditions. Holland [Holl86] refers to this phenomena in expert systems as *brittleness*:

> The systems are *brittle* in the sense that they respond appropriately only in narrow domains and require substantial human intervention to compensate for even slight shifts in domain.

An operational view of the brittleness problem can be seen as the inability of an intelligent system to cope with inexact, incomplete or inconsistent knowledge. Causes of this brittleness problem are twofold — inadequate representation structures and reasoning mechanisms. In expert systems, knowledge is represented as discrete symbols and reasoning consists of *logical* operations on these constructs.

In contrast, reasoning in neural networks involves the numeric aggregation of representations over the whole network [Beal90]. As Smolensky [Smol87] points out,

> Each connection represents a soft constraint; the knowledge contained in the system is the set of all such constraints. If two units have an inhibitory connection, then the network has the knowledge that when one is active the other ought not be. Any of the soft constraints can be overridden by others, they have no implications singly; they only have implications collectively.

This distributed representation and reasoning allows these systems to deal with incomplete and inconsistent data and also allow the systems to *gracefully degrade* [Rume86]. That is, even if some parts of a neural network are made non-operational, the rest of the neural network will function and attempt to give an answer.

This type of inherent fault tolerance contrasts strongly with expert systems which usually fail to function even if a single processing part is non-operational. This manifests itself in many ways, particularly as problems in the maintenance of large, complex knowledge bases [Mcde83].

Rule induction systems such as ID3 are also susceptible to the problems of brittleness. They have difficulties in inducing relationships in data which has contradictory and incomplete examples [Fors86]. However, recent enhancements of the ID3 algorithm have overcome some of the limitations with respect to its handling of noisy data [Quin93].

Fuzzy logic deals with the problem of brittleness by adopting novel knowledge representation and reasoning methods. Fuzzy sets, the form in which knowledge is represented, diffuse the boundaries between concepts. There are no sharp divisions where one concept ends and the next begins. This fuzzy data representation, in conjunction with fuzzy reasoning mechanisms, allows the processing of data which are non-exact or *partially correct* [Zade88].

The key argument for genetic algorithms being able to cope with brittleness is put forward by Holland [Holl86]. Holland argues that it is the maintenance of a *population of solutions* which makes genetic algorithms and classifier systems non-brittle. Each rule in the classifier system population contains a relationship describing the system being modelled. The system's flexibility arises from the rules representing a wide range of competing, conflicting hypotheses. The selection of the appropriate rule to fire is dependent on its past performance — a statistical aggregation of its correct performance. Similar to neural networks, it is this *statistical reasoning* property, based on past performance that gives genetic algorithms their ability to cope with brittleness.

1.2.3 Higher and Lower Level Reasoning

For a complete theory of cognition, there need to be explanations of how we can do "low-level", parallel, pattern recognition tasks as well as how we perform sequential, "high-level" cognitive tasks.

Although expert systems and related methods have provided plausible models to describe "high-level" cognitive tasks such as language generation and comprehension [Bode88], and goal directed reasoning [Bode88], their ability to explain "low-level" pattern recognition type tasks has been limited. In contrast, neural networks offer the exact complementary ability. They are very good at modelling pattern recognition tasks such as visual processing and motor control but are not well equipped to model sequential, high-level cognitive tasks. Because of these limitations, some researchers have criticised neural networks as inadequate models of cognition [Fodo88].

An analogue to cognitive modelling can be found in complex real world industrial and commercial problems. That is, most complex real world tasks are easily decomposed into logical, sequential operations and parallel, pattern recognition operations [Hand89]. For example in industrial robotics, neural networks are good candidates for performing low-level signal processing tasks such as the processing of sensory data while expert systems are good candidates for high-level path planning tasks [Thor90].

Similar to neural networks, fuzzy logic systems are also good at low-level signal processing tasks such as industrial control. They have recently been applied to hand-written character recognition, blast furnace control, robotic-arm control and camera auto-focus control [Mcne93].

Other tasks which have combined signal processing and high-level reasoning subproblems can be found in financial data analysis, medical diagnosis, and autonomous vehicle navigation.

1.2.4 Explanation

The ability to provide users with explanations of the reasoning process is an important feature of intelligent systems [Clan83]. Explanation facilities are required both for user acceptance of the solutions generated by an intelligent system, and for the purpose of understanding whether the reasoning procedure is sound [Davi77]. Good examples of this requirement can be found in medical diagnosis, loan granting, and legal reasoning. There have been fairly successful solutions to the explanation problem by expert systems, symbolic machine learning and case-based reasoning systems. In expert systems and rule induction systems, explanations are typically provided by tracing the *chain of inference* during the reasoning process [Sout91].

In a fuzzy logic system the final decision is generated by *aggregating* the decisions of all the different rules contained in the fuzzy rule-base. In these systems a chain of inference cannot be easily obtained, but the rules are in a simple to understand "IF-THEN" format which users can easily inspect.

Genetic algorithms, especially in the form of *classifier systems*, can build reasoning models in the form of rules [Holl87]. As in the case of expert systems, it is possible to trace a chain of inference and provide some degree of explanation of the reasoning process.

In contrast, in neural networks it is difficult to provide adequate explanation facilities. This is due to neural networks not having explicit, declarative knowledge representation structures but instead having knowledge encoded as *weights* distributed over the whole network [Died89]. It is therefore more difficult to find a *chain of inference* which can be used for producing explanations. This can be a particular problem in certain business and medical applications. For example, the provision of details about the mortgage lending decision process to potential borrowers, is now a legal requirement in certain countries [Hugh92]. This suggests that neural networks cannot be used. Similarly, in medical diagnosis physicians want to be absolutely sure about the detailed reasoning procedures used by an automated diagnostic system before a particular treatment is prescribed.

There is now a small but growing number of researchers who are attempting to provide neural networks with explanation facilities by extracting rules from their internal weight structures [Gall88]. Additionally there has also been progress in using "feature significance estimation" methods that identify the most important variables in a given neural network model [Intr92, Jabr92].

Table 1.1, shows a property assessment table that summarises the above computational properties with respect to individual intelligent techniques.

	Properties				
Technologies	**Automated Knowledge Acquisition**	**Coping with Brittleness**	**High-level Reasoning**	**Low-level Reasoning**	**Explanation**
Expert Systems	✓	✓	✓✓✓✓✓	✓	✓✓✓✓✓
Rule Induction	✓✓✓✓	✓✓	✓✓✓	✓✓	✓✓✓
Fuzzy Systems	✓	✓✓✓✓✓	✓✓✓	✓✓✓✓✓	✓✓✓✓
Neural Networks	✓✓✓✓✓	✓✓✓✓✓	✓	✓✓✓✓✓	✓
Genetic Algorithms	✓✓✓✓✓	✓✓✓	✓✓✓	✓✓✓	✓✓✓

Table 1.1 — Property assessment of different intelligent techniques.

1.3 A Hybrid Systems Classification Scheme

From the above description it is evident that each intelligent technique has particular strengths and weaknesses and that they cannot be applied universally to every problem. There are three main reasons for creating hybrid systems: *technique enhancement*, the *multiplicity of application tasks* and *realising multi-functionality*. Firstly we summarise these reasons, and then put forward a classification scheme directly based on these factors.

(1) *Technique enhancement.* This is the integration of different techniques to overcome the limitations of each individual technique. Here the aim is to take a technique that has weaknesses in a particular property and combine it with a technique that has strengths in that same property.

(2) *Multiplicity of application tasks.* In this instance a hybrid system is created because no single technique is applicable to the many subproblems that a given application may have. Most real world domains have both "logical", static components (that can be easily handled by, say, expert systems) and "fuzzy", dynamic, poorly understood components (which can be handled by, say, neural networks).

(3) *Realising multi-functionality.* The motivation in this instance is to create hybrids that can exhibit multiple information processing capabilities within one architecture. In other words these systems *functionally mimic* or emulate different processing techniques. The prime example of such systems is the use of neural networks for symbol processing, demonstrating the cognitive principle that connectionist networks can be used for serial reasoning [Ajja91, Sun92].

1.3.1 Three Hybrid Classes

In this section, we propose a new classification scheme for intelligent hybrid systems (see Figure 1.1). This classification scheme takes into account factors such as functionality, processing architecture and communication requirements. Although in the following sections we have limited our discussion to hybrids of intelligent techniques, the proposed hybrid classification is also applicable to hybrids of other techniques (e.g. numeric techniques such as statistical clustering, regression techniques, linear programming etc.).

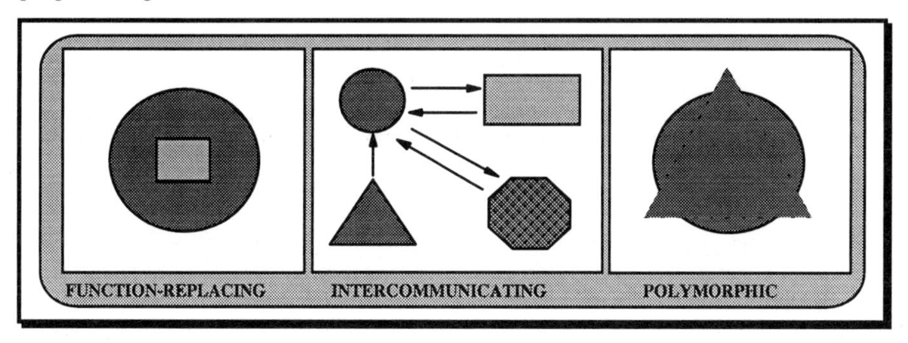

Figure 1.1 — Three proposed hybrid classes.

A classification scheme specifically for hybrids of neural networks and expert systems has been proposed by Medsker [Meds92]. While Medsker's scheme differentiates systems on the basis of implementation issues, our scheme is a more generic scheme that differentiates systems on the basis of functionality [Goon92].

1.3.2 Function-Replacing Hybrids

Function-replacing hybrids address the functional composition of a *single* intelligent technique. In this hybrid class, a *principal function* of the given technique is replaced by *another* intelligent processing technique. The motivation for these hybrid systems, is the *technique enhancement* factor discussed above.

The motivation for replacing these *principal functions* could either be to increase execution speed or enhance reliability. Examples of *principal functions* include pattern matching in an expert system, changing weights in a neural network and crossover operations in a genetic algorithm.

Montana and Davis [Mont89] provide an example of a *function-replacing hybrid*. They replace the backpropagation weight changing mechanism of a neural network with genetic algorithm operators. The genetic algorithm takes the existing weights of the neural network and then applies mutation and crossover operators on these weight values to obtain new values. The weights of the network are encoded in a list structure (see Figure 1.2). Crossover and mutation operators are performed on these lists representing the weights. The performance of this genetic weight updating method was compared with backpropagation [Rume86] on a sonar pattern recognition task. The hybrid method outperformed backpropagation by taking a much smaller number of iterations to converge to a good solution [Mont89].

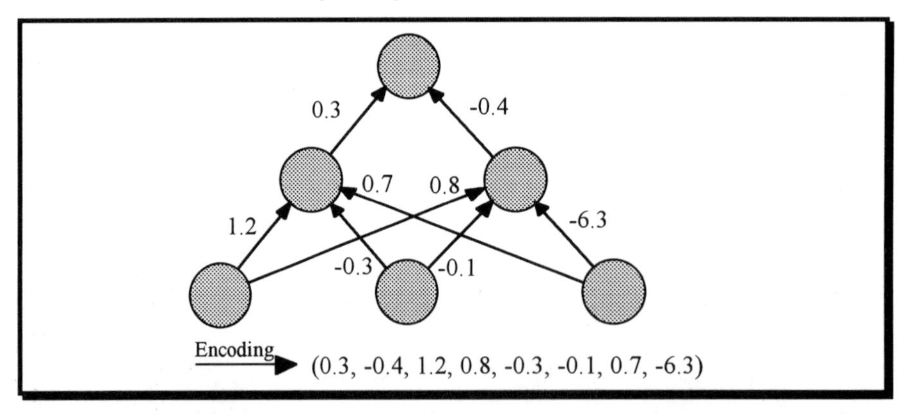

Figure 1.2 — Neural network chromosome encoding.

Another good example of a function replacing hybrid is Karr's [Karr91] work on using a genetic algorithm for replacing the task of manually defining fuzzy membership functions. The definition of a membership function is usually done by a domain expert who uses his or her subjective judgement to specify the shapes and values that the membership function may take. Typically this is a very time consuming trial and error process that takes the largest amount of time in the development of fuzzy systems.

Karr uses a genetic algorithm to design fuzzy membership functions in a fuzzy system for controlling cart-pole balancing. There are four condition variables each having three linguistic variables. Karr uses triangular shapes in the fuzzy membership definition (see Figure 1.3), and the aim of the genetic algorithm is to find the optimal *anchor points* in these triangles. Bit strings were used to represent the possible positions of these anchor points. After having viewed only a small portion of the search space (approximately 32000 of the 2^{132} possible points), the genetic algorithm was able to learn membership functions that were much better at controlling the cart-pole than the membership functions defined by Karr.

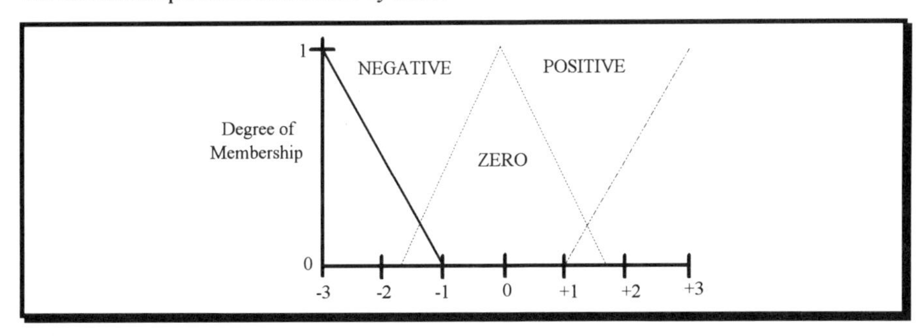

Figure 1.3 — Fuzzy membership definitions.

1.3.3 Intercommunicating Hybrids

Intercommunicating hybrids are *independent*, self-contained, intelligent processing modules that exchange information and perform separate functions to generate solutions.

If a problem can be subdivided into distinct processing tasks, then different independent intelligent modules can be used to solve the parts of the problem at which they are best. These independent modules which collectively solve the given task are co-ordinated by a *control mechanism*.

For example, if a particular task has sub-tasks of pattern recognition, serial reasoning and optimisation, then a neural network, an expert system and a genetic algorithm can be made to perform these respective tasks. These independent modules can process sub-tasks either in a sequential manner (where the control can be provided in the rules of the expert system) or in a competitive/co-operative framework such as a blackboard problem solving architecture (where the control is in a distinct control component).

An example of an *intercommunicating hybrid* is Schreinemaker's and Touretzky's system for veterinary diagnosis [Schr90]. In their system, a rule-based expert system performs serial inferences, and calls neural networks to find patterns in the data as part of the diagnosis. The system uses a particular set of OPS5 data structures called Working Memory Elements (WMEs) for communication between the neural network and the expert system.

For example a typical rule in the system is,

```
(p FEED-COW-TO-MASTITIS-NETWORK
    ( subgoal (goal type diagnose-cow number n )
    ( cow-wme (cow-descriptor number n)
    ⇒(call invoke-network mastitis-net cow-wme)
    (remove subgoal))
```

If the neural network mis-diagnoses a particular case, then the expert system acts as *knowledge manager* by allowing a domain expert to inspect the mis-classification and to change the associated training data if such data is found to be incorrect. The authors claim that the system achieves a classification rate of 87%, which is comparable to the performance of an expert veterinarian.

A further example of this type of intercommunicating hybrid is given by *Dunker et al* [Dunk92] who describe a co-operative environment for integrating neural networks and knowledge-based systems. The principal aim of this system is to decompose large complex reasoning tasks into manageable subproblems and to process them via a blackboard architecture. The agents of the blackboard system are several *neural problem solvers*, each concerned with modelling a particular aspect of the problem domain. A neural problem solver consists of a neural network and an expert system control mechanism. Each of these neural problem solvers pass and receive messages to and from the central blackboard. If all the conditions for a neural problem solver are met (checked by the expert system control mechanism) then that particular neural network agent will be used to process the data. Once the processing has been completed, it will pass information to the blackboard indicating that the sub-task is finished and will post the relevant processed results. Another neural problem solver can now use these results for further processing. Dunker *et al* are planning to use this system in banking for the analysis of investment opportunities.

1.3.4 Polymorphic Hybrids

Polymorphic hybrids are systems that use a *single processing architecture* to achieve the functionality of *different* intelligent processing techniques. The broad motivation for these hybrid systems is *realising multi-functionality* within particular computational architectures. These systems can *functionally mimic* or emulate different processing techniques. These might be viewed as being "chameleon-like" systems which can change their functional form.

Examples of polymorphic hybrids are neural networks that attempt to perform symbolic tasks such as step-wise inferencing and also neural networks which function as if they were doing genetic search.

Ajjanagadde and Shastri [Ajja91] demonstrate a *Polymorphic hybrid system* where a neural network achieves the functionality of a symbolic reasoning system. This system addresses the symbolic variable binding problem (how to dynamically represent variables and their role bindings) by propagating *rhythmic* patterns of activity wherein dynamic bindings are represented as *in-phase* firings of appropriate nodes in the network. This allows a rule-like chain of inference and reasoning to be present within a neural architecture. The primary motivation for this model is to demonstrate the cognitive principle of how serial processing can be achieved with a neural processing mechanism.

A general comment regarding the proposed classification scheme is the limited number of examples in the polymorphic category. The majority of the polymorphic hybrids are examples of neural networks used for symbolic reasoning although there are examples of using neural networks for imitating genetic search procedures [Ackl85]. It is likely that many more examples of such polymorphic hybrids will be developed as more researchers studying different intelligent techniques begin to understand the parallels between the different approaches.

1.4 A Hybrid Systems Development Cycle

While the above classification scheme provides a method of categorising *existing* hybrid systems, it does not provide any guidance in the development of *new* hybrid systems. In this section we draw upon our classification scheme and attempt to describe a typical development cycle in the implementation of hybrid systems. The hope is that a relatively structured approach to the development of these systems may reduce development times and costs.

There are the following stages for the construction of intelligent hybrid systems: problem analysis, property matching, hybrid category selection, implementation, validation and maintenance (see Figure 1.4).

The first step in developing intelligent hybrid systems is the ***problem analysis*** stage which involves two distinct steps.

- The first step is to identify any existing sub-tasks of the problem. A complex problem such as retail distribution, for example, may have both a product demand prediction sub-task and a distribution optimisation sub-task. However, there may also be problems which do not decompose into sub-tasks (e.g. a simple prediction task).

- The second step is to identify the properties of the problem. If the problem has sub-tasks, this involves identifying properties of the sub-tasks. Examples of the identified properties may include the need for automated knowledge acquisition, coping with brittleness and explanation, etc.

The next step, ***property matching***, involves the matching of properties of available techniques against the requirements of the identified tasks. To assess the computational properties of the available techniques, a *property assessment table* similar to Table 1.1 should be constructed. This type of table provides a rating scheme that grades the competence of each intelligent technique in performing tasks such as learning or providing explanations.

The ***hybrid category selection*** phase selects the type of hybrid system required (intercommunicating, function-replacing, or polymorphic) for solving the problem. This phase uses the results of the previous problem analysis and property matching stages. If the developer finds that there is one intelligent technique which has high ratings in all the desired properties, then there is obviously no need for a hybrid systems solution.

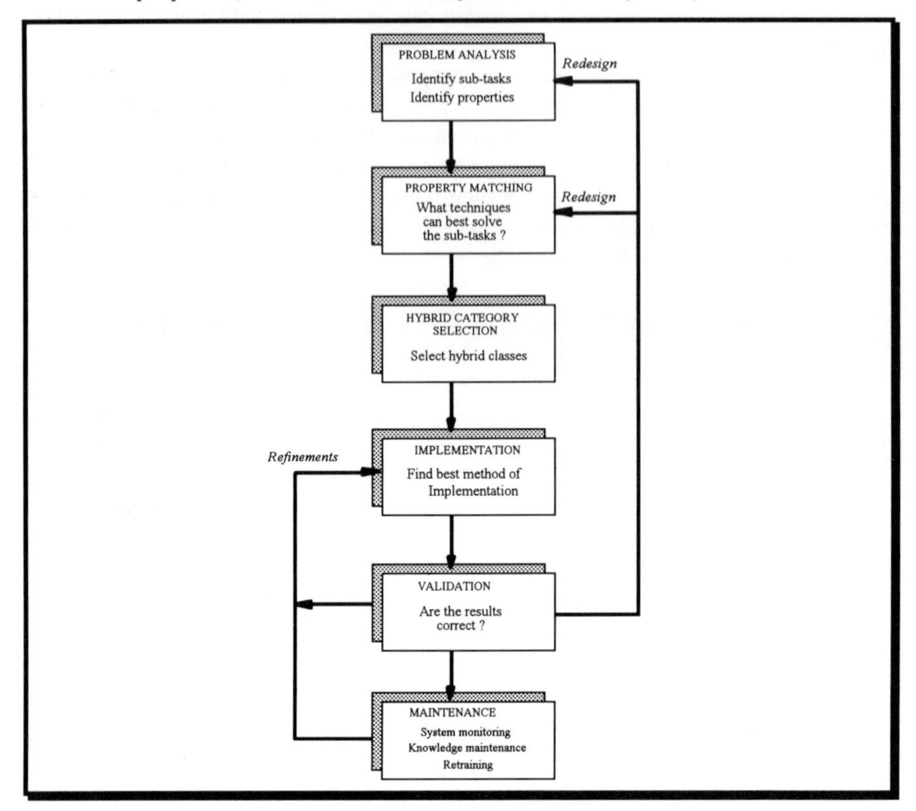

Figure 1.4 — Hybrid system development cycle.

If the problem is composed of distinct, non-overlapping sub-tasks and if there are techniques which have been successfully matched in the previous stage, then an *intercommunicating hybrid* approach can be taken. For example, assume that there is a problem that consists of a prediction task (e.g. energy prediction) followed by a distribution task (e.g. optimisation of energy distribution). Here a neural network, which scores highly in its ability to predict, could be used for the prediction task. This can be followed by a genetic algorithm, which scores highly for its ability to optimise, for the distribution task.

If the problem does not contain distinct sub-tasks that can be solved using different intelligent techniques, then one has to use a *function-replacing hybrid* approach.

The first step is to list the *candidate techniques* which score highly on the different properties of the problem (obtained from the property matching stage).

Then the task is to find a mechanism to combine the different techniques in a manner so that the resulting hybrid has the desired high ratings. Usually such an improvement can be achieved by replacing a *principal function* of one technique by another technique. For example, the neural networks can be used to induce the decision-making knowledge for the fuzzy system, thus *replacing the function* of a domain expert manually specifying knowledge.

A *polymorphic hybrid* is appropriate in situations where the desired functionality dynamically changes, this requires the ability to switch from one style of processing to another. For example, in a robot vision task, a polymorphic hybrid system maybe used to dynamically switch to either a rule-based or pattern-recognition processing style depending on incoming information.

The developer will be now be in a position to start ***implementation*** of the system and will need to select the programming tools and environments needed to implement the hybrid systems. Object-oriented programming offers a natural model for implementing hybrid systems. Object-oriented applications represent the problem domain as a set of objects that interact by passing messages. These messages not only contain data but also the operation to be executed by the receiving object. Because objects have a well defined message interface, they are highly interchangeable. Also, as objects are independent entities that interact through message passing, they are ideally suited for distribution across a network of computers.

Using object-oriented programming to represent different processing techniques enables one to mix several processing styles within the same application. However, to ensure that these techniques can communicate and exchange information also requires a *data exchange protocol*. Currently there are many emerging communication protocols but probably the most widely accepted and used are Dynamic Data Exchange (DDE) [Clar92] and Object Linking and Embedding (OLE) [Micr91]. DDE offers a low-level transport mechanism for "fine grain" information (small chunks of data), whereas OLE is a high-level protocol for linking and exchanging "coarse grain" information (e.g. objects) between different applications. As most commercial applications now comply to these standard protocols, developing new intelligent systems components within this framework would ensure the rapid development of intelligent hybrid systems.

An example of an intelligent hybrid system built using object-oriented methods and protocols is an object-oriented neural network tool used to classify data that is stored within a spreadsheet. The input data for the neural network is sent via a DDE or OLE link from particular cells of the spreadsheet. The neural network classifies the input data and places the results back into the spreadsheet via another DDE or OLE link. These results can then be further used by a graphics package or another intelligent technique (e.g. a genetic algorithm) via another data-link. (Chapter 14, expands on this subject by reviewing current object oriented tools and environments for building hybrid systems.)

The ***validation*** phase is used to test and verify the functioning of the individual components of the application and the hybrid system as a whole. If any malfunctions are discovered, either at the component level or at the integrated system level, then the whole application should be redesigned or refined depending on the severity of the problem.

The purpose of the ***maintenance*** phase is to periodically evaluate the hybrid system's performance, and to refine it as the need arises. Maintenance is particularly important for adaptive components (e.g. neural networks and genetic algorithms) in rapidly changing environments where performance can easily degrade if they are not continually *re-trained* with recent domain data.

An aspect that is not covered in the development cycle is the involvement of domain expertise. A domain expert can be invaluable if available during the problem analysis and property matching stages. The expert's knowledge of the domain helps to identify the sub-tasks and also assess different intelligent techniques.

1.5 Future Directions

We believe that hybrid systems will play an increasingly important role in the development of future intelligent systems. We anticipate a significant growth in the application of hybrid systems, especially in areas such as machine perception, intelligent signal processing, financial decision making, insurance risk assessment and natural language understanding.

It is also likely that there will be a larger repertoire of available techniques for creating hybrids, especially new combinations with conventional statistical methods, numerical analysis methods, case-based reasoning [Kolo93] and distributed artificial intelligence methods [Bond88].

We also foresee an increase in the availability of commercial software tools and environments for the development of hybrid systems. Software support environments such as blackboard systems [Cork91], and distributed computing environments will facilitate the rapid prototyping and development of hybrid systems. Communication protocols such as Object Request Broker (ORB) [OMG92] and intelligent agent languages such as Telescript [Wayn94] will also enable complex hybrid systems to be built from processing components that reside across world-wide networks.

A recent development in intelligent systems is the *silicon compilation* of techniques such as neural networks [Nigr91, Mira93, Trel89]. In this process neural network architectures are mapped directly onto silicon chips allowing the efficient execution of these algorithms. Possible uses for these chips include *silicon retinas* that mimic the function of human eyes. A natural evolutionary progression is the creation of a new generation of highly adaptive hardware systems that combine several intelligent techniques. As the field of intelligent hybrid systems matures and expands, this will fuel growth in hardware systems that genuinely replicate the intricate complexity of natural information processing.

1.6 Organisation of the Book

The book is organised in accordance with the hybrid systems classification scheme detailed previously. Part 1 describes representative systems of the Function-replacing hybrid category. Intercommunicating hybrids and polymorphic hybrids are represented in Parts 2 and 3 respectively. Part 4 surveys current tools and environments for implementing hybrid systems. In addition, Part 4 also provides a general resource guide for intelligent techniques used in hybrid systems.

Part 1 — Function-Replacing Hybrids

There are four chapters in this section which discuss combinations of fuzzy logic, neural networks, genetic algorithms, and expert systems. All the chapters are concerned with the function-replacing hybrid class, where the limitations of one particular technique are overcome by the use of another intelligent technique.

Chapter 2 by Wendy Foslien and Tariq Samad describes the use of neural networks for constructing fuzzy logic controllers. They also assess the use of genetic algorithms to further optimise the resulting hybrid controllers.

Chapter 3 by Henry Tirri details a method of improving expert systems by replacing the pattern matching mechanisms by neural networks. It is argued that this approach enables expert systems to reason in a more robust manner.

Chapter 4 by Charles Karr discusses the use of genetic algorithms for defining fuzzy logic membership functions. This is demonstrated firstly in the control task of the cart-pole balancing problem and secondly in an industrial control task.

Chapter 5 by David Montana proposes the use of genetic algorithms as an alternative method for weight adjustments in neural network learning.

Part 2 — Intercommunicating Hybrids

This section consists of four chapters detailing combinations of genetic algorithms, expert systems, symbolic clustering, rule induction, neural networks and fuzzy logic. All hybrids in this section are intercommunicating hybrids which are composed of independent, self-contained, intelligent processing modules.

Chapter 6 by Powell *et al* describe a radically new approach to engineering design optimisation where genetic algorithms, expert systems and numerical optimisation methods are combined in a manner that exploits the strengths of each approach.

Chapter 7 by Kerber *et al* presents a database mining framework that combines statistical analysis, symbolic clustering, rule induction and visualisation methods. The application of the approach is discussed within the context of discovering patterns in financial trading data.

Chapter 8 by Casimir Klimasauskas describes the use of applying fuzzy logic transformations on input data for neural networks. This fuzzy pre-processing method is used for solving complex chemical process diagnostic problems that are not amenable to "pure" neural network approaches.

Chapter 9 by Andreas Scherer and Gunter Schlageter detail the use of distributed artificial intelligence approaches for combining neural networks and expert systems. The approach is based on a blackboard architecture and is demonstrated in the domain of economic forecasting.

Part 3 — Polymorphic Hybrids

This section reviews Polymorphic hybrids which are systems that use a single processing architecture to achieve the functionality of different intelligent processing techniques. There are four chapters in this section which are concerned with neural network, genetic programming and symbolic reasoning hybrids.

Chapter 10 by Vasant Honavar and Leonard Uhr proposes a framework for the synthesis of symbolic processing and connectionist networks by generalising the standard definition of connectionist models to incorporate symbolic processing mechanisms.

Chapter 11 by Venkat Ajjanagadde and Lokendra Shastri presents a neural network architecture for serial reasoning that solves the variable binding problem. This allows a system to have expert system-like reasoning while retaining the advantages of distributed processing found in neural networks.

Chapter 12 by Khai Minh Pham details a polymorphic approach where *neural agents* have both neural network and symbolic reasoning capabilities. This *neurOagent* approach and an associated knowledge modelling methodology is discussed with reference to industrial problems.

Chapter 13 by Frederic Gruau describes the use of a novel encoding scheme for neural networks. He uses this scheme to synthesise neural networks which have functional similarities to genetic algorithm search procedures — an approach referred to as the genetic programming of neural networks.

Part 4 — Developing Hybrid Systems

This last section is included to give developers a practical guide to implement hybrid systems.

Chapter 14 by Sukhdev Khebbal and Danny Shamhong surveys the currently available software tools and environments for the development of hybrid systems. Special emphasis is placed on object-oriented techniques and communication protocols.

A glossary of terms and a resource guide for different intelligent techniques is also provided in this section.

Acknowledgements

We would like to thank Mike Luck of the University of Warwick, Konrad Feldman of University College London and Steve Green for constructive comments on earlier drafts of this paper.

References

[Ackl85] Ackley, D.H., Hinton, G.E., and Sejnowski, T.J., "A learning algorithm for boltzmann machines", *Cognitive Sci.*, vol. 9, pp. 147-169, 1985.

[Ajja91] Ajjanagadde, Venkat and Shastri, Lokendra, "Rules And Variables In Neural Nets ", *Neural Computation*, vol. 3, no. 1, pp. 121-134, 1991.

[Beal90] Beale, R. and Jackson T., "Neural Computing - an introduction", Adam Hilger, Bristol, 1990.

[Bode88] Boden, Margaret, "Computer Models of Mind", Cambridge University Press, 1988.

[Bond88] Bond, A.; Gasser, L.: Readings in Distributed Artificial Intelligence, Morgan Kaufmann Publishers, San Mateo, CA., 1988.

[Clan83] Clancy, W.J, "The Epistemology of a Rule-based Expert System: A Framework For Explanation", *Artificial Intelligence* , vol. 20, pp. 215-251, 1983.

[Clar92] Clark, J. D., "Windows Programmers Guide to OLE/DDE", SAMS/Prentice Hall Computer Publishing, 1992.

[Cork91] Corkill, Daniel, "Blackboard Systems", *AI Expert*, pp. 41-47, Sept 1991.

[Davi77] Davis, R., Buchanan, B., and Shortcliffe, E., "Production Rules as a Representation For a Knowledge-Based Consultation Program", *Artificial Intelligence*, vol. 8, no. 1, pp. 15-45, 1977.

[Died89] Diederich, Joachim, "Explanation and Artificial Neural Networks", GERMAN National Research Centre for Computer Science(GMD), W.Germany, 1989.

[Dunk92] Dunker, Jurgen, Scherer, Andreas, and Schlageter, Gunter, "Integrating Neural Networks Into Distributed Knowledge Based Systems", *Proc of 12th Conf on AI, Expert Systems and Natural Language, Avignon, France*, 1992.

[Feig79] Feigenbaum, E., "Themes and Case Studies of Knowledge Engineering", in *Expert Systems in the micro-technology age*, ed. D. Mitchie, Edinburgh University Press, 1979.

[Fodo88] Fodor, Jerry A. and Pylyshyn, Zenon W., "Connectionism and Cognitive Architecture : A Critical Analysis ", *Cognition*, vol. 28, pp. 3-71, 1988.

[Fors86] Forsyth, R., "Machine Learning: Applications in Expert Systems and Information retrieval", Ellis Horwood Ltd, 1986.

[Gall88] Gallant, Stephen I., "Connectionist Expert Systems", ACM, *Communications of the ACM*, vol. 31, no. 2, pp. 152-169, Feb 1988.

[Goon92] Goonatilake, Suran and Khebbal, Sukhdev, "Intelligent Hybrid Systems ", *Proceedings of First International Conf on Intelligent Systems, Singapore*, pp. 207-212, Sept 1992.

[Hand89] Handelman, David A., Lane, Stephen H., and Gelfand, Jack J., "Integrating Knowledge-Based System and Neural Network Techniques for Robotic Skill Acquisition", Morgan Kaufmann, *Proceedings of IJCAI, Detroit*, pp. 193-198, August 1989.

[Holl86] Holland, J., "Escaping Brittleness", in *Machine Learning*, ed. R. Michalski, J. Carbonell, and T. Mitchell, Morgan Kaufmann, vol. 2, 1986.

[Holl87] Holland, J., "Genetic Algorithms and classifier systems: Foundations and future directions", In *Genetic Algorithms and their Applications: Proceedings of the 2rd International Conference on Genetic Algorithms,* 1987.

[Hugh92] Hughes, C., "Neural Networks and Expert Systems - A Partnership". In the Proceedings of the IEE Colloquium on the Integration of Neural Networks and Knowledge Based Systems, IEE London, 1992.

[Intr92] Intrator, N. "Feature Extraction Using an Unsupervised Neural Network", *Neural Computation*, vol. 4 no. 1, pp. 98-107, 1992.

[Jabr92] Jabri, M. and Flower, B. "Weight Perturbation: an Optimal Architecture and Learning Technique for Analog VLSI Feedforward and Recurrent Multilayer Networks", *IEEE Transactions on Neural Networks,* vol. 3. no. 1, pp. 154-157, 1992.

[Jack86] Jackson, P., "Introduction to Expert Systems", Addison Wesley, 1986.

[Karr91] Karr, Charles L., "Design of an Adaptive Fuzzy Logic Controller Using a Genetic Algorithm", *Proceedings of Fourth Genetic Algorithms Conference*, pp. 450-457, 1991.

[Kidd87] Kidd, A., "Knowledge Acquisition for Expert Systems: A practical handbook", Plenum Press, 1987.

[Kolo93] Kolodner, J., "Case-Based Reasoning", Morgan Kaufman Publishers, San Matereo, CA, 1993.

[Mcde83] McDermott, D., "Extracting Knowledge From Expert Systems", *Proceeding of 8th International Joint Conference on Artificial Intelligence*, pp. 100-107, 1983.

[Mcne93] McNeill and Frieberger, "Fuzzy Logic", Simon and Schuster, New York, 1993.

[Meds92] Medsker, L.A, and Bailey, D.L, "Models and Guidelines for Integrating Expert Systems and Neural Networks", In *Hybrid architectures for intelligent systems*, ed. Kandel, A, and Langholz, G, CRC Press, 1992.

[Mich80] Michalski, R. and Chilansky, R. L., "Learning by Being Told and Learning From Examples", *Journal of Policy Analysis & Information Systems*, vol. 4, 1980.

[Micr91] Microsoft Corporation, "Object Linking and Embedding", 10 June 1991.

[Mira93] Mira, E J., Cabestany, E J, Prieto, E A., M.E. Nigri, and Treleaven,P.C, "High-Level Synthesis of Neural Network Chips", In *New Trends in Neural Computation*, Proceedings of the International Workshop on Artificial Neural Networks, IWANN'93, Sitges, Spain, pp 448-453, June 9-11, 1993.

[Mont89] Montana, D. and Davis, L., "Training Feedforward Neural Networks using Genetic Algorithms", *Proceedings of 11th International Joint Conference on Artificial Intelligence*, pp. 762-767, 1989.

[Nigr91] Nigri, M.E, Treleaven, P.C., and Vellasco, M.M.B.R., "Silicon Compilation of Neural Networks, Proceedings of the IEEE CompEuro'91, Bologna, Italy, pp 541-546, May 13-16, 1991.

[OMG92] Object Management Group (OMG), "The Common Object Request Broker: Architecture and Specification", Object Management Group, 1992.

[Otsu92] Otsuka, Y., Hanaoka, K., and Konishi, M., "Applications of Neural Network Models to Iron and Steel Making Process", *Proceedings of the 2nd International Conference on Fuzzy Logic & Neural Networks*, pp. 815-818, Iizuka, Japan, July 1992.

[Part86] Partridge, Derek, "Is Intuitive Expertise Rule-Based?", University of Exeter, UK, 1986.

[Quin86] Quinlan, J. R., "Induction of Decision Trees", *Machine Learning*, vol. 1, pp. 81-106, 1986.

[Quin93] Quinlan, J. R., "C4.5: A Machine Learning Paradigm", Morgan Kaufmann, 1993.

[Rume86] Rumelhart, David E., and McClelland, James L., "*Parallel Distributed Processing - Explorations in the Microstructure of Cognition*", ed. David E. Rumelhart and James L. McClelland, MIT Press, Cambridge, MASS, vol. 1, 1986.

[Schr90] Schreinemakers, J.F. and Touretzky, D., "Interfacing a Neural Network with a Rule-Based Reasoner For Diagnosing Mastitis", *Proceedings of International Joint Conference of Neural Networks, Washington*, pp. 487-490, July 1990.

[Seve92] Severwright, J., "Neural Networks for Direct Marketing", In the *Proceeding of the IBC conference on Data Mining in Finance and Marketing.* IBC Technical Services Ltd., 1992.

[Smol87] Smolensky, P., "Connectionist AI, symbolic AI, and the Brain", Artificial Intelligence Review, 1:95-109, 1987.

[Sout91] Southwick, Richard, "Explaining Reasoning: An Overview of Explanation in Knowledge-Based Systems", *The Knowledge Engineering Review*, vol. 6, no. 1, pp. 1-19, 1991.

[Sun92] Sun, Ron, "On Variable binding in connectionist networks", *Connection Science: Journal of Computing, Artificial Intelligence and Cognitive Research,* 4(2): 93-124, 1992.

[Thor90] Thorpe, C., "Vision and Navigation"*: The Carnegie Mellon NAVLAB*, Kluwer Academic Publishers, 1990.

[Trel89] Treleaven, P.C, Pacheco, M.A.C, and Vellasco, M.M.B.R, "VLSI Architecture for Neural Networks", *IEEE Micro,* vol. 9, no. 6, pp. 8-27, December 1989.

[Wayn94] Wayner, P, "Agents Away", *BYTE,* pp. 113-118, May 1994.

[Wien88] Wiener, Richard S. and Pinson, Lewis J., "An Introduction to Object-Oriented Programming and C++", Addison-Wesley Publishing Company, 1988.

[Weis91] Weiss, S.M. and Kulikowski, C.A., "Computer Systems that Learn: Classification and Prediction Methods From Statistics, Neural Nets, Machine Learning and Expert Systems", Morgan Kaufmann Publishers Inc., San Mateo, California, 1991.

[Zade88] Zadeh, L.A., "Fuzzy Logic", *IEEE Computer*, pp. 83-93, April 1988.

Part One

Function-replacing Hybrids

2

Fuzzy Controller Synthesis with Neural Network Process Models

Wendy Foslien and Tariq Samad

2.1 Introduction

A leading application domain for intelligent hybrid systems is process control. Control systems based on fuzzy logic and neural networks are no longer just research topics. Indeed, their popularity has redefined the field of "intelligent control", much activity in which is now devoted to the investigation of hybrid architectures that integrate neural networks, fuzzy logic, and other novel (or newly resurgent) technologies.

Neural networks and fuzzy logic have in common two features that distinguish them from traditional control. First, they both provide non-linear processing capabilities that allow them to tolerate the non-linear nature of real-world systems in ways that the prevailing theory and practice in process control — rooted as it is in linear systems theory — simply cannot. Second, they do not demand control-theoretic sophistication of the user, permitting instead other types of information as substitutes — heuristic knowledge in the case of fuzzy logic and experimental data for neural networks.

This second feature also hints at the need for hybrid fuzzy/neural control. In many cases, neither heuristics nor process data is of sufficient quality by itself to allow the true potential of non-linear process control to be realised. However, an appropriate integration of the two technologies can enable a synergism of these two complementary types of information.

Intelligent Hybrid Systems, Edited by S. Goonatilake and S. Khebbal. ©1995 John Wiley & Sons Ltd.

We propose such an integration in this chapter. In particular, we show how a neural network process model, developed from process data, can be used to optimise a fuzzy controller. The hybrid system thus combines the intuitive representation of a fuzzy rule set and the empirical modelling capabilities of neural networks. This approach requires the solution of a complex optimisation problem, one for which conventional gradient-based optimisation algorithms are ill suited. We have addressed this issue by adopting a genetic algorithm as the optimisation method. Thus our overall control synthesis scheme is a hybrid system that integrates three technologies: fuzzy logic, neural networks, and genetic algorithms.

This chapter elaborates and extends work previously reported in [Fosl93].

2.2 Overview of Technologies

The novelty of our research lies not in any extensions to fuzzy logic, neural network, or genetic algorithm methods — we have adopted standard approaches for each of these — but in their integration. In this section we provide a brief overview of fuzzy logic and control, and we provide capsule introductions to our use of neural networks and genetic algorithms — technologies that are well represented in several other chapters of this book.

2.2.1 Fuzzy Logic

Fuzzy control is based on the concepts of fuzzy logic, which were formalised in the mid-1960s by Dr. Lotfi Zadeh [Zade65]. Zadeh developed the mathematics of fuzzy logic to provide a tool for reasoning in a mode which is approximate, rather than exact. Fuzzy logic differs from binary or Boolean logic because fuzzy logic allows a statement to be partially true (or partially false), rather than requiring all statements to be *either* true or false. The first use of fuzzy logic for control was by Mamdani and Assilian, who developed a fuzzy controller for regulation of a laboratory scale steam engine [Mamd74, Mamd75].

Fuzzy logic is fundamentally based on the concept of a fuzzy set. In conventional set theory, an element is either a member of a set, or not a member of a set. Zadeh proposed a generalisation of set theory in which some elements are more thoroughly members of a set than others. The degree of membership in a particular set may take on various values between zero and a maximum value, where a zero value indicates complete exclusion and the maximum value indicates complete membership. Thus, in fuzzy logic, a statement such as "The current temperature is warm" actually is an evaluation of the degree to which the temperature fits in to the fuzzy set "warm". Membership in a fuzzy set is defined using a membership function. Typically the membership functions are defined to have a range between 0 and 1 or alternatively between 0 and 100. Figure 2.1 demonstrates a definition of membership functions for the sets "cool", "warm" and "hot", where complete membership in each set is defined as a degree of membership equal to 100. The set of possible temperatures is referred to as the universe of discourse — here the universe of discourse ranges from 50 to 100.

Fuzzy logic also uses a set of operators which are similar to those used for Boolean logic — i.e. AND, OR, NOT. However the interpretation of each of these operators is different from the Boolean interpretation. Although many different interpretations for these operators have been proposed, the most commonly used is a definition proposed by Zadeh. In this definition, (A AND B) corresponds to the minimum of the degrees of membership of A and B. Similarly, (A OR B) corresponds to the maximum of the degrees of membership of A and B.

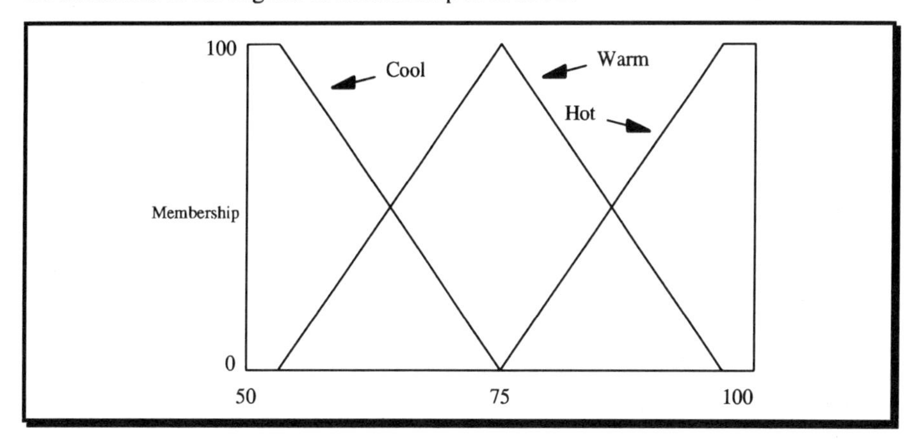

Figure 2.1 — Membership functions for temperature in °F.

Given the concepts of fuzzy sets and fuzzy operators, we can form IF-THEN rules in a form similar to that of Boolean logic. However, in fuzzy logic, the conclusion specified by the THEN portion of a rule is based on the degree of truth of the IF portion of the rule. Thus, if a condition is true to degree 0.5, the conclusion will be executed to degree 0.5. There are several methods for accomplishing the inferencing step. One of the commonly used methods is known as max-product inferencing. In max-product inferencing, the membership function for the conclusion of a rule is scaled by the degree of membership of the rule's premise.

If there are a number of rules, one or more of which apply to a given process state at a given point in time, they are all applied by taking the fuzzy *union* of the fuzzy sets corresponding to the rules. The fuzzy union is represented by the *max* operator. There is an important point to be made about how the fuzzy set which results from the fuzzy union is used: the result of a fuzzy union is not a crisp number but a *fuzzy set.*

If we are going to use the output of the knowledge base for control, we obviously cannot send a fuzzy set out to an actuator. Thus, we must select a representative value from the action fuzzy set. This step is called defuzzification. One popular approach to defuzzification is *centre of gravity defuzzification,* which uses the centroid of the membership function as the representative value.

For further information on fuzzy sets, see [Klir92].

2.2.2 Neural Networks

Neural networks are now a popular approach for modelling non-linear processes. Several architectures have been successfully used for this task, and our work is based on one of the more common ones. We have used multilayer feedforward networks trained with standard backpropagation [Rume86, Werb74]. The input to the network consists of a historical trace of the process input — the "tapped-delay-line" approach to modelling dynamical systems with non-dynamical (algebraic) structures. In the experiments discussed later, 100 or 150 past samples of the process input are used. The neural network outputs the predicted value of the process output at some future time, such as the next time-step.

This neural network modelling architecture is a non-linear finite impulse response filter, a non-linear extension of FIR models often used in conventional system identification (in a FIR model, the process response is modelled as a linear combination of some number of past process input samples).

2.2.3 Genetic Algorithms

Two broad classes of optimisation algorithms can be distinguished: gradient-based and non-gradient-based algorithms. The former — examples of which include steepest descent, conjugate gradient, Newton, and quasi-Newton algorithms — can be effective when the problem permits gradient computation (i.e., when the partial derivatives of the optimisation criterion with respect to the search variables can be feasibly computed). With a non-differentiable criterion, non-gradient-based methods must generally be used. Genetic algorithms constitute a class of non-gradient-based optimisation algorithms that are inspired by biological evolution [Holl75, Gold89].

We have used the GENESIS genetic algorithm system for our experiments [Gref84]. GENESIS implements a standard function-optimisation GA with mutation and two-point crossover operators. The continuous-valued search variables of our optimisation problem were discretized into 8 bit vectors for GA manipulation. All experiments were conducted on HP/Apollo workstations.

2.3 A Brief Review of Control Fundamentals

Although the discipline of control systems is broad and multifaceted, at its most fundamental it is concerned with a single specific problem: regulating the behaviour of a dynamical system — a system whose output at any time is a function not only of its current input but also of its history.

The notion of feedback plays a central role in control. A feedback controller uses information about the current state of a system and it exploits knowledge (often implicit) about the nature of the system to control its behaviour Figure 2.2.

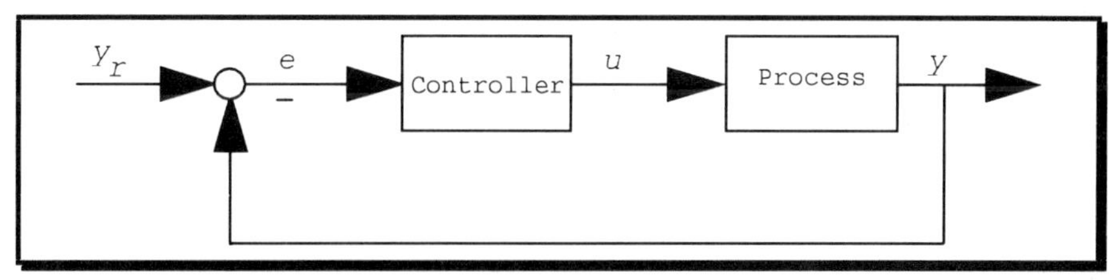

Figure 2.2 — Feedback control system.

A simple feedback controller is the proportional controller. The controller output u is a function of the error $y_r(t)–y(t)$ between the desired process output y_r (referred to as the process "setpoint") and the actual process output y at time t. Proportional controllers are sometimes used in simple applications, but they have a significant shortcoming: they do not allow zero error responses in plants which do not contain an integrating element. An asymptotic offset error will always exist. (To see this, note that the controller output is a function only of the difference between the desired and actual output, not of the desired output itself. When this difference is zero the controller output will provide the same input to the process regardless of the setpoint, which cannot therefore be achieved.)

A higher level of sophistication in feedback control is the proportional-integral (PI) controller. Now the control output is a linear combination of the error and the integral (or summation) of the error over the past history of controller operation. The integral action can eliminate the offset error. PI controllers are in widespread use in the process industries (e.g., oil refineries, chemical plants, pulp and paper manufacturing plants).

Both proportional and PI controllers are linear controllers. The controller output is a linear function of the error signals:

Proportional controller: $$u(t) = K_c\, e(t)$$

PI controller: $$u(t) = K_c\, e(t) + K_i \sum_{k=0}^{k=t} e(k)$$

K_c and K_i are referred to as the proportional and integral "gains" respectively. Equivalent incremental or "velocity" forms, shown below for PI controllers, are also common:

Incremental PI controller: $$\Delta u(t) = K_c \Delta e(t) + K_i e(t)$$

Where $\Delta u(t)$ and $\Delta e(t)$ are the change (or, under an appropriate normalisation, the rate of change) in u and e respectively from time $t–1$ to time t. Linear controllers are analytically convenient and continue to be the focus of the vast majority of theory and practice in control systems. However, this convenience is at the expense of performance. For example, even for linear processes (and all real-world processes are non-linear to some extent[1]) it is only for fairly specific performance criteria that the optimal controller is known to be linear [Khar87, Pool87].

Neural networks and fuzzy control can help realise effective non-linear control systems. Non-linearities can be beneficial even in the context of PI control structures: there is no theoretical basis for assuming that the best function of proportional and integral error is linear. Many fuzzy control systems in fact can be considered non-linear PI controllers. One such system that forms the basis for our work will be discussed in section 2.6.

2.4 Rationale for Hybrid Processing

The quality of a control system depends largely on the degree to which it is informed by knowledge about the process to be controlled. We have already noted above that fuzzy control allows control system development in the absence of analytic first-principles process models. The knowledge source is instead the experience-based heuristic rules of process operators.

The encoding of this heuristic knowledge is not an exact science, and the performance of a fuzzy control system is critically dependent on its design parameters, such as membership function and rule definitions. These are usually defined with the help of human operators of the target process. This is a manual, trial-and-error procedure that is time-consuming and is typically completed with little attempt at optimising performance.

An additional source of information that still does not rely on a mathematically detailed understanding of the process is data gathered from the process. In particular, such data can be used to generate empirical (as distinct from first-principles) models, thereby complementing the expert-defined knowledge bases. Neural networks are now a preferred approach for empirical non-linear process modelling.

The issue that arises is how can we utilise the non-linear modelling capabilities of neural networks for optimising fuzzy control design. Our approach is inspired partly by recent work in the neural network community, in which neural network feedback controllers have been optimised with neural network models [Nguy90]. The fact that the controller is a neural network is incidental in these cases; the essence of the concept is the use of a neural network model to optimise, through off-line closed loop simulation, some non-linear control structure. This concept can be applied to a fuzzy controller in much the same way as to a neural network controller.

There is one significant complicating factor: most work in neural network controller optimisation has relied on the differentiability of the neural network models for computing gradients of the cost function (the control system performance measure) with respect to the design parameters (the neural network controller weights). "Backpropagation through time" is a popular method for gradient computation [Werb90]. With gradients available, steepest descent or more sophisticated search

[1]There is a fundamental exception to this "fact": the quantum mechanical Schrödinger wave function (which determines the probability of a particle being at a specific position at a specific time) is a linear system.

methods can be employed. However, conventional fuzzy control architectures are not differentiable, and optimisation algorithms commonly used in the neurocontrol community are therefore not readily applicable. We have thus been led to consider non-gradient-based optimisation methods, and, in particular, genetic algorithms.

Figure 2.3 shows how the fuzzy controller, the neural network process model, and the genetic algorithm are integrated for controller optimisation.

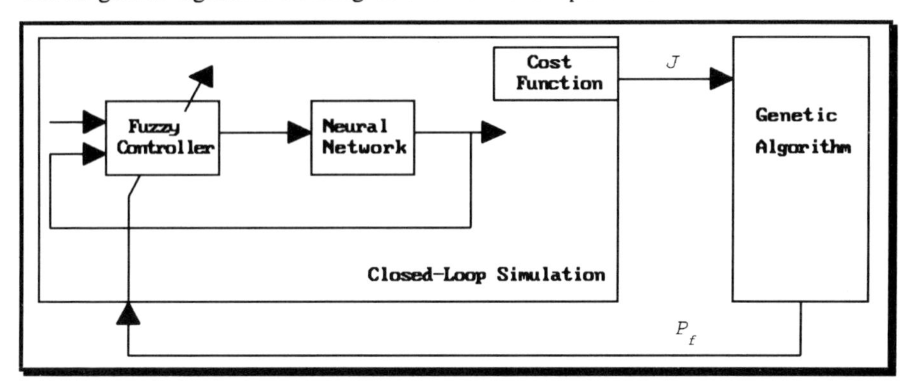

Figure 2.3 — Fuzzy controller optimisation schematic.

The optimisation is conducted in simulation with the neural network process model and the fuzzy controller. Design parameters p_f of the fuzzy controller are modified to minimise a predefined criterion $J(p_f)$. We note that this is an off-line controller synthesis procedure. No learning or adaptation occurs on-line.

2.5 Survey of Related Hybrid Systems

Neural networks, fuzzy logic, and genetic algorithms are all enjoying a resurgence of popularity, and the present volume is only the latest indication of the increasing interest in the integration of these technologies. Fuzzy control systems have been a particularly fertile area for intelligent hybrid systems. Although we know of no previous work that has effectively integrated all three of these technologies, there is already a significant literature in the applications of neural networks or genetic algorithms for fuzzy control.

Some authors have noted that the execution of a (previously trained) neural network is computationally cheaper than undertaking the calculations involved in fuzzy control. Thus one popular application of neural networks is as models of fuzzy controllers [Iwat90, Patr90] where data from a fuzzy controller serves as a training set for a neural network.

A side-effect of neural network modelling of fuzzy control systems is that the resulting controller is differentiable everywhere. Gradient-based algorithms can then be used for adapting the neuro-fuzzy controller [Lin91, Lee91, Ishi93]. This requires that a process model be available, and the availability of first-principles models is generally assumed in this line of research. [Werb92] makes the general point that if differentiable fuzzy models are adopted, then a variety of existing techniques in neurocontrol are readily applicable.

In the absence of gradient information, non-gradient-based learning methods can be applied. Reinforcement learning techniques inspired by the [Bart83] scheme have been used by Lee [Lee91] and Berenji and Khedkar [Bere92]. The peak values of membership functions are adapted in Lee. Berenji and Khedkar model the fuzzy controller using a neural network and use an action-state evaluation network to determine weight updates for the fuzzy controller using a weak reinforcement signal. Genetic algorithms have been used by [Wigg92] and [Karr93]. In the former, peak values of the membership functions are evolved for solving the truck-backer-upper problem [Nguy90]. Karr and Gentry search a space of trapezoidal membership functions for both condition and action variables for a pH control problem. Again, a first-principles model of the chemical process is presumed available.

The integration of fuzzy control with a neural network process model is investigated by [Mccu92]. A conventional fuzzy controller is used to calculate an initial control action which is then given as input to the pretrained neural network model which predicts its effect. If the effect is not as desired, additional rules are applied to modify the control action.

2.6 Case Studies

In this section we describe experimental results on two applications: a simple non-linear process model and a reversible continuous stirred tank reactor (CSTR) model. For each, we describe the rule base for the fuzzy controller, the results of neural network modelling, and the results of the genetic optimisation experiments. The same optimisation criterion was used for both applications, as discussed below.

Ultimately, controller synthesis is a matter of explicitly or implicitly solving an optimisation problem:

$$\min_{p_f} J(p_f)$$

Here p_f is the vector of fuzzy controller parameters and $J(p_f)$ is the cost function derived from the user's specification. Because of its reliance on gradient-based optimisation algorithms, conventional control science has typically been concerned with quadratic criteria. With optimisation algorithms such as GAs, other criteria are also feasible. For our experiments, the integral of the absolute value of the setpoint error (IAE) was selected as the optimisation criterion. The IAE is defined as:

$$J(p_f) = \int_{t=0}^{t=T_f} \left| y(t, p_f) - y_r(t) \right| dt \tag{2.1}$$

Here y is the process output and y_r the setpoint. The cost function is computed over a fixed time interval T_f.

2.6.1 Case Study 1: A Simple Non-linear Process

To assess the feasibility of using a neural network model in the fuzzy controller synthesis process, a simple process model was selected. This was a non-linear model in which the dynamics are governed by a linear second-order transfer function:

$$\frac{Y(s)}{U(s)} = \frac{K(T_n s+1)}{(T_{p1}s+1)(T_{p2}s+1)}$$

The non-linearity is embedded in the process gain, which is dependent upon process input via the following relationship:

$$K = K_{nom} + 0.1 * u$$

where u is the process input. Nominal process parameters were: T_n=2.5, T_{p1}=5.5, T_{p2}=8.9, K_{nom}=1.2.

2.6.1.1 Fuzzy Control System Formulation

The fuzzy controller for this application has two inputs, error and rate of change of error; and one output, the change in process input. Thus the controller is an incremental form of a proportional-integral controller. Note that in the incremental form, the error term is tied to integral action, and the change of error term is tied to proportional action.

For the fuzzy controller, we defined five membership functions for error, five membership functions for rate of change of error, and five membership functions for controller output. The universe of discourse for each input membership function was normalised to a range of 0-100, and a scaling factor was applied to measured inputs to adjust measurements into the normalised universe of discourse. The universe of discourse for the controller output was also scaled, but the range of the universe of discourse was expanded to –150 to +150.

Five membership functions were used for each of the two inputs and for the output. These membership functions defined fuzzy sets for Negative Big, Negative Small, Zero, Positive Small, and Positive Big. Although similarly labelled, the membership functions for the variables were independently specified and adaptable. The controller thus comprised a total of 15 linguistic variables.

Thus, the rule base consists of 25 rules, which are summarised in Figure 2.4. For example, the top right-hand entry specifies the fuzzy rule: IF the Error is Positive Big and the Rate of Change of Error is Negative Big, THEN the Change in Controller Output is Negative Small. Max-product inferencing was used, along with centre of gravity defuzzification.

To simplify processing in the fuzzy controller, we restricted the membership functions to be either triangular or trapezoidal. Further restrictions were placed on the output membership functions — all output membership functions were symmetric triangles.

ERROR

ERROR RATE	NB	NS	Z	PS	PB
NB	PB	PB	PS	PS	NS
NS	PB	PS	PS	NS	NS
Z	PB	PS	Z	NS	NB
PS	PS	PS	NS	NS	NB
PB	PS	NS	NB	NB	NB

NB = Negative Big
NS = Negative Small
Z = Zero
PS = Positive Small
PB = Positive Big

Figure 2.4 — Fuzzy controller rule base.

As a baseline, all membership functions were evenly spaced over their respective input spaces, and nominal values were selected for the scaling factors. Three setpoint changes were made, and the response of the controller was observed. The performance of the baseline fuzzy controller is shown in Figure 2.5. Note that the process output does not reach set point for the first two set point changes. It is not obvious why the controller is performing poorly — membership functions could be defined improperly, rules could be incorrect, or scaling factors could be poorly selected. In the work reported in this chapter, we investigated the optimisation of membership functions and scaling factors; the rule base was fixed. Figures 2.6, 2.7, and 2.8 show the symmetric membership functions which were defined for the nominal controller. Figure 2.9 shows the resulting three-dimensional output map.

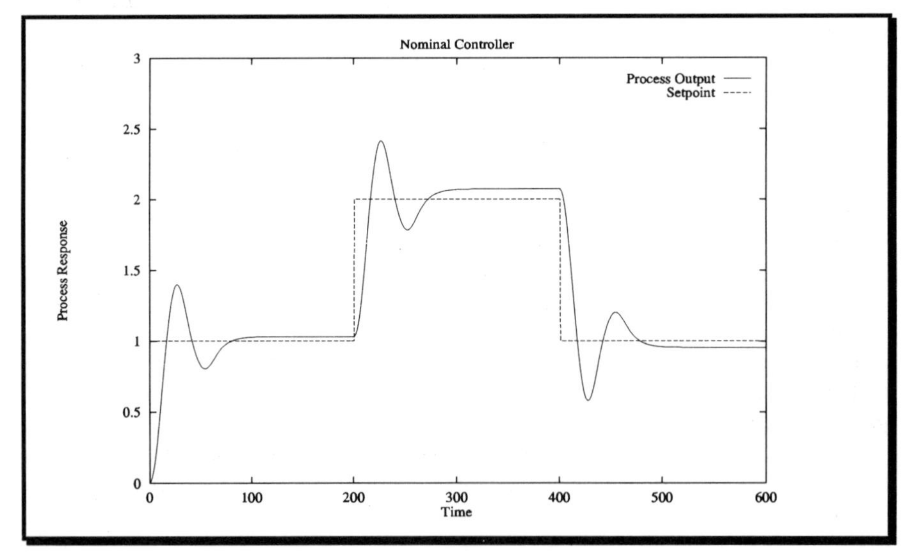

Figure 2.5 — Nominal controller performance.

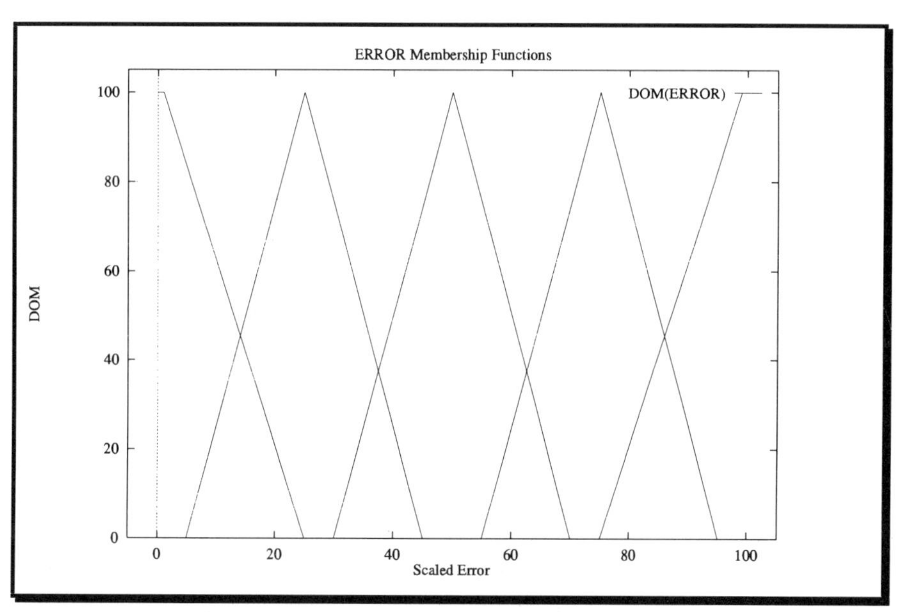

Figure 2.6 — Nominal error membership functions.

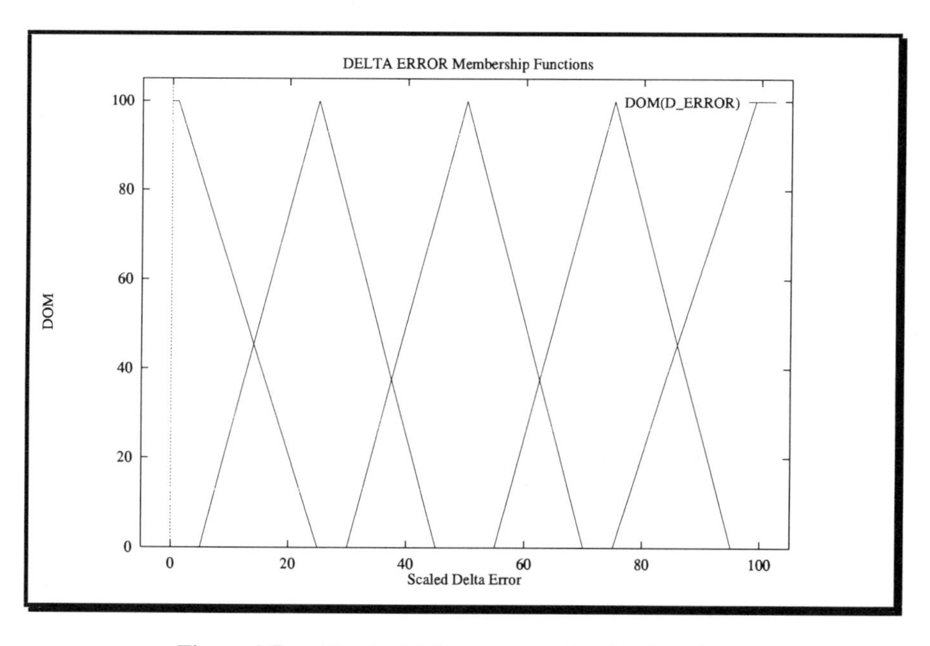

Figure 2.7 — Nominal delta error membership functions.

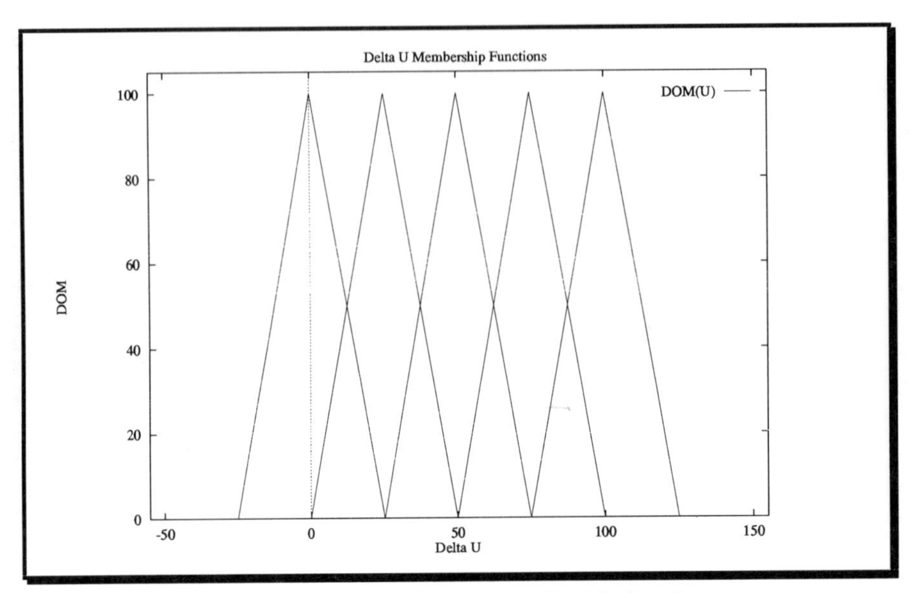

Figure 2.8 — Nominal output membership functions.

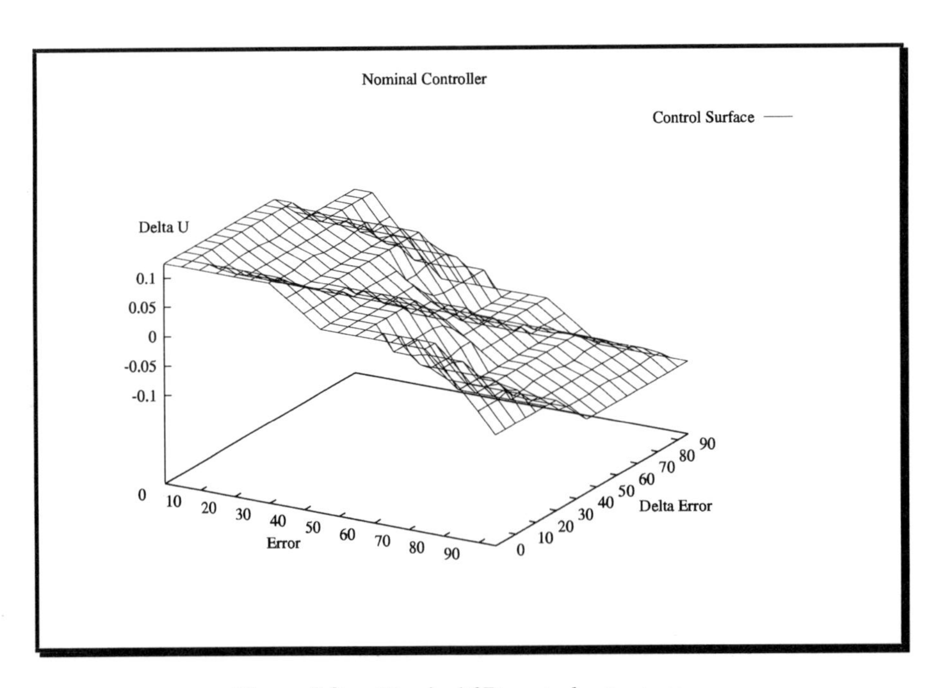

Figure 2.9 — Nominal 3D control output map.

2.6.1.2 Neural Network Modelling

To generate training data for the neural network model, the process model was exercised with a randomly generated piecewise constant input signal. Random white noise was also applied to the process, with a standard deviation of 0.05. At every sample time-step (0.5 seconds in our simulations) the input signal was maintained at its current value with a fixed probability of 0.95; otherwise it was changed to a randomly generated new value between limits of 0 and 2.

One hundred samples (i.e. 50 simulated seconds) of the process input was given as a parallel input vector to a feedforward neural network; thus the neural network was a non-linear finite impulse response filter. The conventional fully connected feedforward network structure was used with two hidden layers of 10 and 5 sigmoidal units each. The output unit — an estimate of the process output one time-step in the future – was linear. Weight values were determined during the learning process with standard backpropagation. The response of the trained network to a test input stimulus is shown in Figure 2.10.

2.6.1.3 Experiments and Results

The primary objective is to determine if the neural network model can be used effectively to synthesise a fuzzy controller for the process. As noted earlier, the IAE was selected as the optimisation criterion (Eq. 2.1). For our experiments, T_f was 600 seconds, and the reference signal or setpoint $y_r(t)$ was fixed at 1.5 for the first 200 seconds, stepped to 2.0 from 200 to 400 seconds, and stepped down to 1.0 from 400 to 600 seconds.

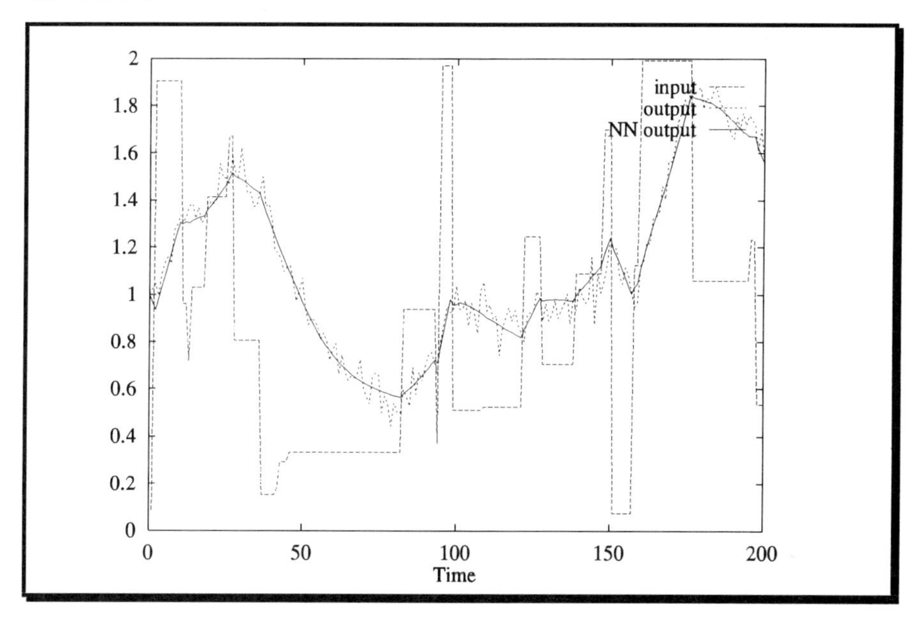

Figure 2.10 — Neural network model response to time-varying input.

We determined appropriate values for GA parameters based on several preliminary experimental runs with the process (as opposed to the neural network) model. All subsequent experiments with the neural network model used these parameter values: a population size of 50, crossover probability of 0.9, and mutation rate of 0.001. For each experiment, we ran the GA for 10000 trials (a trial being the evaluation of one fuzzy controller).

The next step in the synthesis procedure is to select a set of controller parameters to modify. For the initial set of experiments, we modified the centres of membership functions and scaling factors for universes of discourse for input and output parameters. Thus, the initial set of experiments allowed 18 parameters to be modified simultaneously. A second set of experiments extended the set of modifiable parameters to include the slopes of the piecewise linear membership functions as well as centres and scaling factors. This modification expanded the search space to 39 parameters.

To establish a baseline for the synthesis procedure, the process model was used for an initial experiment to verify the performance of the GA for the desired optimisation criterion and set of modifiable controller parameters. The result of the GA search using the process model is shown in Figure 2.11 The process now reaches setpoint for all setpoint changes, and the IAE is reduced from 78.15 for the untuned controller to 59.01 for the optimised controller, which is a reduction of 24%. This experiment established the synthesis procedure using the process model. The next experiment used the neural network model for the process, modifying the same set of controller parameters. Since the neural network model is an approximation of the process model, we would expect improved performance with respect to the nominal set of controller parameters, but not as large an improvement as that observed using the process model in the synthesis procedure.

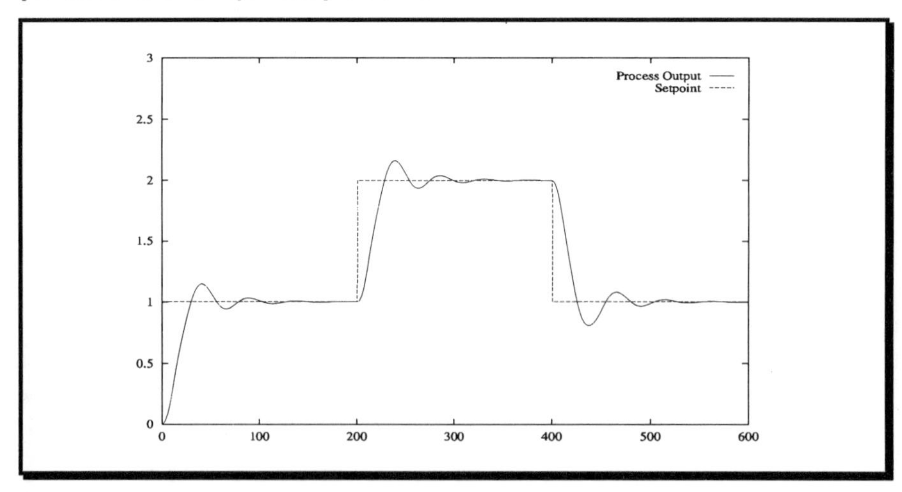

Figure 2.11 — Time response of controller synthesised using process model, modifying centres of membership functions and scaling factors.

Figure 2.12 shows the response of the process model to the set of controller parameters optimised using the neural network model of the process. As expected, the performance is considerably better than the nominal controller, with the process reaching all setpoints. The IAE is reduced from 78.15 for the untuned controller to 64.69 for the optimised controller, which is a 17% reduction.

In the next set of experiments, we allowed the slopes of the piecewise linear membership functions to change, as well as the centres of the membership functions and the scaling factors for the universes of discourse. Once again, the optimal controller synthesis procedure was used on the process model, and the synthesis procedure was then repeated using the neural network model.

The results of the optimisation using the process model are shown in Figure 2.13. Once again, all setpoints are reached. However, the response is much less oscillatory than that obtained using the controller synthesised by modifying a smaller set of controller parameters. The IAE is actually slightly higher than the previously synthesised controller — 59.84 as opposed to 59.01. The decrease in IAE over the untuned controller is 23%. Results of the synthesis procedure using the neural network process model are shown in Figure 2.14. In this case, we do not see the decrease in oscillation, but the IAE does decrease (from 64.69 to 59.25) over that obtained using the neural network process model and modifying only the centres of membership function and scaling factors. The decrease in IAE over the untuned controller is 24%.

The membership functions after optimisation are shown in Figures 2.15, 2.16 and 2.17. Note that in several cases, there is considerable overlap between optimised membership functions. This tends to suggest that we could redesign the controller to use fewer membership functions without a significant effect on controller performance. The 3D map for the optimised controller is shown in Figure 2.18.

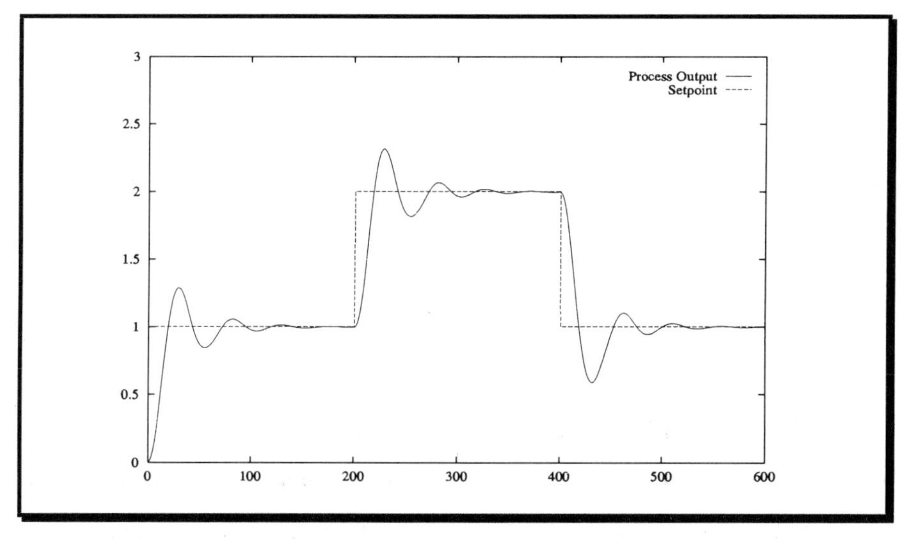

Figure 2.12 — Time response of controller synthesised using neural network model, modifying centres of membership functions and scaling factors.

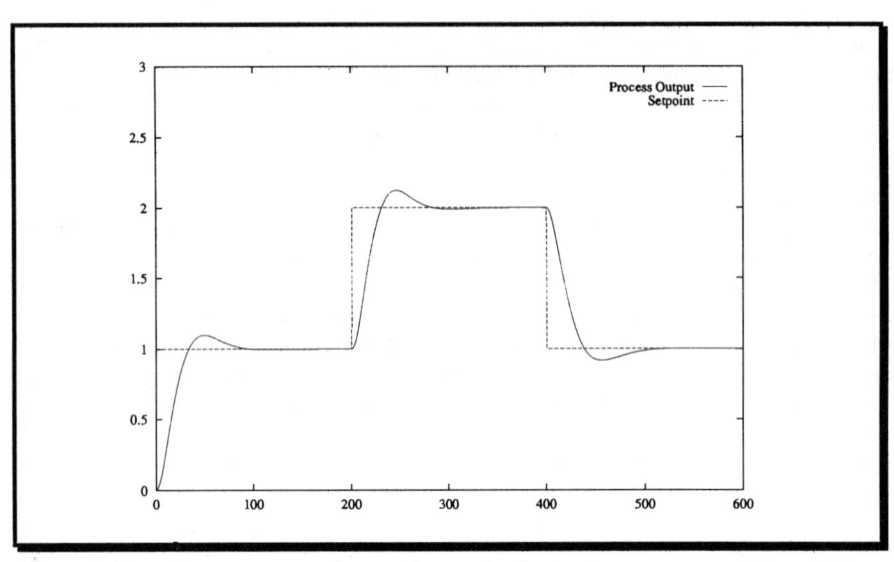

Figure 2.13 — Time response of controller synthesised using process model, modifying centres and slopes of membership functions and scaling factors.

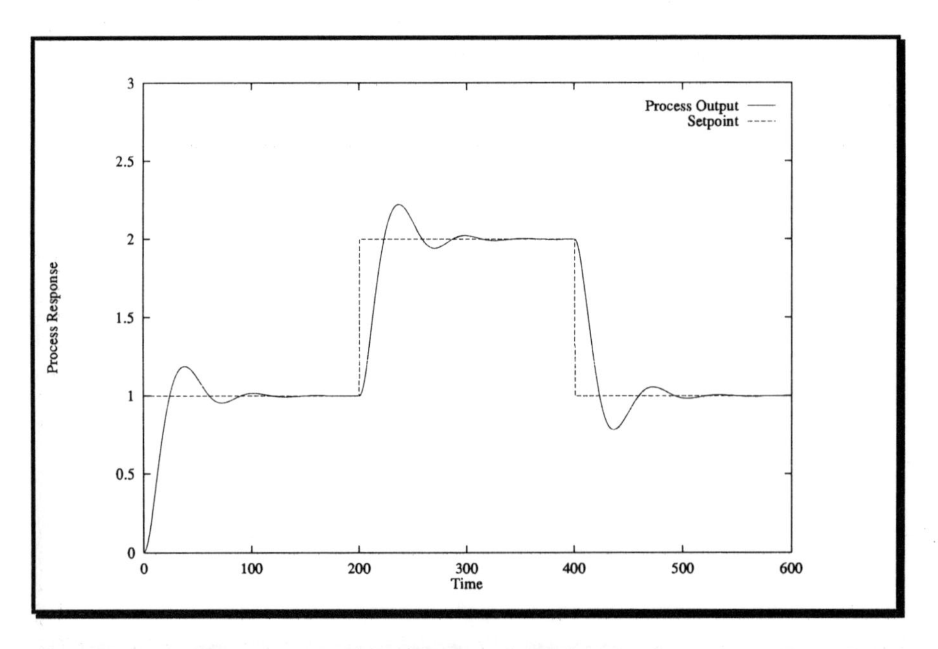

Figure 2.14 — Time response of controller synthesised using process model, modifying centres and slopes of membership functions and scaling factors.

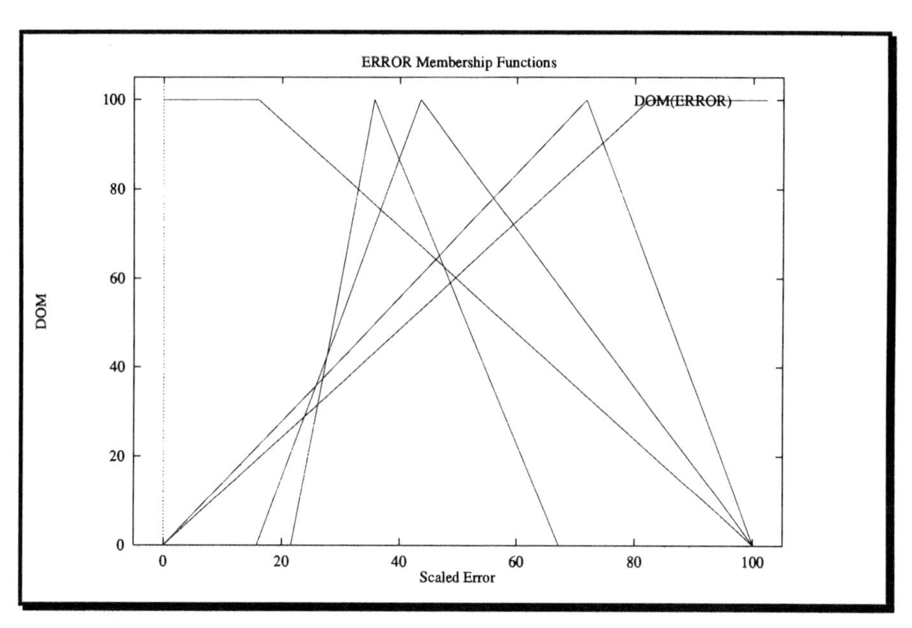

Figure 2.15 — Error membership functions of controller synthesised using neural network model.

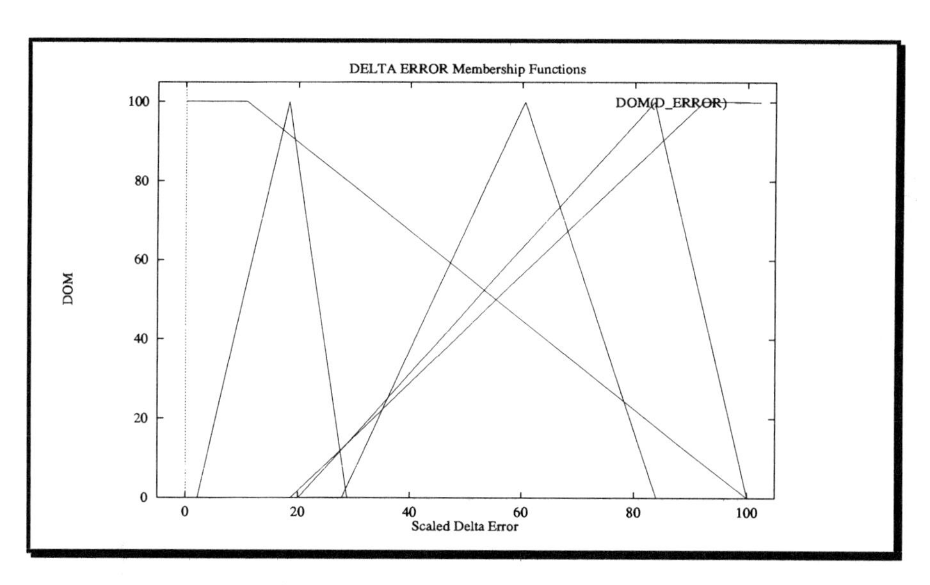

Figure 2.16 — Delta error membership functions of controller synthesised using neural network model.

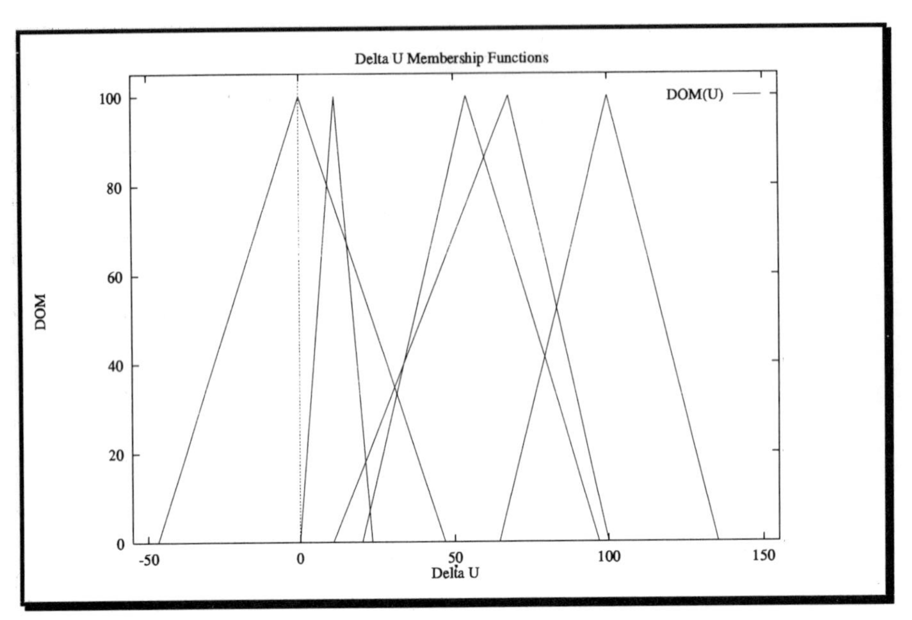

Figure 2.17 — Output membership functions of controller synthesised using neural network model.

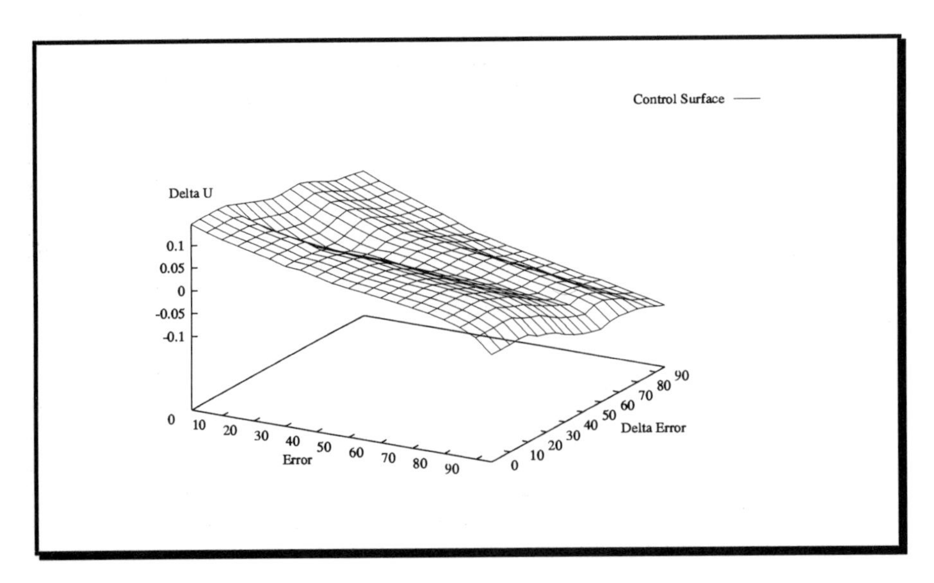

Figure 2.18 — 3D control output map for controller synthesised using neural network model.

2.6.2 Case Study 2: Non-linear CSTR Model

For a second application of our approach, we used a model for a continuous stirred tank reactor (CSTR), a common piece of equipment in chemical plants. Figure 2.19 depicts the particular CSTR we used. Two reactants A and R, with input concentrations A_i and R_i flow into the reactor vessel. A reversible chemical reaction between A and R takes place in the reactor and output concentrations A_o and R_o are produced. The rate of reaction is a function of the input temperature T_i. Furthermore, the reaction is exothermic so that the output temperature T_o is higher than T_i. A three-state system of coupled non-linear differential equations constitutes the CSTR model and is described in [Econ86].

We assume that A_i and R_i are fixed, and the problem of interest is to control T_i to produce a desired R_o. There is in fact an optimum temperature T_{opt} at which R_o reaches a maximum. Because of the non-monotonicity of the R_o/T_i relationship, very different control actions are required for $T_i<T_{opt}$ and $T_i>T_{opt}$. We thus included T_i as a third input to the fuzzy controller: For simplicity, three membership functions, NEGATIVE, ZERO, and POSITIVE were used for the error and error rate inputs, and three membership functions, BELOW_TOPT, NEAR_TOPT, and ABOVE_TOPT were used for T_i. The fuzzy rule base thus consisted of 27 rules, shown in Figure 2.20.

As in the previous example, we restricted the membership functions to be either triangular or trapezoidal. We allowed the genetic algorithm to modify centres of membership functions, slopes, and scaling factors for each input and output. This choice gave the genetic algorithm a total of 35 parameters to modify. The objective function was selected as the integral of the absolute value of error, and the IAE was also normalised using the nominal setpoint of 0.505. Our simulation ran for 4000 seconds, holding the setpoint for R_o at 0.505 for the first 500 seconds, with a step change to a setpoint of 0.502 from 500 seconds to 1800 seconds, and a step back to a setpoint of 0.505 for the duration of the simulation.

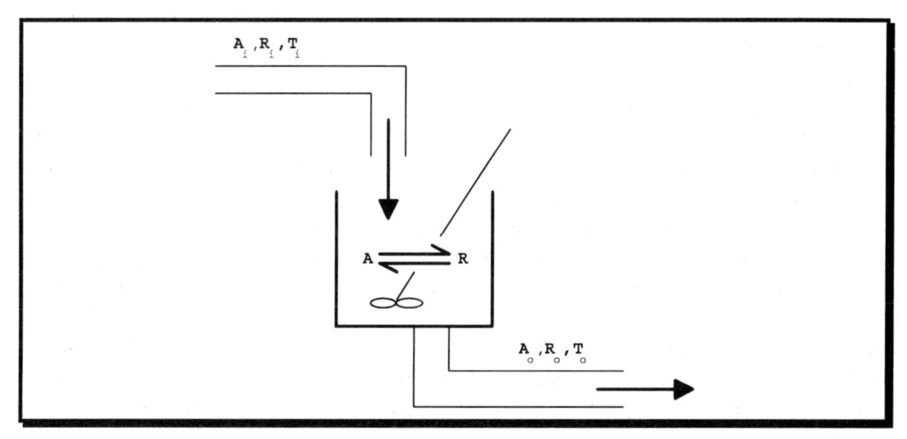

Figure 2.19 — Schematic of continuous stirred tank reactor.

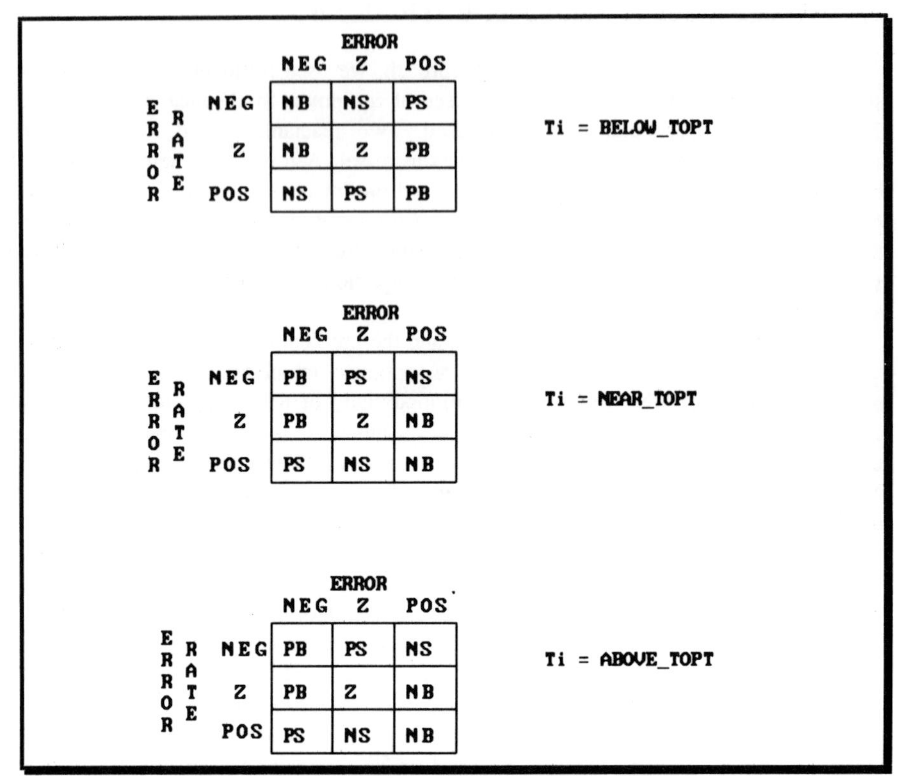

Ti = BELOW_TOPT

ERROR RATE \ ERROR	NEG	Z	POS
NEG	NB	NS	PS
Z	NB	Z	PB
POS	NS	PS	PB

Ti = NEAR_TOPT

ERROR RATE \ ERROR	NEG	Z	POS
NEG	PB	PS	NS
Z	PB	Z	NB
POS	PS	NS	NB

Ti = ABOVE_TOPT

ERROR RATE \ ERROR	NEG	Z	POS
NEG	PB	PS	NS
Z	PB	Z	NB
POS	PS	NS	NB

Figure 2.20 — Fuzzy rules for the CSTR example.

Figure 2.21 shows the performance of the fuzzy controller tuned using the process model in the evaluation function for the genetic algorithm. The resulting controller has a normalised IAE of 0.9476, and we can see that the control has some overshoot, with quick settling after the overshoot, and no steady state offset.

A 150-5-5-1 feedforward neural network was used to model the process dynamics. The training input was again a piecewise constant input stream. Every other input sample was presented to the network, which was trained to predict one time-step ahead. Figure 2.22 shows the prediction accuracy of the trained network.

In Figure 2.23, we see the performance of the fuzzy controller tuned using the neural network model in the evaluation function for the genetic algorithm. This controller has a normalised IAE of 1.1615, 1.33 times greater than the IAE obtained using the process model. However, if we look at the process response, we see that the performance is similar to the controller optimised using the process model. This controller actually has a slightly smaller overshoot, but has a longer settling time. The lengthened settling time is the primary contributor to the increased IAE.

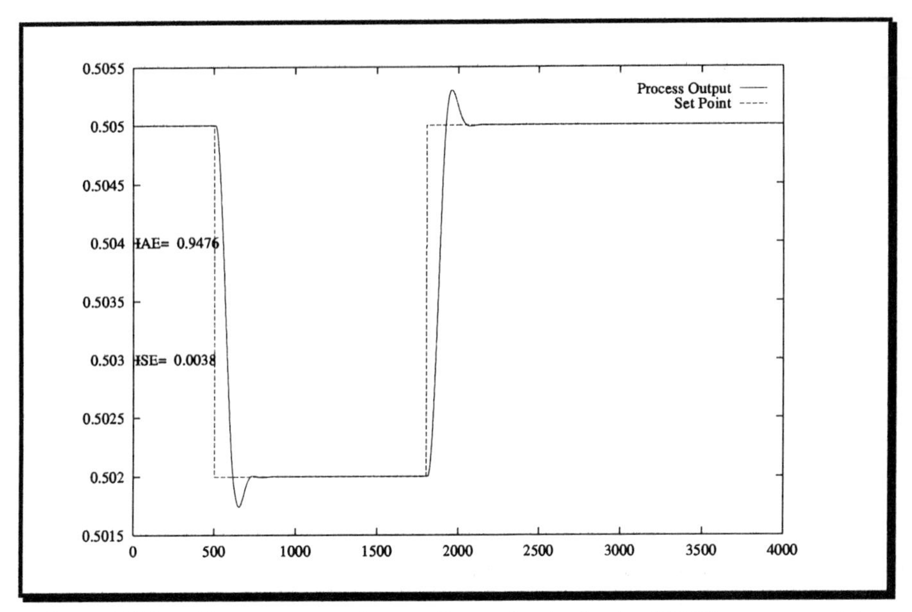

Figure 2.21 — Time response of controller synthesised using CSTR process model.

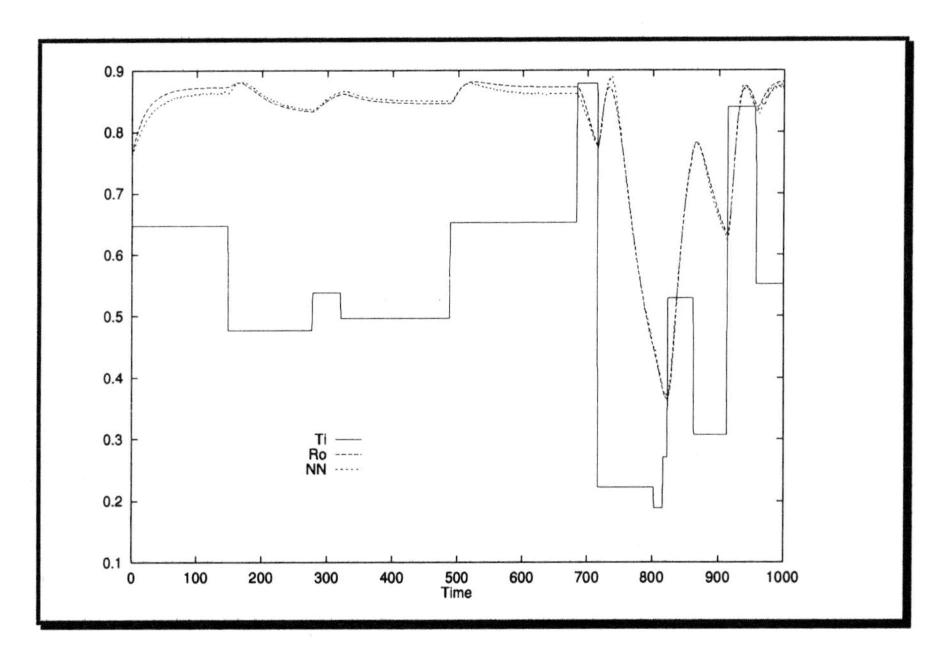

Figure 2.22 — Neural network modelling of CSTR output concentration.

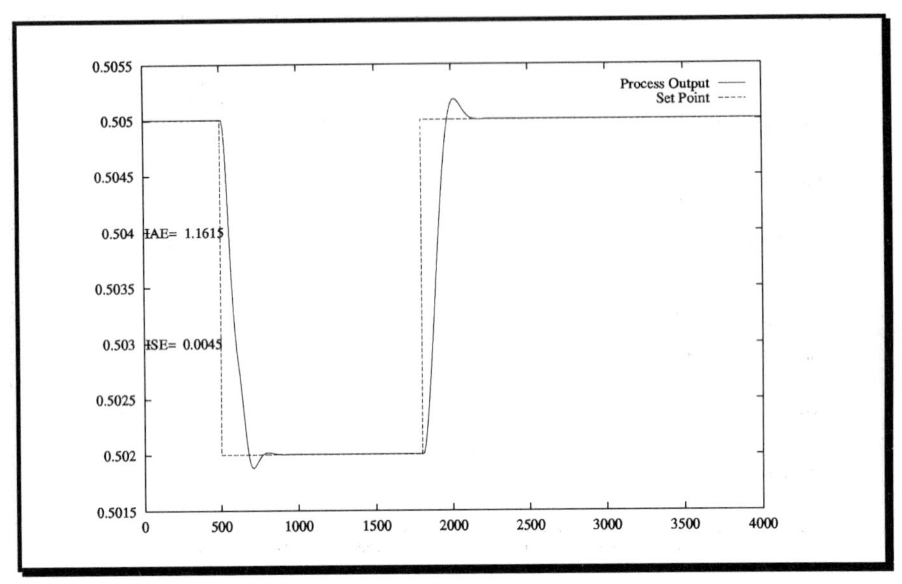

Figure 2.23 — Time response of controller synthesised using neural network model.

2.7 Future Directions

In addition to the ever-present need to validate initial research results on larger, more complex applications, there are three avenues for future work that appear to us to be particularly promising.

The power of genetic algorithms arises in large part from the absence of constraints it imposes on its application: virtually any optimisation problem is fair game.

Our research has demonstrated that this is an effective approach for designing membership functions of a fuzzy controller, but for our initial research we have deliberately excluded other design choices that, from a performance aspect, are no less important. One topic for future research is the extension of our approach to include the fuzzy rules within the design space addressed by the genetic algorithm. The genetic algorithm would simultaneously generate fuzzy rules and select appropriate membership functions.

The motivation for our choice of a genetic algorithm as the optimum derives from the need for a robust non-gradient-based method. There are other algorithms that share this feature with genetic algorithms. In particular, we are interested in exploring the use of chemotaxis [Brem91] for fuzzy controller optimisation. Chemotaxis has one advantage over a conventional GA: it searches a continuous-valued space (however, GA variants are available that do not require discretization of the search dimensions). Another non-gradient-based optimisation algorithm of interest is simulated annealing.

We have stressed the off-line nature of our approach, but it may be feasible to extend it to include an on-line learning and adaptation capability. Processes are seldom truly stationary and, as they drift over time, an initially optimal controller may no longer be even adequate. With a neural network process model, process drift can be tracked and an incremental re-optimisation conducted to ensure that control performance does not degrade. Adaptive control must always be approached with caution — inappropriate adaptation can have catastrophic consequences — but this is an intriguing and important area for future research.

References

[Bart83] Barto, A.G., Sutton, R.S. and Anderson, C. "Neuronlike elements that can solve difficult learning control problems", *IEEE Trans. on Systems, Man, and Cybernetics, 13*, 1983.

[Bere92] Berenji, H. and Khedkar, P. "Learning and Tuning Fuzzy Logic Controllers Through Reinforcements.", *IEEE Trans. on Neural Networks, 3* pp. 724-740, 1992.

[Brem91] Bremermann, H.J. and Anderson, R.W. "How the brain adjusts synapses—maybe.", In *Automated Reasoning: Essays in Honour of Woody Bledsoe*, R.S. Boyer (Ed.), Kluwer, 1991.

[Econ86] Economou, C.G. Morari, M. and Palsson, B.O. "Internal model control. 5. Extension to non-linear systems.", *Ind. Eng. Chem. Process Des. Dev., 25*, pp. 403-411, 1986.

[Fosl93] Foslien, W. and Samad, T. "Fuzzy controller synthesis with neural network process models." *Proc. IEEE International Symposium on Intelligent Control*, Chicago, 1993.

[Gold89] Goldberg, D.E. "*Genetic Algorithms in Search, Optimisation, and Machine Learning*. Addison-Wesley, 1989.

[Gref84] Grefenstette, J.J. "GENESIS: a system for using genetic search procedures." *Proc. of the 1984 Conference on Intelligent Systems and Machines*, pp. 161-165, 1984.

[Holl75] Holland, J.H. "*Adaptation in Natural and Artificial Systems.*", Ann Arbor: The University of Michigan Press, 1975.

[Ishi93] Ishibuchi, H., Fujioka, R. and Tanaka, H. "Neural networks that learn from fuzzy if-then rules.", *IEEE Trans. on Fuzzy Systems, 1*, pp. 85-97, 1993.

[Iwat90] Iwata, T., Machida, K. and Y. Toda, "Fuzzy control using neural network techniques.", *Proc. International Joint Conference on Neural Networks, III*, pp. 365-370, 1990.

[Karr93] Karr, C.L. and Gentry, E.J. "Fuzzy control of pH using genetic algorithms.", *IEEE Trans. on Fuzzy Systems, 1*, pp. 46-53, 1993.

[Khar87] Khargonekar, P., Georgiou, T. and Pascoal, A. "On the robust stabilizability of linear time invariant plants with unstructured uncertainty.", *IEEE Trans. on Automatic Control. AC-32*, 1987.

[Klir92] Klir, G.J. and Folger, T.A. "*Fuzzy Sets, Uncertainy and Information.*", Prentice-Hall, 1992.

[Lee91] Lee, C.-C., "A self-learning rule-based controller employing approximate reasoning and neural net concepts.", *International Journal of Intelligent Systems*, *6*, pp. 71-93, 1991.

[Lin91] Lin, C.-T., and Lee, C.S.G. "Neural-network-based fuzzy logic control and decision system.", *IEEE Trans. on Computers*, *40*, pp. 1320-1336, 1991.

[Mccu92] McCullough, C.L., "An anticipatory fuzzy logic controller utilising neural net prediction. Simulation," 58, pp. 327-332, 1992.

[Mamd74] Mamdani, E.H., "Application of fuzzy algorithms for control of a simple dynamic plant.", Proc. IEEE, Vol. 121, pp. 1585-1588, 1974.

[Mamd75] Mamdani, E.H., and Assilian, S. "A fuzzy logic controller for a dynamic plant.", *International Journal of Man-Machine Studies*, Vol. 7, No. 1, pp. 1-13, 1975.

[Nguy90] Nguyen, D. and Widrow, B. "The truck backer-upper: An example of self-learning in neural networks.", In *Neural Networks for Control*, W.T. Miller, R.S. Sutton, and P.J. Werbos (Eds.). MIT Press, 1990.

[Patr90] Patrikar, A. and Provence, J. "A self-organising controller for dynamic processes using neural networks.", Proc. *International Joint Conference on Neural Networks*, III, pp. 359-364, San Diego, 1990.

[Pool87] Poolla, K. and Ting, T. "Non-linear time-varying controllers for robust stabilisation.", IEEE Trans. on Automatic Control, AC-32, 1987.

[Rume86] Rumelhart, D.E., Hinton, G.E. and Williams, R.J. "Learning internal representations by error propagation.", In *Parallel Distributed Processing*, Vol. 1, D.E. Rumelhart and J.L. McClelland (Eds.). MIT Press, 1986.

[Werb74] Werbos, P.J., "Beyond Regression: New Tools for Prediction and Analysis in the Behavioural Sciences.", Ph. D. Dissertation, Harvard University Committee on Applied Mathematics, Harvard University, 1974.

[Werb90] Werbos, P.J., "Backpropagation through time: what it is and how to do it.", *Proceedings of the IEEE*, 78, pp. 1550-1560, 1990.

[Werb92] Werbos, P.J., "Neurocontrol and fuzzy logic: connections and designs." *International Journal of Approximate Reasoning*, 6, pp. 185-219, 1992.

[Wigg92] Wiggins, R., "Docking a truck: a genetic fuzzy approach.", *AI Expert*, May, 1992.

[Zade65] Zadeh, L. A., "Fuzzy Sets", *Information and Control*, Vol. 8, pp. 338-353, 1965.

3

Replacing the Pattern Matcher of an Expert System with a Neural Network

Henry Tirri

3.1 Introduction

Expert systems are usually designed based on a particular knowledge representation scheme, typically a rule-based, a frame-based or an object-oriented framework. This raises the question how much of the expert system's problem-solving performance is restricted by the limitations of the representation technique chosen. It is evident that many of the fundamental problems in expert system design are due to imperfect "knowledge translation" — part of the domain knowledge is difficult or impossible to express in the format defined by the representation framework.

A traditional expert system performs inferences with symbols, assuming that all the interesting computation occurs at this level. However, in many applications the reasoning with concepts only covers part of the knowledge. Quite often, at least as much of the knowledge is in the correct coding of input information to a symbolic form. However the research on expert systems has focused almost solely on the properties of the inference engine and the related data structures [Jack90]. In spite of the valid critic of symbolic expert system technology (see for example the very recent book by Kelly [Kell93]) there has been only isolated attempts to address questions related to the coding process (see e.g., MACIE in [Gall88]). Implementing semantic questions such as "Is the traffic pattern showing congestion?" or "Is the concentration of sulphuric acid in

Intelligent Hybrid Systems, Edited by S. Goonatilake and S. Khebbal.

the process within tolerance bounds?" require intricate pattern matching mechanisms, which in fact may require more complex calculations than the symbolic reasoning task which uses that information. Implementing such pattern matching modules imposes a new challenge to expert system technology: how to compress complex high-dimensional sensory data into symbolic form. If such a translation can be achieved, this symbolic information can then be used by the inference engine, not unlike the way in which information received by human expert's sensory systems can be utilised at higher cognitive levels.

In spite of the apparent maturity of the symbolic level expert system technology, one can identify several areas where programs have had little success of showing similar performance to human experts [Wate86, Kell93]. Many of these areas are studied intensively in the artificial intelligence literature, but here we will focus on two problems related to the general theme of building hybrid AI systems: allowing direct use of complex sensory input by implementing rule predicates with neural networks, and attribute selection by connectionist learning, i.e., how to "ground" symbols to the statistical information available to the application.

3.2 Overview of the Neural Networks Used

Recent years have seen rapid growth in neural computing research and many monographs and text books have been published on the topic. As it is outside the scope of this chapter to review all the various theoretical models and their properties appearing in the literature, we suggest for an uninitiated reader to study, for example, the two-volume set by the PDP-group [Rume86], the monograph by Kohonen [Koho88], or the text book by Hertz, Krogh and Palmer [Hert91]. Here we will only briefly introduce the concepts that are necessary to understand the neural network approach for implementing pattern recogniser predicates and attribute selection.

Formally one can define a neural network *N* as the dynamical system which has a topology of a directed graph and which carries out information processing by means of its state response to (continuous) input [Hech87a]. One class of neural networks suitable for the translation of sensory data into symbols are networks that directly approximate the target function $g: S \subseteq R^n \longrightarrow S' \subseteq R^m$ after self-adjustment in response to a finite descriptive set of example mapping pairs $(i_1, o_1), \ldots, (i_k, o_k)$ (where $o_j = f(i_j) + \varphi$, φ is a stationary noise process). Such layered networks are discussed for example in [Hech87b], [Pogg90] and [Werb88].

The nodes in the networks are simple computational elements: a node sums *k* weighted inputs and passes the result through a non-linearity *f*. Much of the interest in such feedforward networks is due to the observation that they can be used as pattern classifiers (see e.g., [Duda73]) in the *d*-dimensional feature space defined by the network inputs [Rume86, Lipp87a]. The decision boundary in the feature space is determined by the state of the network defined by the weight matrix *W*, activation function fa and the threshold value vector Θ. Although in principle the position of this decision boundary in the feature space could be "programmed" directly by setting arc weight values, the values of the weights are difficult to determine *a priori*. Hence the correct positioning of the boundary is approximated by a training process, where a set of examples of input instances $(\mathbf{i}_1,...,\mathbf{i}_k)$ and a set of correct classifications $(\mathbf{o}_1,...,\mathbf{o}_k)$ are presented to the network, and an algorithm called a *learning rule* is used to calculate weight changes depending on network's performance with the current weights. The network input also includes the correct classifications, hence this type of training is called supervised learning [Duda73]. Finally, the complexity of the shape of the boundary that can be realised is dependent on the number of layers [Lipp87a]. However, for our purposes it is enough to know that any realisable shape can be produced by a three-layer network of the above elements — at least in principle.

For the above layered networks many learning algorithms have been suggested [Lipp87a, Werb88, Pogg90], for example backpropagation and its variants. In our case the details of the learning algorithms are not important, it is sufficient to know that, for example, backpropagation is an iterative gradient descent algorithm designed to minimise the mean square error between the actual and desired output of the multilayer network. In the case study discussed below the networks used were variants of the so-called radial basis functions [Pogg90] in which the function approximation is achieved by a mixture of units with Gaussian activation functions.

If a neural network N solves a two-class classification problem, for example the question whether an input $\mathbf{I}$ is regular/irregular, in our knowledge representation formalism the neural network corresponds to a single data predicate P_I. However, in many cases a single network is capable of solving m-class problems and hence implements a set of mutually excluding predicates $P_I = \{P1_I,...,Pm_I\}$.

For attribute selection unsupervised networks can be used, in particular the self-organising maps (SOM) by Kohonen. In SOM networks the learning algorithm generates a mapping of a higher dimensional input space S onto discrete lattice M of output nodes. The map is generated by establishing a correspondence between the inputs in S and output nodes in M, such that the topological neighbourhood relationships among the input instances are reflected as closely as possible in the arrangement of the corresponding nodes in the lattice. As a result of this process, a non-linearly reduced two-dimensional version of the input space is found. This data structure can be used to cluster input attributes.

This correspondence is obtained as follows. Each input instance is represented by a vector $s \in S$. For each training cycle an input instance $s \in S$ is chosen randomly according to a probability distribution $Pr(s)$. Each location $m \in M$ has an associated vector $\mathbf{w}_m \in S$. These vectors $\mathbf{w}_m$ map lattice locations m to points in the space S. For each training cycle the mapping is modified according to the following abstract algorithm:

[A1] Determine lattice location c for which $\|\mathbf{w}_c - \mathbf{s}\| = \min_{m \in M}\|\mathbf{w}_m - \mathbf{s}\|$ where s is the input chosen for the training cycle.

[A2] For all nodes m in the neighbourhood of c modify $\mathbf{w}_m(t+1) = \mathbf{w}_m(t) + \alpha\delta_{mc}(\mathbf{s}, \mathbf{w}_m(t))$. Here $0 \le \delta_{mc} \le 1$ is the adjustment function for the distance $\|\mathbf{w} - \mathbf{s}\|$ and α is the learning step size.

By decreasing the step size α and the width of δ_{mc} slowly during training, the algorithm gradually yields values for the vectors w_m which define a discretized neighbourhood conserving mapping between lattice nodes m and points of the input space S [Koho88].

3.3 Pattern Matching in Expert Systems

For pragmatic reasons in the following we will restrict ourselves to represent symbolic knowledge in the form of rules [Wate86]. The rule-based representation is still the most common means to give rigorous formal expressions of knowledge about the problem domain. However, applying the ideas presented below is by no means restricted only to rule-based systems.

Rule-based expert systems manipulate symbols that represent ideas and concepts. However, in many application areas complex sensory data such as images, auditory data (speech), spectral data etc., have to be first transformed into symbols manipulated by the inference mechanism. In pure rule-based systems this translation process inevitably loses information, and depending on the application, that information may be crucial to the successful operation of the overall system.

To illustrate the problems involved let us suppose that we aim at building a circuit board verification expert system. One part of the system performs non-destructive testing on components to identify pieces that may lead to malfunctions caused by thermal dissipation. In our simple example case we assume that the test can be based on two parameters: thermal dissipation T and the location $L = (L_1,L_2)$ of the component in question. Here the location L is expressed as a point on a finite plane. Due to the electrical rules and other design factors, the tolerance for thermal dissipation is a function of the location. Potential malfunctions can be recognised by checking the actual measurements t and $l = (l_1,l_2)$ against the "thermal dissipation tolerance profile" of the design.

Assume that Figure 3.1 represents the relationship between the tolerance for thermal dissipation and location L for a particular circuit board.

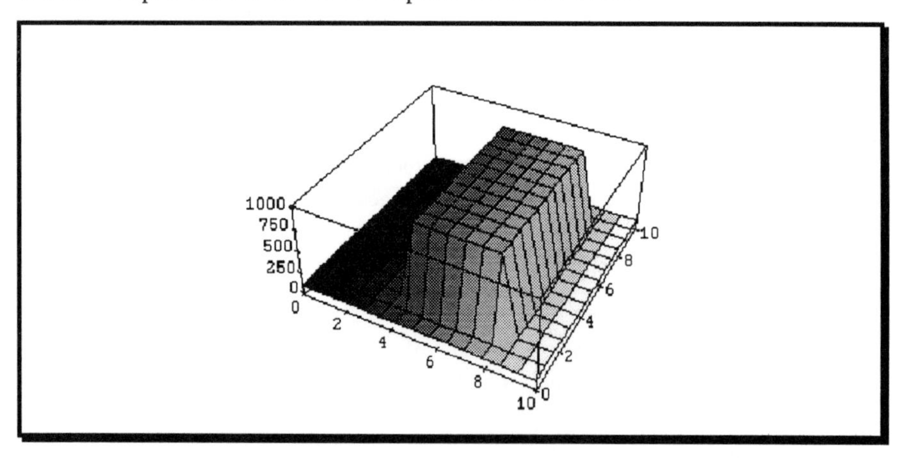

Figure 3.1 — A thermal dissipation profile that is easily describable as a collection of rules.

In this case the predicate of "acceptable thermal dissipation" can easily be described with a rule, i.e.,

IF $4 < l_1 < 8$ AND $1 < l_2 < 9$ AND $t < 1000$ OR (($0 < l_1 < 4$ OR $8 < l_1 < 10$) AND ($0 < l_2 < 1$ OR $9 < l_2 < 10$) AND $t < 100$)

THEN th-dissipation = acceptable

In this case we have a compact description of the predicate as the above expression can be coded as a simple rule. However, is this a realistic situation?

Very probably the thermal dissipation profile is more like the surface in Figure 3.2 if modelled with real world data. For this profile there is no simple set of rules that describes the predicate, in fact if one tries to describe it by rules the description will be approximately as long as giving the list of locations with acceptable dissipation values. Things get even worse if we add more dimensions to our example, since we need to construct high dimensional models to understand the interrelationships that define our predicate. In many realistic domains this type of "translation" from high-dimensional inputs to a discrete symbolic representation involves predicates that need to be implemented with pattern recognisers such as neural networks. As we will see, for this type of a translation, i.e. recognition of whether the "pattern condition predicate" is true or not, the predicate can be modelled as a characteristic mapping describing whether or not an input is an instance of the set defined by the predicate. For real concepts, an exact description of such a function can be intractably complex in practice, however it can be well approximated with a function realised as a neural network.

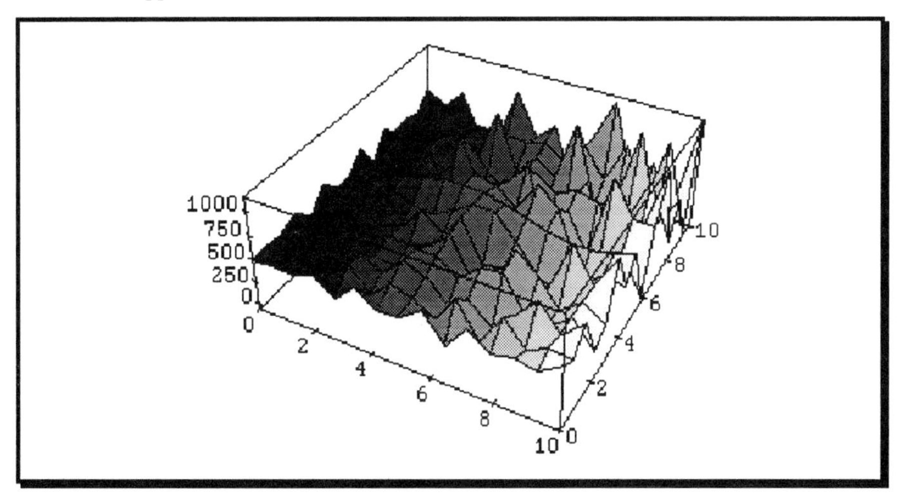

Figure 3.2 — A realistic thermal dissipation profile that is *not* easily describable as a collection of rules.

To illustrate that the above example is not an artificial construction, let us study some typical rules from existing expert systems.

SPE [Weis84] is a typical example of an interpretive expert system whose purpose is to infer situation descriptions from sensory data, in this case analyse waveforms from a scanning densitometer to distinguish between different causes of inflammatory conditions in medical patients.

IF the tracing pattern is asymmetric gamma and the gamma quantity is normal

THEN the concentration of the gammaglobulin is in normal range.

The example rule from TATR [Call84] is an example of rules in planning expert systems.

IF the airfield does not have an exposed aircraft and the number of aircraft in the open at the airfield is greater than 0.25 times the total number of aircraft at that airfield

THEN let excellent be the rating for aircraft at that airfield

REACTOR [Nels82] falls into the category of monitoring systems that compare actual system behaviour to expected behaviour and perform control actions based on the differences observed.

IF the heat transfer from the primary coolant system to the secondary coolant system is inadequate and the feedwater flow is low

THEN the accident is loss of feedwater.

It should be evident that for all of the example rules from these different types of systems there exists a need for translation from sensory data to a symbolic representation, and the correctness of this translation is essential to the viability of the rule.

Although all the above notions "tracing pattern is asymmetric gamma", "does not have exposed aircraft" and "heat transfer ... coolant system is inadequate" have a precise interpretation, problems arise when one attempts to give a formal definition to describe the condition. The situation is further complicated by the fact that the sensory data used as input is of poor quality: noisy, sparse or incomplete, or in the worst case, all of them. In the literature there exist many studies on fuzzy logic [Zade83] for approximate reasoning strategies which aim at a good estimate for uncertain data and imperfect rules. Instead of describing the certainty factor of the truth value of a rule R, our approach is much more pragmatic, we simply implement a reliable detector for the pattern matching predicate P in the condition part of the rule. This approach is based on the observation that very seldom the rules themselves are fuzzy, but in many cases the pattern matchers for the condition predicates are hard to describe with logical primitives. Thus the problem is how to preserve the relevant information when changing information into a symbolic form rather than the impreciseness of the rules themselves.

3.4 The Benefits Of Hybrid Systems

As discussed above, one possible approach to this "translation" from sensory information into symbols is to use neural networks. These neural networks act as computing modules that perform noise-tolerant pattern recognition of input information by providing a realisation of the approximate function needed for the pattern recognition. Although we adopt a purely engineering approach to expert system design (as opposed to for example the competence modelling approach of [Kera86]), and do not try to mimic cognitive behaviour of human experts, we still find an interesting starting point in studying areas where human expertise is clearly superior to the capabilities of modern expert systems.

Why not use only neural networks and omit the symbolic level totally? During the last few years there has been an intense debate between the proponents of connectionism and symbolic artificial intelligence, on the relative strengths and weaknesses of both worlds [Pink88]. Although this debate still continues, it seems to be commonly accepted that both the symbolic and the subsymbolic approaches (i.e., neural networks) has its advantages. We have already discussed many of the problems when purely symbolic representations are used. The other extreme of using only neural networks also suffers from serious disadvantages.

First, if the constraints are well known and can be described by rules there should not be a need to go through a lengthy training process with a neural network, only to achieve an approximation of the function already known. Second, in spite of various attempts no one has been able to find a general mechanism to incorporate an explanation facility to neural networks (for partial success see [Gall93]), which leaves their use at the "black box" level — unacceptable in many application domains. Finally, once reasoning at the symbolic level can be applied, there is no substitute in the efficiency of manipulating discrete symbols. (Thus replacing the rules using the pattern recognition predicates would in most cases result in a more complex and difficult to maintain implementation).

Our use of neural computing principles differs considerably from the usual studies of neural networks as expert systems see e.g., [Beck87, Boun88, Dutt88, Gall88, Gall93] which concentrate solely on acquiring knowledge by learning processes. Our approach to expert system design is a hybrid one, and we describe the architectural framework to integrate neural networks and the higher levels of reasoning, which are always described within the rule-based paradigm.

3.5 The Architecture of an Expert System with Pattern Matcher Predicates

In our case the knowledge base consists of three (sub)knowledge bases: a rule base R_b, a fact base F_b and a neural network base N_b. Following the common expert system terminology a rule is understood as a condition-action-statement rule i:

IF C_i THEN A_i ;

with the obvious semantics (if the condition part C_i is evaluated to be true, the action part A_i will be performed). A condition C_i is an expression containing one or more predicates P_{ik}. For our purposes it is sufficient to make a distinction between a pattern matcher (a detector) and a non pattern matcher predicate. The former predicates (denoted by P_d) are implemented by neural networks in N_b.

An action A_i is defined as a set of operations $\{o_{ij}\}$, each operation being either

- internal, i.e. it modifies the fact base F_b and/or internal variables,
- external, i.e. call to an external procedure (e.g. for an alarm signal) or
- adaptive, i.e. call to a neural network in N_b in training mode.

In this architecture internal operations allow the storing of deduced knowledge for further use and external operations provide the interface to the environment where the expert system is functioning. These two types of operations are usually found in all monitoring, planning or interpretive expert systems, where the software is embedded in a larger system. The adaptive operations, which are related to the adaptiveness of the pattern matcher predicates, will be discussed in more detail in the context of the dynamic behaviour of this expert system architecture.

3.5.1 Fact Base

The fact base F_b consists of facts stating that a particular predicate P holds for certain objects s_i in the object domain S of the expert system. In addition to the normal facts, predicates appearing in the fact base F_b can also be detector predicates. In this case the detector predicate on current input **I** has already been evaluated and stored. Hence a fact base acts as the "memory" of the expert system. In addition to the static part that states the universal facts about the problem domain, it also contains dynamically changing knowledge about a particular execution that can be erased later on.

3.5.2 Neural Network Base

The neural network base N_b contains a set of neural networks $\{N_i\}$, each of which corresponds to one or more pattern matcher predicates. Analogously to the rules in the rule base, neural networks are active components of the system. They perform a classification operation that implements a predicate test on their input data. They also keep the latest data stored in an associated buffer. This process is called knowledge translation, as the statistical relationships within the input data are translated into symbolic facts. In this sense N_b is analogous to the rule base R_b; the latter contains the knowledge for inference with discrete symbols at the symbolic level, the former the knowledge for reasoning at the subsymbolic, pattern data level.

3.5.3 Dynamic Behaviour of the System

The dynamic behaviour of the hybrid system is illustrated in Figure 3.3. The inference engine performs the normal backward/forward-chaining of rules. However, the inference engine may encounter a rule which has detector predicates $P_{d1}(l,x)$, $P_{d2}(j,y)$,... in its condition part, and if it cannot decide the value of a predicate from the facts in fact base F_b, it performs a call for the corresponding neural network(s) with predicate arguments as parameters to the network. It should be observed that these parameters are not the actual input values to the neural network N in question. They define what classification result (l) is significant to the condition (since a multi-class classifier corresponds to a set of predicates $\{P_{di}\}$) and the address of the input device (x). The actual input values the neural network receives come from the sensory data equipment directly, and the arrival of the parameters acts only as a trigger to the classification process.

In its simplest form the network N returns a boolean predicate value. If the predicate is true, the action part A of the rule in question has an operation o_i that stores the corresponding fact in the fact base. If it is evaluated false, the system automatically stores the negative fact. This is necessary in order to prevent the subsequent encounters with the predicate in some other condition parts from invoking the knowledge translation process again, i.e., the system "memorises" the fact resulting from the application of the pattern predicate. Naturally this stored information is query-dependent, and is removed after the query is completed. If the detector predicates use graded values, the output value itself is stored in the fact base. This means that a graded predicate always "succeeds".

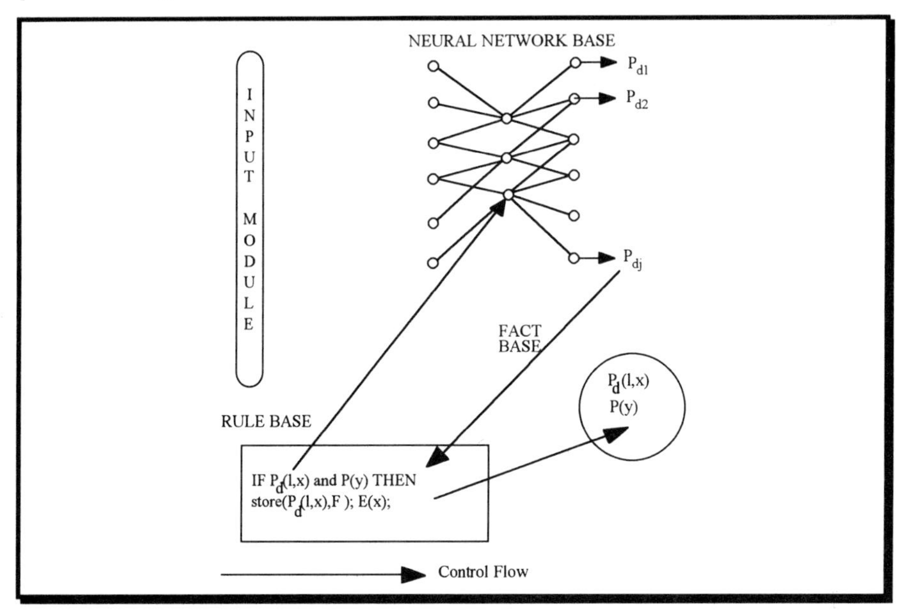

Figure 3.3 — The execution of a rule with a boolean detector predicate.

3.5.4 Adaptation to Refine the Pattern Matcher Predicates

Layered neural networks, such as the ones we are using, often require an extensive training period, with a sample set adjusting the weights to reflect a good approximation of the decision boundary in the feature space. Obviously the larger and statistically more representative this training sample set is of the whole input space S, the better the accuracy of the approximation is. Unfortunately in most cases it is not possible to gather enough data samples in advance to create a truly representative set. Therefore the system described above must be prepared to continue adjusting the network weight matrix while already functioning by using the adaptive operations o_a in the action parts. If during the inference process there is substantial evidence of the fact $P_d(l,x)$ in the fact base being incorrect, deduced either automatically or by human intervention, an adaptive operation is performed.

The adaptive operation is implemented as a call to the corresponding neural network N with the correct classification and the request to train so as to perform this classification. Observe that we required our N modules to be able to store the latest input data whose classification was triggered. This makes the data transparent to the symbolic reasoning module, i.e., it does not have to deal with the actual data at all. This adaptation mechanism gives a way to gradually improve the accuracy of the detector predicate implementations with real input data.

This training process need not necessarily be slow, the speed is dependent on the network type in question. In particular one of the benefits of the RBF-networks is the existence of fast learning algorithms, i.e., usually an addition of a new node and few iterations to relocate by means of the Gaussian bases is sufficient [Pogg90].

3.5.5 Self-organisation for Attribute Selection

In the above discussion it has been assumed that the relevant data attributes on which the pattern matcher predicates can be based have already been identified. In practice one of the most difficult issues in the design of an expert system is the question of attribute selection for knowledge representation. As any real world process has infinitely many attributes, the real problem is how to choose such a small attribute set for the knowledge base so that it would be descriptive enough for the proper functioning of the system. This problem is present especially when machine learning methods (either neural or symbolic, decision tree based ones such as ID3 [Quin79]) are used for knowledge acquisition.

According to results in neurobiology, one important organising principle of sensory pathways in the brain is that the placement of neurons is orderly and often reflects some characteristic of the external stimulus being sensed [Kand85]. Inspired by this biological fact some of the neural network models and their associated learning algorithms promote self-organisation [Koho88, Gros88]. Since the Kohonen networks can also be used directly as classifiers, they are especially suitable for our purposes.

We now turn to the problem of using the self-organisation process for attribute selection. Let us assume that our input space S is d-dimensional, i.e., each input instance is a vector $\mathbf{v} = (v_1, v_2, \ldots, v_d)$. Let T be the training set, i.e., a set of such vectors. Further assume that the output nodes M are arranged as a grid (size k^2).

In the training process an input instance enforces the sensitivity of the most responsive node c (closest in d-space) and the nodes in its immediate neighbourhood defined by δ_{mc}, hence the resulting network has a tendency to form clusters of nodes that are sensitive to similar inputs. After the completion of the training process, each cluster C_i is labelled with a meaningful attribute name B_i semi-automatically by finding an example set of vectors T_i from the training set such that the nodes in C_i are sensitive to these input instances. This example vector set T_i helps giving a meaningful interpretation for the clusters C_i. The convergence for this process has an effect on the quality of the clustering — too fast a decrement of the step size will result in a poor solution. The resulting clustering resembles multivariate methods such as factor analysis, but differs from it due to the non-linearities used in the process.

As the output of the cluster nodes c_i is graded, these attributes could directly be used in relational expressions to form detector predicates P_{di} discussed in the previous sections. For example, "sensor object s is a submarine" if detector predicate *is-sub(s)* holds, where *is-sub(i)* = SUB > .7 (SUB is an attribute defined by the clustering process). However, better results are obtained if this self-organisation process is used as a pre-processing step for a more sophisticated classifier. The clustering and labelling process gives the number and type of the classes after which a neural implementation of a nearest-neighbour method such as LVQ [Koho88] can be used to tune the classifier with the same training set T. This approach also provides an alternate neural implementation of a set of detector predicates to those based on layered networks discussed above, and is more viable in the cases where the structure of the expert system input space is not well-understood in advance.

3.6 Example: Expert System for Automatic Inspection

The general ideas of pattern matcher predicates discussed above can be demonstrated by applying them to the design of an expert system for automatic inspection of circuit packs. The overall system performs several tests on the produced circuit pack including possible wiring problems, chip connectivity and chip orientation. Here our focus is the last one of the sub-tasks: chip orientation. We will only focus on showing the benefits of the overall hybrid approach in this particular application, readers interested in the details of the problem and a comparison of the different solution methods adopted should consult [Morr90].

Computer vision systems are becoming more and more important in assuring the quality of manufacturing processes by allowing low cost, reliable inspection. A computer vision system placed on-line after the placement operation can catch errors before the soldering process, thereby also reducing the repair cost. One special problem that arises in electronic assembly is the component orientation error. In many cases hundreds of components are placed on a single circuit pack, but the components must be loaded into their hoppers manually and the symmetry of the component allows orientation errors. An easy solution to the problem would be to force a standard orientation mark on all chips. However, presently there is no standardisation of the orientation marks (some of them are notches, some of them dots etc.) and even if such marks are used, they are often very hard to detect with a computer vision system because of the poor contrast. Consequently the only starting point for the orientation detection is the information printed on the electronic component, as each of them has at least type information printed on the surface.

To solve this particular detection problem in the inspection expert system, the system must have a rule with a pattern matcher:

IF orient($chip_i$) = DESIGN_DB(orient($chip_i$))
THEN check-pins($chip_i$)

This rule checks the component orientation before initiating the inspection of component pins. Here the value DESIGN_DB(orient(chip$_i$)) is the orientation value stored in the design database. In principle one could use backward-chaining to solve the value of orient(chip$_i$) and then compare it to the value accessed from the design database. In the traditional approach one would try to use a set of rules that detect the orientation based on the features extracted from the image produced by the machine vision system. However, the viability of this solution is questionable if we consider more closely some of the key requirements for this text orientation detection problem:

- There is no advance knowledge of font style or size.
- As opposed to optical character recognition there is no opportunity to use contextual information (dictionaries) to resolve difficult-to-detect characters.
- The printing is often poor quality, e.g., characters are touching, misaligned and may contain non-character symbols.
- Many characters are invariant or almost invariant to a 180 degree rotation (or when rotated resemble some other character without rotation). Hence the system must be prepared to output also an "indeterminate" response.
- Detection must be carried out very quickly (typically up to 50 characters/second).

Our solution to this problem is to implement a detector predicate orient() as a RBF network, which can then be called when executing the corresponding rule. In fact in this particular application an alternative implementation for the detector predicate was based on the LVQ-method described above, and these two approaches were compared.

The feedforward neural network model used in this application is a variant of RBF-networks [Pogg90]. These networks structurally resemble the ones used with gradient descent based learning methods such as backpropagation. However, it should be pointed out that with this variant of RBF neural networks there is no iterative learning process. Computation in these RBF networks are based on a set of reference vectors $(up_1, up_2, \ldots, up_s)$ and $(do_1, do_2, \ldots, do_s)$ (numeric vectors representing symbols in normal orientation and rotated, respectively), and the weights are set only once when the reference vector set is stored. Thus each of the nodes in the middle layer represents a particular example in the training set. Adaptation is achieved by incrementally adding new nodes if new reference vectors become available. One should also notice that the effect of adding new reference vectors to the computed function is only gradual. Thus the undesirable interference properties frequently observed with many of the neural network learning algorithms can be avoided.

Figure 3.4 shows a discrete RBF network within which the first layer consists of "pattern nodes" which compute Hamming distances $h(x,up_i)$ and $h(x,do_i)$. The second layer consists of "summation nodes" which form the conditional probabilities as weighted sums. Finally the "output units" give the final decision by thresholding. In this particular case computation of the conditional probabilities in the discrete RBF networks is based on calculating Hamming distances, therefore the RBF neural network model can also be understood as a generalisation of the "Hamming Network" [Stei61]. It is well known [Lipp87b] that under the assumption of independent bit errors, the optimal minimum error classifier calculates the Hamming distance to a single reference instance in each concept and selects the vector with the minimum distance as the decision. Our RBF variant can be viewed as a natural generalisation of the idea. Rather than taking the nearest reference instance, the likelihood ratio from families of reference instances is computed.

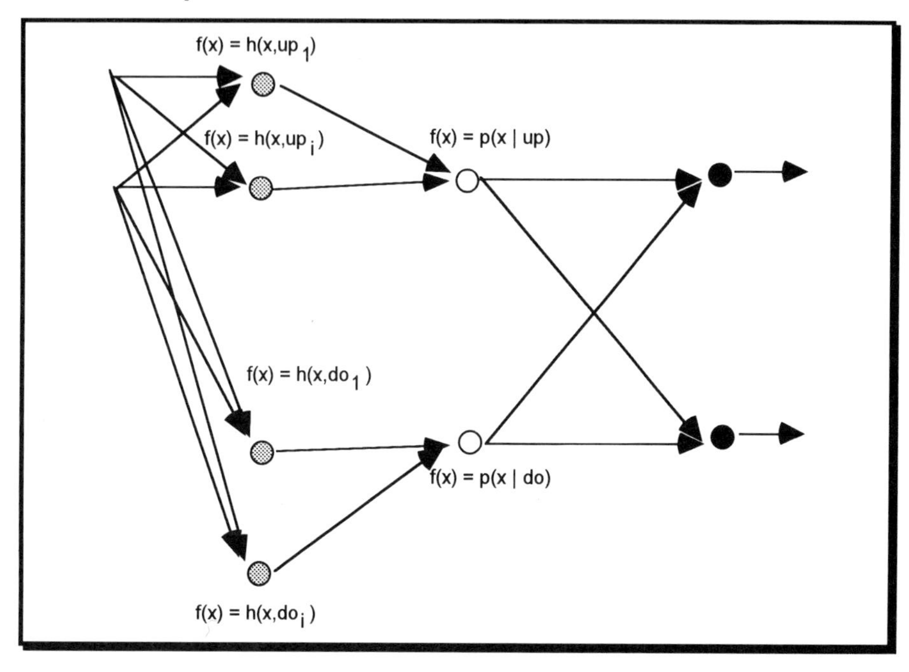

Figure 3.4 — The RBF neural network implementing the pattern matcher predicate. The gray nodes are pattern nodes, the white summation nodes and the black ones threshold nodes. The activation functions are given above the nodes.

This case study illustrates a typical case where neural networks can be successfully used to implement the translation from stochastic data to symbols manipulated by the inference engines. The most obvious application areas for this pattern matcher predicate approach are expert systems with direct use of complex sensory input. However, the ideas presented are in fact much more general. We have already mentioned the notion of self-organisation as a means to determine properties of data that reflect meaningful statistical relationships in the expert system input space. We believe that such mechanisms will play an increasingly important role in several areas including robotics control, image processing and autonomous vehicle navigation.

References

[Beck87] Becker, L.E. and Peng, J., "Using activation networks for analogical ordering of consideration: one method for integrating connectionistic and symbolic processing." *Proceedings of the IEEE International Conference on Neural Networks*, San Diego, pp.367-371, 1987.

[Boun88] Bounds, D.G, Lloyd, P.J., Matthew, B. and Waddell,G. "A multilayer perceptron network for the diagnosis of low back pain." *Proceedings of the IEEE International Conference on Neural Networks*, San Diego, pp.481-489,1988.

[Call84] Callero, M., Waterman, D.A. and Kipps, J. "TATR: a prototype expert system for tactical air targeting." Rand Report R-3096-ARPA, The Rand Corporation, Santa Monica, CA, 1984.

[Duda73] Duda, R.O. and Hart, P.E., *Pattern Classification and Scene Analysis*, John Wiley & Sons, 1973.

[Dutt88] Dutta, S. and Shekhar, S., "Bond rating: a non-conservative application of neural networks.", *Proceedings of the IEEE International Conference on Neural Networks*, San Diego, pp.443-450, 1988.

[Gall88] Gallant, S.I., "Connectionist expert systems." *Communications of the ACM* (31), pp. 152-169, 1988.

[Gall93] Gallant, S.I., "Neural Network learning and expert systems.", *A Bradford Book*, MIT Press, 1993.

[Gros88] Grossberg, S.(Ed.), *Neural Networks and Natural Intelligence,* The MIT Press, 1988.

[Hech87a]Hecht-Nielsen, R., "Neurocomputer applications.", In R. Eckmiller and C.v.d.Malsburg (eds.), *Neural Computers*, Springer-Verlag, pp. 445-451, 1987.

[Hech87b]Hecht-Nielsen, R., "Counterpropagation networks" To appear in *Applied Optics*, 1987.

[Hert91] Hertz, J., Krogh, A. and Palmer, R., "Introduction to the theory of neural computation.", *Lecture Notes Volume I*, Santa Fe Institute. Addison-Wesley, 1991.

[Jack90] Jackson, P. *Introduction to Expert Systems*, 2nd ed. Addison-Wesley, Reading, MA, 1990.

[Kand85] Kandel, E.R. and Schwartz, J.H., *Principles of Neural Science*, Elsevier, 1985.

[Kera86] Keravnou, E.T. and Johnson, L. *Competent Expert Systems - A Case Study in Fault Diagnosis*, McGraw-Hill, 1986.

[Kell93] Kelly, J., *Artificial Intelligence*, Ellis Horwood, 1993.

[Koho88] Kohonen, T., *Self-organisation and associative memory*, 2nd Edition. Springer-Verlag, 1988.

[Lipp87a] Lippmann, R.P., "An introduction to computing with neural nets", *IEEE ASSP Magazine*, pp. 4-22, 1987.

[Lipp87b] Lippmann, R.P., Gold, B. and Malpass, M.L., "A comparison of Hamming and Hopfield neural nets for pattern classification", Technical report 769, MIT Lincoln Laboratory, MIT, pp. 463-502, 1987.

[Morr90] Morris, R.J.T., Rubin, L. and Tirri, H. "Neural network techniques for object orientation detection: solution by optimal feedforward network and learning vector quantization approaches", *IEEE Trans. on Pattern Analysis and Machine Intelligence*, 12, pp. 1107-1115, 1990.

[Nels82] Nelson, W.R., "REACTOR: an expert system for diagnosis and treatment of nuclear reactor accidents", *Proceedings of AAAI*, 1982.

[Pink88] Pinker, S. and Mehler, J. (eds.), "Connections and symbols", *A Bradford Book*, MIT Press, 1988.

[Pogg90] Poggio, T. and Girosi, F., "Networks for approximation and learning." *Proceedings of the IEEE* 78, pp. 1481-1497, 1990.

[Quin79] Quinlan, R., "Discovering rules from large collections of examples. A case study" In D.Mitchie (ed.), *Expert Systems in the Microelectronic Age*. Edinburgh University Press, 1979.

[Rume86] Rumelhart, D.E. and J.L.McClelland (Eds.), *Parallel Distributed Processing: Explorations in the Microstructures Of Cognition*, Vol 1,2. The MIT Press, 1986.

[Stei61] Steinbuch, K., "Die Lernmatrix.", *Kybernetik* 1, pp. 36-45, 1961.

[Wate86] Waterman, D.A., *A guide to expert systems*, Addison-Wesley, 1986.

[Weis84] Weiss, S.M. and Kulikowski, C.A., *A practical guide to designing expert systems*, Rowman & Allanheld, NJ, USA, 1984.

[Werb88] Werbos, P., "Backpropagation: past and future", *Proceedings of the IEEE International Conference on Neural Networks*, San Diego, pp. 343-353, 1988.

[Zade83] Zadeh, L.A., "The role of fuzzy logic in the management of uncertainty in expert systems", *Fuzzy Sets and Systems* 11, pp. 199-227, 1983.

4

Genetic Algorithms and Fuzzy Logic for Adaptive Process Control

Charles L. Karr

4.1 Introduction

Researchers at the U.S. Bureau of Mines have developed adaptive process control systems in which genetic algorithms (GAs) are used to augment fuzzy logic controllers (FLCs). GAs are search algorithms that rapidly locate near-optimum solutions to a wide spectrum of problems by loosely modelling the search procedures of natural genetics. FLCs are rule based systems which efficiently manipulate a problem environment by modelling the rule-of-thumb strategy used in human decision making. Together, GAs and FLCs include all of the capabilities necessary to produce powerful, efficient, and robust adaptive control systems. To perform efficiently, such control systems require a *control element* to manipulate the problem environment, an *analysis element* to recognise changes in the problem environment, and an *adaptive element* to adjust to the changes in the problem environment. This chapter describes the use of a GA and an FLC in the development of a process control system that includes each of the three elements mentioned above. The details of the process control system are presented for a specific cart-pole (broom-balancing) system. Later, the use of such a control system is described for a column flotation unit, a device that is used extensively in mineral beneficiation.

Intelligent Hybrid Systems, Edited by S. Goonatilake and S. Khebbal. ©1995 John Wiley & Sons Ltd.

4.2 Overview of Technologies

Rule-based systems, commonly called expert systems, have proven to be effective tools for solving a wide variety of practical problems [Wate89]. However, their performance in the area of process control problems has lagged because of the uncertainty inherent in measuring and adjusting parameters in most process control environments. The uncertainty associated with process control, and with human decision-making, can be incorporated into expert systems via fuzzy set theory [Zade65]. In fuzzy set theory, abstract concepts can be represented with *linguistic variables* (fuzzy terms); terms like "very high", "fairly low", and "kind of fast". The use of these fuzzy terms provides FLCs with a degree of flexibility generally unattainable in conventional rule-based systems used for process control.

FLCs have been used successfully in a number of control problems [Suge85, Evan89]. These *fuzzy expert systems* include rules to direct the decision process and membership functions to convert linguistic variables into the precise numeric values needed by a computer for automated process control. The rule set is gleaned from a human expert's knowledge which is generally based on his experience. Because linguistic terms are used in their construction, writing the rule set is often a straightforward task. Defining the fuzzy membership functions, on the other hand, is almost always the most time-consuming aspect of FLC design. Although there has been some work in the area of rule modification [Proc78], little has been done to ease the task of defining the membership functions. In fact, this chore is often accomplished via trial-and-error.

In general, the development of methods for choosing membership functions that optimise FLC performance has received little attention. A standard method for determining the membership functions that produce maximum FLC performance would expedite FLC development. However, locating optimal membership functions is a difficult task because the performance of an FLC often changes dramatically due to small changes in either the membership functions or the rule set [Karr91b]. Additionally, this task is even more difficult when it must be accomplished on-line as in an adaptive control system.

A search technique that is finding increasing popularity in the field of optimisation is the genetic algorithm (GA) [Holl75]. GAs are search algorithms based on the mechanics of genetics; they use operations found in natural genetics to guide their trek through a search space [Gold89]. Empirical investigations by Hollstien [Holl71] and De Jong [DeJo75] have demonstrated the technique's efficiency in function optimisation. De Jong's work, in particular, establishes the GA as a robust search technique, one that is efficient across a broad spectrum of problems as compared to several traditional schemes. Subsequent applications of GAs to the search problems of pipeline engineering [Gold89], very large scale integration microchip layout [Davi85b], structural optimisation [Gold86], job shop scheduling [Davi85a], and equipment design [Karr89], add considerable evidence to the claim that GAs are broad based.

The robust nature and simple mechanics of GAs make them inviting tools for establishing the membership functions to be used in FLCs. GAs are also potentially beneficial in the design of adaptive FLCs which alter their membership functions on-line to account for changes in the physical environment. In the adaptive process control system presented, a GA is used for several tasks necessary to establish adaptive control. This chapter introduces a problem environment that requires a control system with adaptive capabilities for efficient manipulation. The tasks that must be accomplished by the GA in the overall scheme of the adaptive control system are discussed, and an adaptive process control system is developed for a specific problem environment.

4.3 Problem Domain

The problem of interest in this application is the control of a cart-pole balancing system. A cart is free to translate along a one-dimensional track while a pole is free to rotate only in the vertical plane of the cart and track. A multivalued force, $\vec{F}$, can be applied at discrete time intervals in either direction to the centre of mass of the cart. A schematic of the cart-pole system is shown in Figure 4.1

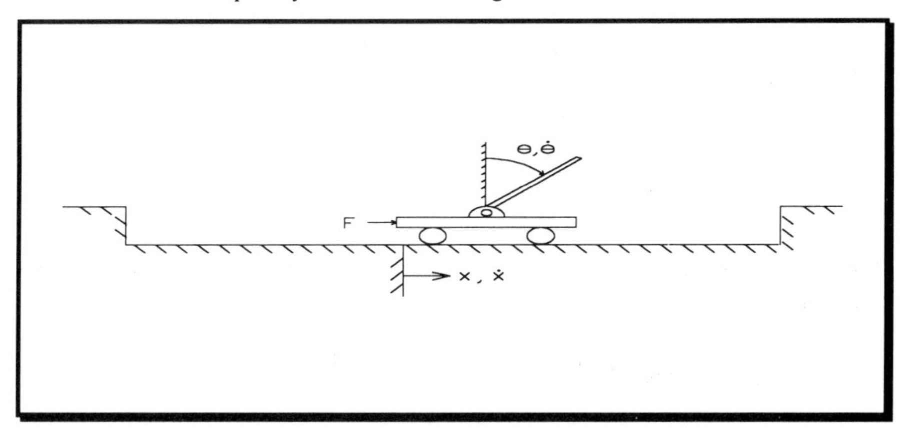

Figure 4.1 — Schematic of the cart-pole balancing system.

The objective of the control problem is to apply forces to the cart until it is motionless at the centre of the track and the pole is balanced in a vertical position. A block diagram of a very basic control loop is shown in Figure 4.2. This task of centring a cart on a track while balancing a pole was established as a benchmark problem for artificial intelligence based control schemes by Barto, Anderson, and Sutton [Bart83]. In their work, this problem was solved using an innovative neural network scheme. However, numerous artificial intelligence techniques have been used to address the cart-pole problem [Thri91, Wu91]. Despite the fact that this problem has been deemed a "toy problem" by some, it is often used as an example of the inherently unstable, multiple-output, dynamic systems present in many balancing situations, e.g., two-legged walking and the aiming of a rocket thruster.

The adaptive control system described later in this chapter relies on a computer simulation of the problem environment. Thus, it is necessary to examine the equations of motion that describe the cart-pole system.

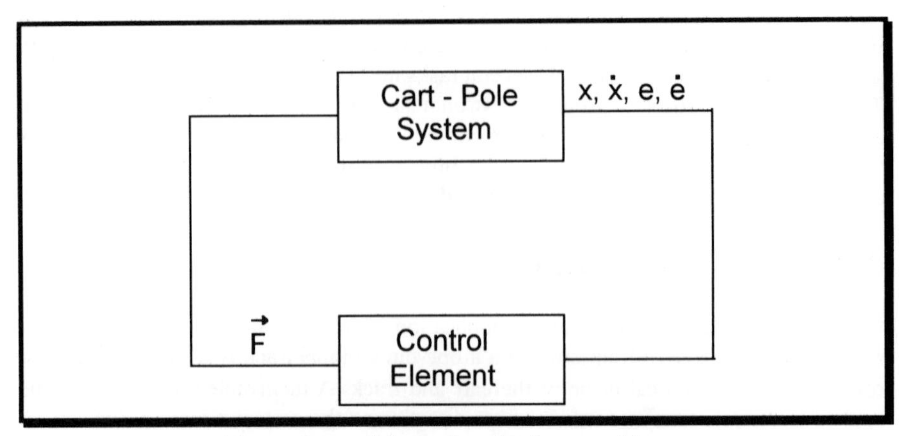

Figure 4.2 — Schematic of basic control loop.

The state of the cart-pole system at any time is described by four real-valued state variables:

x = position of the cart;

$\dot{x}$ = linear velocity of the cart;

θ = angle of the pole with respect to the vertical;

and $\dot{\theta}$ = angular velocity of the pole.

The system is modelled by the following non-linear ordinary differential equations [Bart83]:

$$\ddot{\theta} = \frac{g \sin\theta + \cos\theta \left[\frac{-F - m_p l \dot{\theta}^2 \sin\theta + \mu_c \operatorname{sign}(\dot{x})}{(m_c + m_p)} \right] - \frac{\mu_p \dot{\theta}}{m_p l}}{l \left[\frac{4}{3} + \frac{m_p \cos^2\theta}{(m_c + m_p)} \right]} \tag{4.1}$$

$$\ddot{x} = \frac{F + m_p l \left[\dot{\theta}^2 \sin\theta - \ddot{\theta} \cos\theta \right] - \mu_c sign(\dot{x})}{(m_c + m_p)} \tag{4.2}$$

where:

g = -9.81 m/s^2, acceleration due to gravity;

m_c = 1.0 kg, mass of cart (as will be seen later, this value changes with time);

m_p = 0.1 kg, mass of pole;

l = 0.5 m, length of pole;

μ_c = 0.0005, coefficient of friction of cart on track;

μ_p = 0.000002, coefficient of friction of pole on cart;

and $\quad -10.0\ \mathrm{N} \leq \vec{F} \leq 10.0\ \mathrm{N}$, force applied to cart's center of mass.

The solution of these equations was approximated using Euler's method, thereby yielding the following difference equations:

$$\theta^{t+1} = \theta^{t} + \dot{\theta}^{t}\Delta t \tag{4.3}$$

$$\dot{\theta}^{t+1} = \dot{\theta}^{t} + \ddot{\theta}^{t}\Delta t \tag{4.4}$$

$$x^{t+1} = x^{t} + \dot{x}^{t}\Delta t \tag{4.5}$$

$$\dot{x}^{t+1} = \dot{x}^{t} + \ddot{x}^{t}\Delta t \tag{4.6}$$

where the superscripts indicate values at a particular time, Δt is the time step, and the values $\ddot{x}^{t}$ and $\ddot{\theta}^{t}$ are evaluated using equations (4.1) and (4.2). A time step of 0.02 seconds was used because this time step struck a balance between the accuracy of the solution and the computational time required to find the solution. As will be seen later, the GA's ability to locate effective membership functions depends heavily on the computational time required for a simulation.

The preceding is a description of the classic cart-pole system as it is generally addressed in the literature. The characteristic parameters of the physical system, parameters such as cart mass and pole length, remain constant in the problem traditionally addressed. However, in the current effort, the control problem is made considerably more difficult by considering a system in which the parameters of the system can change with time. This expansion transforms the problem into a time-varying control problem. Note that the parameters can change by large amounts; the cart mass can increase by as much as 500%. These changes significantly alter the response of the cart-pole system to a given force stimulus.

The most straightforward way to address the time-varying nature of the process control problem is to simply make the control action prescribed by the controller a function of the time-varying parameter. In other words, the mass of the cart can be incorporated into the rules that are being used to prescribe the force to be applied to the cart. However, there are some definitive reasons for not adopting this approach. Each time a new parameter is introduced into the conventional FLC, the number of rules increases exponentially; this is of course a problem when the speed of execution of the process control system is to be considered. But far worse than that, the rule set must be re-designed each time a new parameter is considered. Thus, an alternative approach is adopted here. In the process control system being considered, changes in the cart mass are accounted for by altering the membership functions describing the cart position, cart velocity, pole angle, pole angular velocity, and force to be applied to the cart on-line. Such changes are made when the mass of the cart (or any of the other system parameters) changes. In this way, the FLC is able to prescribe a force to be applied to the cart by considering the state of only four parameters; it accounts for changes in any of the other parameters associated with the problem environment by re-tuning its membership functions, as opposed to re-designing the rule set. Thus, an adaptive FLC is produced: one that is able to account for changes in variables that do not explicitly appear in the controller's rule set.

4.4 Why Hybrid Processing?

FLCs have been shown to be effective tools for solving process control problems. However, manipulating a problem environment with rapidly changing process dynamics proves difficult for virtually any static controller. And here, it is important to realise that the changes in the parameters actually alter the way in which the problem environment responds to control actions. When the physical parameters of the cart-pole system vary with time (i.e., cart mass, pole length, frictional coefficients, etc.), some mechanism for adapting the control strategy is necessary. Implementing such an adaptive element to the controller is not a trivial task. Unfortunately, there is even more to this story. Consider the situation in which no feedback regarding the time-varying parameters is available. For the purpose of this study, no information concerning the magnitude of the time-varying parameters is available to the controller.

With the introduction of this new problem constraint, it should be apparent that the process control system faces several problems. First, the controller must efficiently manipulate the cart-pole system by employing an FLC to prescribe values of force to be applied to the cart (control element). Second, the control system must determine when one of the parameters in the problem environment has changed, and further, to what extent the parameter has been changed (analysis element). Third, the process control system must make the necessary alterations to its control strategy as a consequence of the time-varying parameters. Such a complicated process control system must be provided with techniques that are both robust and flexible.

The task of manipulating the cart-pole system can be accomplished with a variety of techniques. From traditional proportional-integral-derivative controllers, to neural networks, to fuzzy logic, numerous techniques have been demonstrated to be effective in this problem domain. An FLC was chosen in this work for a variety of reasons. First, fuzzy logic produces a control system that is akin to the procedure humans use to tackle decision making problems. A rule base is produced that contains rules that incorporate subjective terms that appear in the rules-of-thumb that humans utilise so effectively. Second, fuzzy logic allows for the efficient treatment of the uncertainty associated with most process control problems; uncertainty in both measuring and manoeuvring system parameters such as the cart position, pole velocity, and the force applied to the cart. Third, in the control system outlined later in this chapter, it is imperative that the control strategy can be altered without excessive disruption to the current efforts of the controller. Fuzzy logic, with its flexible membership function definitions of subjective terms, certainly offers this opportunity.

Recognising and quantifying changes in the problem environment can be accomplished in a number of ways. Probably the most direct way to address this problem is to develop efficient sensors that can provide the control system with the information needed. However, just as the approach of incorporating all of the parameters that may change values into the rule base is a never ending task, so too is the task of developing sensors for all of the parameters that may need to be measured. With the elimination of this option, and the realisation that the problem environment being considered can be accurately simulated on a computer, an alternative solution comes to light. A computer model, one that includes all of the parameters that affect the cart-pole response (see equations (4.1) and (4.2) which include cart mass, pole length, and frictional coefficients), can be used to "track" the response of the system. If the computer model is accurate, each time the response of the actual cart-pole system differs from the predicted response of the computer model of the cart-pole system, some parameter in the physical system has been altered. Now, the problem of system characterisation must be solved. This effort may be thought of as solving an inverse problem. The actions (actions prescribed by the FLC are recorded and sent to the analysis element) that produce a particular response on the problem environment (the values of the state variables from the physical cart-pole system are measured and forwarded to the analysis element) are known. Computer simulations can be run with varying cart-pole parameter values until the computer simulated response to the control actions matches the actual response of the physical cart-pole system. At this point, the parameters of the cart-pole system are known. In the process control system presented later in this chapter, a GA is used to solve the system characterisation problem. Certainly, other search algorithms could be used to locate the parameters that produce a particular response. However, the robust nature of a GA makes it a particularly inviting candidate.

Altering the control strategy also requires a search algorithm; the search involves selecting either new rules or new membership functions for the FLC. The GA possesses three qualities that make it attractive for the search at hand. First, a GA requires no derivative information about the search space. This property is particularly desirable in the quest for an improved FLC because the meaning of derivative information is not clear. Even if it were clear, computing derivative information would most likely prove to be an imposing task at best. Second, a GA works with populations of points in the search space, not individual points. Thus, a GA is not easily fooled by discontinuities in the search space, nor is it deceived by problem environments that are highly sensitive to changes. Third, the function evaluations associated with a GA's search of a problem space can be accomplished in parallel. Each of the potential solutions considered in the GA's population approach can be evaluated independently. This quality is important when the search for a new control strategy is being carried out on-line, as is the case in the current research.

The task of designing an FLC can be eased by incorporating a GA for selecting the rules and tuning the membership functions. Above and beyond that, when the problem environment is time-varying, and the control system does not receive adequate feedback concerning the changes to the problem environment, some mechanism is required for enhancing the feedback that is available. The tasks of recognising and quantifying the changes in the cart-pole system, and the task of altering the control strategy implemented by the FLC can all be cast in the light of a search problem. A GA has qualities that are inviting for solving a wide spectrum of search problems, and it rapidly locates near-optimum solutions to difficult search problems.

4.5 Survey of Related Hybrid Systems

The popularity of fuzzy logic systems has increased dramatically in the past several years. With this increase in popularity has come a definite increase in the research associated with fuzzy logic. Although the basic structure of FLCs has remained, for the most part unchanged because they are simple and effective, a good deal of research has been conducted to expedite the process of FLC development. Interestingly enough, most of this research has involved the synergy of artificial intelligence techniques such as neural networks and GAs.

One of the first articles to address the synergy of GAs and fuzzy logic was written by Karr [Karr91b]. This article acknowledged the difficulty of selecting membership functions that allowed for efficient FLC performance. A description was provided of an approach to membership function tuning that involved the use of a GA. The approach described in Karr's original paper is adopted here for the learning element which alters the control strategy in response to changes in the problem environment. In the current effort, however, the membership functions are altered on-line, not in the original development of the FLC.

It did not take long for others to realise that there was a real need for methods of designing FLCs. Thrift [Thri91] also proposed the use of GAs for designing FLCs. In his work, he proposed the use of GAs for both selecting the rule set and for tuning the membership functions. His approach was applied to a computer simulated translating cart (the inverted pendulum). Results indicate that the GA-FLC approached the performance of an optimal controller.

Not surprisingly, several researchers have extended the approaches described by Karr and Thrift. Feldman [Feld93] suggested the use of a GA for the synthesis of a fuzzy network. A fuzzy network is a connectionist extension of a fuzzy logic system allowing partially connected associations, or rules, that incorporate fuzzy terms. Fuzzy networks can be used either to model or control physical systems. In his paper, Feldman developed a fuzzy network to control a computer simulation of the very same translating cart system presented by Thrift. Results indicate that the GA-designed fuzzy network is able to achieve a level of performance that is comparable with that of the FLC developed by Thrift. However, it is important to note that the fuzzy network required only six rules to achieve this performance level whereas, Thrift's FLC utilised eighteen rules. Thus, a GA developed a fuzzy logic based control system that required a third of the rules. This is, of course, important because it substantially reduces the time necessary to arrive at an appropriate control action.

Sun and Jang [Sun93] described the use of a GA to design a fuzzy model of a medical diagnosis problem. The focus of their work was to use fuzzy logic not for a control strategy, but rather for a prediction strategy. In the development of their model, a GA once again played a key role in designing the fuzzy logic system.

Homaifar and McCormick [Homa93] utilised a GA for simultaneously developing the rule set and tuning the membership functions associated with an FLC. In their paper, they argue that the performance of an FLC is dependant on the coupling of the rule set and the membership functions, and that therefore, the two cannot be developed independent of one another. Interestingly enough, their approach was applied to the development of an FLC for a cart-pole system. Their results indicate that a GA is quite capable of generating a rule set while simultaneously tuning membership functions. However, whether or not the simultaneous designing of the rule set and the membership functions is vital, is unclear.

Valenzuela-Rendon [Vale91] proposed an intriguing system in which a classifier system was provided with fuzzy logic capabilities. A classifier system is a rule-based system that generates rules using a GA. The classifier system rewards or punishes its rules based on their past performance. His results indicate that there are situations in which the traditional classifier system needs the capabilities supplied by the inclusion of fuzzy logic. Additionally, the results indicate that there are a number of innovative ways to combine the search capabilities of GAs with the approximate reasoning capabilities of fuzzy logic.

The preceding citations are meant to provide the reader with a sampling of the work that has been done in the area of hybrid fuzzy logic and GA systems. By no means is it meant to be an all inclusive review of the literature. Certainly the above review neglects a portion of work that is being continuously added to the literature. A number of excellent papers have not been mentioned. Two such papers come immediately to mind: (1) one by Surmann, Kanstein, and Goser [Surm93] that address the synergism of GAs and fuzzy logic for a decision-support system, and (2) a paper by Eick and Jong [Eick93] who combined fuzzy logic and GAs to produce a classification system.

It should be pointed out that the current process control system is more ambitious than any of the efforts cited above. In the current system, not only must an FLC be designed, but it must be altered on-line. Additionally, the process control system is charged with the task of recognising changes to and quantifying the extent of, the time-varying parameters of the problem environment. These tasks amplify the importance of works in which GAs are used for system characterisation. Fortunately, some research has been done that addresses the effectiveness of using GAs for system characterisation [Karr91a]. Results of this effort indicate that a GA can be quite effective in the characterisation of system parameters. The approach outlined in this paper is used by the analysis element of the process control system described later in this chapter.

4.6 An Adaptive Ga-Flc

Figure 4.3 shows a schematic of the Bureau's adaptive process control system. The heart of this control system is the loop consisting of the control element and the problem environment. The control element receives information from sensors in the problem environment concerning the status of the *condition variables*, i.e., x, $\dot{x}$, θ, and $\dot{\theta}$. It then computes a desirable state for a set of *action variables*, i.e., force to be added to the cart ($\vec{F}$). These changes in the action variable force the problem environment toward the setpoint ($x = \dot{x} = \theta = \dot{\theta} = 0$). This is the basic approach adopted for the design of virtually any closed loop control system. However, this basic loop includes no mechanism for adaptive control.

The adaptive capabilities of the system shown in Figure 4.3 are due to the analysis and learning elements. In general, the analysis element must recognise when a change in the problem environment has occurred. A "change", as it is used here, consists of any alteration of a parameter that is not included in the rule set that makes up the FLC, i.e., mass of the cart, length of the pole, or one of the frictional coefficients. (Of importance is the fact that all of these changes affect the response of the problem environment, otherwise it has no effect on the way in which the control element must act to efficiently manipulate the problem environment.) The analysis element uses information concerning the condition and action variables over some finite time period to recognise changes in the environment and to compute the new performance characteristics associated with these changes.

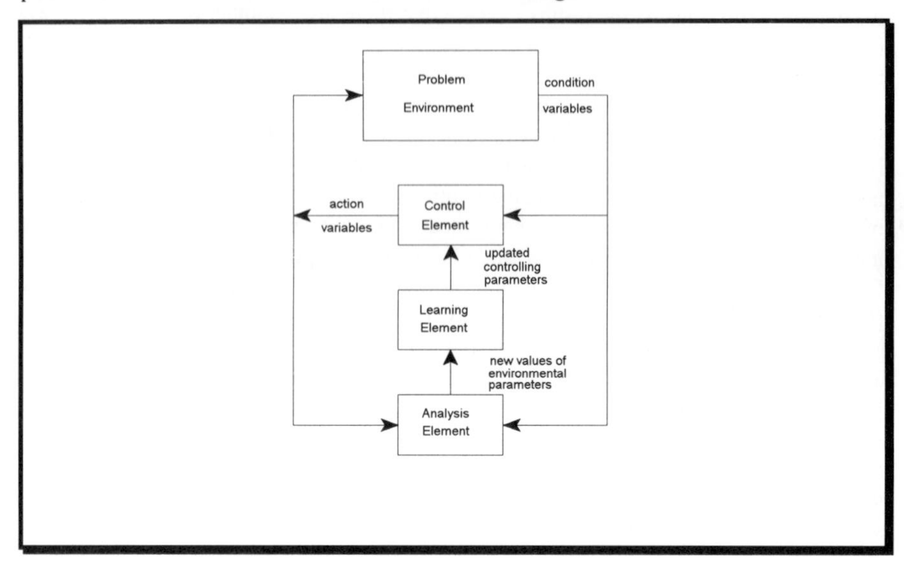

Figure 4.3 — Structure of the adaptive control system.

The new environment (the problem environment with the altered parameters) can pose many difficulties for the control element, because the control element is no longer manipulating the environment for which it was designed. Therefore, the algorithm that drives the control element must be altered. As shown in the schematic of Figure 4.3, this task is accomplished by the learning element. The most efficient approach for the learning element to use to alter the control element is to utilise information concerning the past performance of the control system. The strategy used by the control, analysis, and learning elements of the stand-alone, comprehensive adaptive controller developed by the U.S. Bureau of Mines is provided in the following sections.

4.6.1 Control Element

The control element receives feedback from the cart-pole system, and based on the current state of x, $\dot{x}$, θ, and $\dot{\theta}$, must prescribe appropriate values of $\vec{F}$. Any of a number of closed-loop controllers could be used for this element. However, because of the flexibility needed in the control system as a whole, an FLC is employed. Like conventional rule-based systems (expert systems), FLCs use a set of production rules which are of the form:

IF {*condition*} THEN {*action*}

to arrive at appropriate control actions. The left-hand-side of the rules (the *condition* side) consists of combinations of the controlled variables (x, $\dot{x}$, θ, and $\dot{\theta}$); the right-hand-side of the rules (the *action* side) consists of combinations of the manipulated variables ($\vec{F}$). Unlike conventional expert systems, FLCs use rules that utilise fuzzy terms like those appearing in human rules-of-thumb. For example, a valid rule for an FLC to use for manipulating the cart-pole system is:

IF {x is NEGATIVE BIG and $\dot{x}$ is NEGATIVE and θ is POSITIVE and $\dot{\theta}$ is POSITIVE} THEN {$\vec{F}$ is POSITIVE BIG}.

This rule is self-explanatory and is typical of the rules generally used in FLCs.

The fuzzy terms are subjective; they mean different things to different "experts", and can mean different things in varying situations. Fuzzy terms are assigned concrete meaning via fuzzy membership functions [Zade65]. The membership functions used in the control element to describe cart position appear in Figure 4.4. (As will be seen shortly, the learning element is capable of changing these membership functions in response to changes in the problem environment.) These membership functions are used in conjunction with the rule set to prescribe single, crisp values of the action variable ($\vec{F}$). Unlike conventional expert systems, FLCs allow for the enactment of more than one rule at any given time. The single crisp action is computed using a weighted averaging technique that incorporates both a *min-max* operator and the *centre-of-area* method [Karr91b].

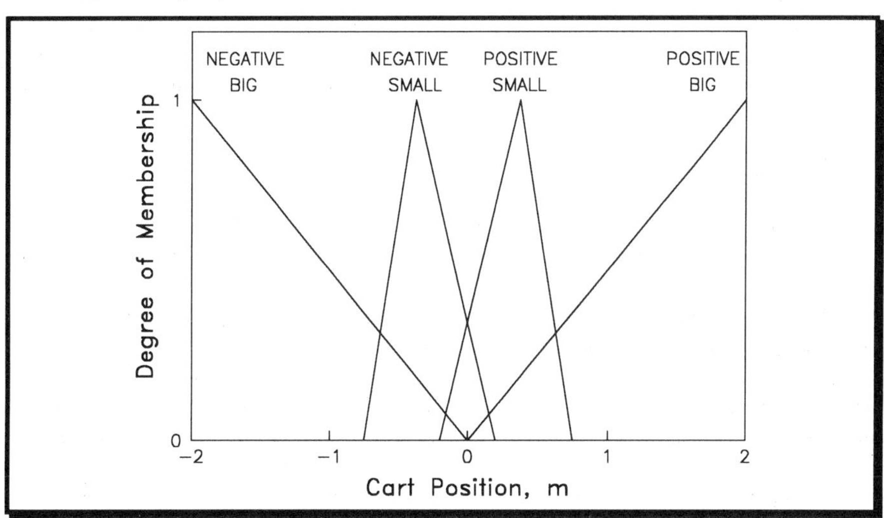

Figure 4.4 — Cart position membership functions.

The following fuzzy terms were used, and therefore "defined" with membership functions, to describe the significant variables in the cart-pole system:

x NEGATIVE BIG (NB), NEGATIVE SMALL (NS), POSITIVE SMALL (PS) and POSITIVE BIG (PB);

$\dot{x}$ NEGATIVE (N), NEAR ZERO (NZ), and POSITIVE (P);

θ NEGATIVE (N), NEAR ZERO (NZ), and POSITIVE (P);

$\dot{\theta}$ NEGATIVE (N), NEAR ZERO (NZ), and POSITIVE (P);

$\vec{F}$ NEGATIVE BIG (NB), NEGATIVE MEDIUM (NM), NEGATIVE SMALL (NS), ZERO (Z), POSITIVE SMALL (PS), POSITIVE MEDIUM (PM) and POSITIVE BIG (PB).

Although the cart-pole system is quite complex (especially when the time-varying parameters are considered), some basic common sense rules can be written for manipulating the system. An effective FLC for manipulating the cart-pole system can be written that contains only 108 rules. The 108 rules are necessary because there are 4 fuzzy terms describing x, 3 fuzzy terms describing $\dot{x}$, 3 fuzzy terms describing θ, and 3 fuzzy terms describing $\dot{\theta}$ (4*3*3*3 = 108 rules to describe all possible combinations that could exist in the cart-pole system as described by the fuzzy terms represented by the membership functions selected). Now, the rules selected for the control element are certainly inadequate for effectively manipulating the full-scale cart-pole system; the one that includes the time-varying parameters. However, the performance of an FLC can be dramatically altered by changing the membership functions. This is equivalent to changing the definition of the terms used to describe the variables being considered by the controller. As will be seen shortly, GAs are powerful tools capable of rapidly locating efficient fuzzy membership functions that allow the controller to accommodate changes in the value of system parameters; even though, those parameters do not explicitly appear in the rule base.

4.6.2 Analysis Element

The analysis element recognises changes in parameters associated with the problem environment not taken into account by the rules used in the control element. In the cart-pole system, these parameters include: (1) the mass of the cart (m_C), (2) the length of the pole (l), and (3) the frictional coefficients (μ_C and μ_P). Changes to any of these parameters can dramatically alter the way in which the cart-pole system responds to applications of a force to the cart, thus forming a new problem environment requiring an altered control strategy. Recall that the FLC used for the control element presented includes none of these parameters in its 108 rules. Therefore, some mechanism for altering the prescribed actions must be included in the control system. But before the control element can be altered, the control system must recognise that the problem environment has changed, and compute the nature and magnitude of the changes.

The analysis element recognises changes in the system parameters by comparing the response of the cart-pole system being manipulated by the control element to the response of a model of the cart-pole system used in the analysis element. In general, recognising changes in the parameters associated with the problem environment requires the control system to store information concerning the past performance of the problem environment. This information is most effectively acquired through either a database or a computer model. Storing such an extensive database can be cumbersome and requires extensive computer memory. Fortunately, the dynamics of the cart-pole system are well understood, and can be modelled using the two differential equations, equations 4.1 and 4.2, yielding values of x, $\dot{x}$, θ, and $\dot{\theta}$. In the approach adopted here, a computer model predicts the response of the cart-pole system. This predicted response is compared to the response of the physical system (which is actually itself a simulated system). When the two responses differ by a threshold amount over a finite period of time, the physical cart-pole system is considered to have been altered.

When the above approach is adopted, the problem of computing the new system parameters becomes a curve fitting problem [Karr91a]. The parameters associated with the computer model produce a particular response to changes in the action variables. The parameters must be selected so that the response of the model matches the response of the actual problem environment.

An analysis element has been forged in which a GA is used to compute the values of the parameters associated with the cart-pole system. When employing a GA in a search problem, there are basically two decisions that must be made: (1) how to code the parameters as bit strings and (2) how to evaluate the merit of each string (the fitness function must be defined). The GA used in the analysis element employs concatenated, mapped, unsigned binary coding [Gold89]. The bit-strings produced by this coding strategy were of length 60: 4 bits for each of 15 parameters that had to be represented. For a complete definition of the set of triangular membership functions for the cart-pole balancer as described above, 30 points had to be specified. Certainly, a 30 parameter search problem is challenging to say the least. Fortunately, the search space associated with the selection of membership functions for the cart-pole balancer can be pruned. Notice that the rule set presented above is symmetric because every condition wherein the cart is to the left of the track's centre has an analogous condition wherein the cart is to the right of the track's centre. Therefore, NEGATIVE BIG should be "opposite" of POSITIVE BIG, NEGATIVE SMALL should be "opposite" of POSITIVE SMALL, and so on for all of the membership functions. Thus, instead of finding 30 parameters, the GA is faced with the task of finding only the 15 points shown in Figure 4.5. Realise that even though the original search space has been reduced by a factor of 2, a 15 parameter search problem is still of some consequence. The 4 bits associated with each individual parameter were read as a binary number, converted to decimal numbers (0000 = 0, 0001 = 1, 0010 = 2, 0011 = 3, etc.,), and mapped between minimum and maximum values according to the following:

$$C = C_{min} + \frac{b}{\left(2^m - 1\right)}\left(C_{max} - C_{min}\right) \tag{4.7}$$

where C is the value of the parameter in question, b is the binary value, m is the number of bits used to represent the particular parameter (4 bits), and C_{min} and C_{max} are minimum and maximum values associated with each parameter that is being coded.

A fitness function has been employed that represents the quality of each bit-string; it provides a quantitative evaluation of how accurately the response of a model using the new model parameters matches the response of the actual physical system.

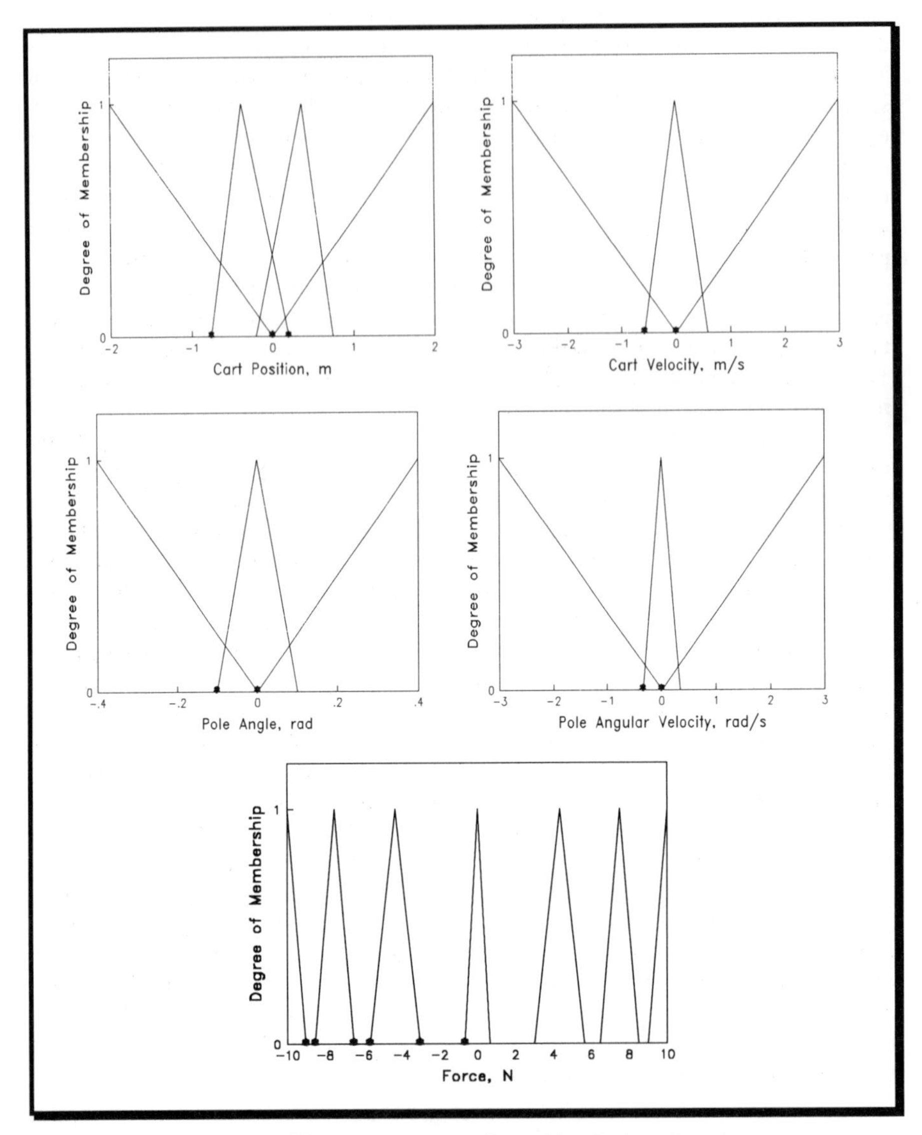

Figure 4.5 — Fifteen parameters adjusted by the learning element.

The fitness function used in this application is:

$$f = \sum_{i=0 \text{ sec}}^{i=30 \text{ sec}} \left(w_1 |x_i| + w_2 |\theta_i| \right) \tag{4.8}$$

where $w_1 = 1.0$ and $w_2 = 10.0$ are weighting constants. With this definition of the fitness function, the problem becomes a minimization problem: the GA must minimize f, which as it has been defined, represents the difference between the response predicted by the model and the response of the physical cart-pole system.

Figure 4.6 compares the response of the cart-pole system being manipulated by the control element to the response of the cart-pole simulation in the analysis element using the parameters determined by a GA. This figure shows that the responses of the two simulated systems are virtually identical, thereby demonstrating the effectiveness of a GA in this application. The GA was able to locate the correct parameters after only 500 function evaluations, where a function evaluation consisted of simulating the cart-pole system for 30 seconds. Locating the correct parameters took approximately 20 seconds on a 386 personal computer. Industrial systems may mandate that a control action be taken in less than 20 seconds. In such cases, the time the GA is allotted to update the model parameters can be restricted. Once new parameters (and thus the new response characteristics of the problem environment) have been determined, the learning element must alter the control element.

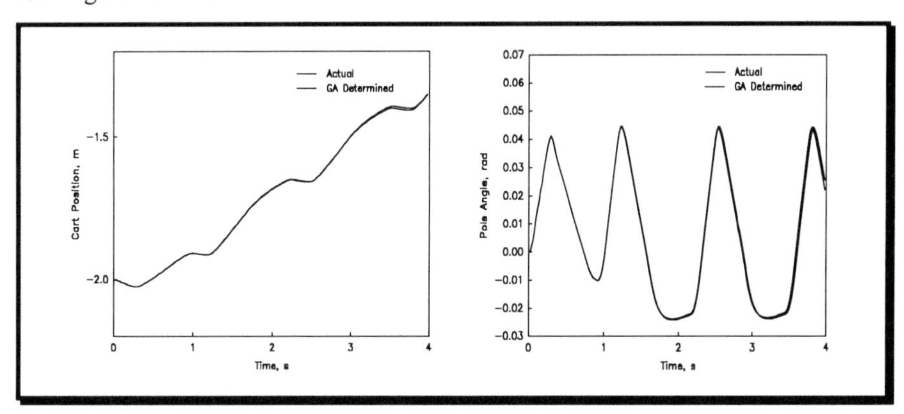

Figure 4.6 — Performance of an analysis element.

4.6.3 Learning Element

The learning element alters the control element in response to changes in the problem environment. It does so by altering the membership functions employed by the FLC of the control element. Since none of the randomly altered parameters appear in the FLC rule set, the only way to account for these conditions (outside of completely revamping the system) is to alter the membership functions employed by the FLC. These alterations consist of changing the location of the triangles used to define the fuzzy terms.

Altering the membership functions (the definition of the fuzzy terms in the rule set) is consistent with the way humans control complex systems. Quite often, the rules-of-thumb humans use to manipulate a problem environment remain the same despite even dramatic changes to that environment; only the conditions under which the rules are applied are altered. This is basically the approach that is being taken when the fuzzy membership functions are altered.

The current adaptive control system uses a GA to alter the membership functions associated with an FLC, and this technique has been well documented [Karr91b]. A learning element that utilises a GA to locate high-efficiency membership functions for the time-varying cart-pole system has been designed and implemented.

Figure 4.5 shows the results of an adaptive GA-FLC that accounted for changes in the cart mass. This adaptive GA-FLC was able to avoid the catastrophic failures of the pole falling over or the cart striking a wall, despite dramatic changes in cart mass by altering its membership functions on-line. Every time a change in cart mass was made, the analysis element recognised and quantified the changes, and the learning element employed a GA to locate new membership functions that were effective for the current state of the system. As can be seen in Figure 4.7, the adaptive GA-FLC outperformed a non-adaptive FLC which allowed the pole to fall over.

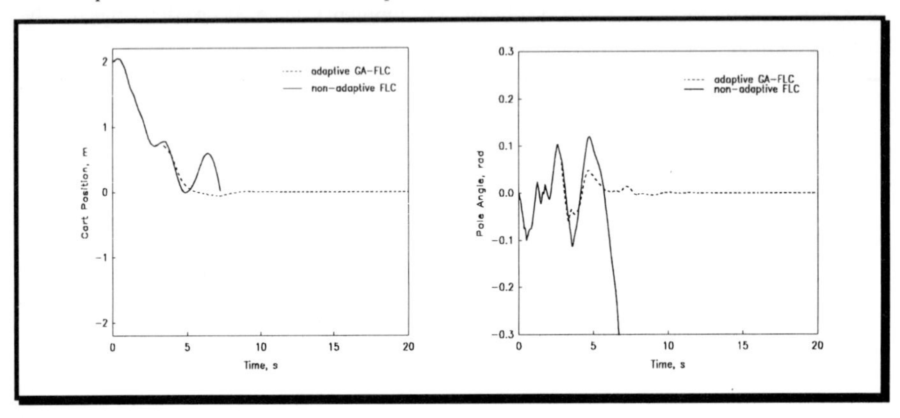

Figure 4.7 — Performance of an adaptive process control system.

4.6.4 A "Real-World" System

The previous sections provide discussion of the adaptive process control system developed and implemented by researchers at the U.S. Bureau of Mines. The controller was applied to a cart-pole system which, although it has been used for a model of the inherently unstable, multiple-output, dynamic systems present in many balancing situations, is not a full-scale, real-world situation. The adaptive process control system presented, however, is currently being applied to industrial systems. This section describes one such implementation to a flotation column.

Flotation is an established method for concentrating minerals [Sast78]. Separation occurs when selected particles are rendered hydrophobic (water repellent) with chemical agents (called collectors) and are attached to air bubbles moving upward through a slurry of mineral ore and water. Hydrophilic (water attracted) particles are unaffected by the air and settle separately. Conventional flotation cells have been a cornerstone of the mineral processing industry for a number of years. However, column flotation is becoming increasingly popular as an alternative to conventional flotation cells because of its efficiency and mechanical simplicity. Several copper-molybdenum operations in the United States and Canada are currently operating flotation columns, while numerous other ventures are running pilot-scale tests [Agar91]. The effectiveness of these separators should continue to improve as a better understanding of the underlying mechanics of the operation is gained.

Figure 4.8 shows a schematic of a column flotation unit. A chemically treated mineral slurry (feed) is fed into the column approximately one-third of the way down from the top of the column. Bubbles are introduced at the bottom of the column. As the feed particles sink and the bubbles rise, collisions occur. Due to the chemical treatment of the feed, some particle species attach to the bubbles and rise to the top, while others are repelled by the air and sink to the bottom; thus, a separation occurs. This has been a brief description of column flotation, but even when considered in its entirety, the mechanics of column flotation are not difficult.

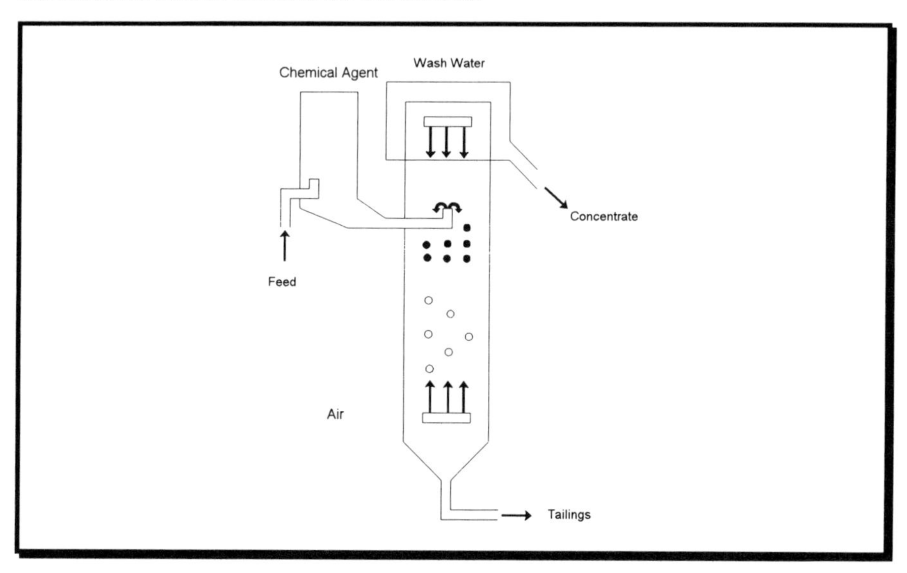

Figure 4.8 — A schematic of a column flotation unit.

Despite the fact that the mechanics of column flotation are not difficult, and that column flotation units offer distinct advantages for improving the metallurgical efficiency of a flotation circuit, there is still a lot of room for improvement in the area of controlling the operation of a column flotation unit. Process variables (also known as condition variables) such as pulp level (the height of the froth in the column), gas hold-up, bias rate (the net flow into the column), bubble size, slurry feed rate, and overflow rate must constantly be measured and monitored if maximum performance is to be achieved. Furthermore, controlled variables (also known as action variables) such as grade, recovery, and system capacity must be closely monitored so that manipulated variables such as flow rates, pH, aeration rate, and chemical agent addition can be adjusted to provide for optimum process efficiency. The objective of the current research effort is to produce an adaptive control system that is capable of efficiently manipulating a column flotation unit in an industrial environment.

The adaptive process control system relies heavily on a computational model of the problem environment, in this case a column flotation unit. In recent years, numerous researchers have worked to develop an accurate model of column flotation. Some of the more noteworthy research has been performed by Finch and Doby [Finc90], Herbst and Rajamani [Herb89], Luttrell, Adel, and Yoon [Lutt87], Sastry and Lofftus [Sast88], and Ynchausti, Herbst, and Hales [Ynch88]. Despite these efforts in developing a suitable computer model of column flotation, there is still no single model that is suitable for the control system that is being proposed. Therefore, the current research effort includes the development of a neural network model of a column flotation unit. Neural networks are techniques originating in the field of artificial intelligence that are capable of modelling the performance of a system simply by being presented with examples of that systems response [Wass89]. Previous efforts have demonstrated the ability of neural networks to accurately model conventional flotation units [Karr92], and therefore should be effective in the modelling of column flotation units.

Quality performance in a flotation operation is not difficult to quantify. Generally, the intent of a control system is to adjust the process parameters so that recovery is maximised subject to a constraint on grade. Recovery, which is a measure of the material separated, is the percentage of the desired mineral that is successfully removed from the ore sample. Grade, which is a measure of the degree of the separation, is the percentage of the desired mineral relative to the total material removed. An effective control strategy can be implemented via an FLC once a basic understanding has been achieved of the relationship between the process variables and grade and recovery. Fortunately, such strategies appear in the literature [Ynch88, Finc90]. These strategies can be transformed into the typical production rules used in an FLC.

An adaptive process control system becomes necessary when the ore sample that is being fed into the column flotation unit is changing with time. In this case, the chemistry of the separation process is changed, and the control strategy must be altered if maximum performance is to be achieved. Thus, the format used for the cart-pole system is applicable in the column flotation environment. First, a control element will be needed to adjust the process parameters to achieve some performance level (maximise recovery subject to a constraint on grade). Second, when the concentration of the feed changes, an analysis element is needed to recognise and to quantify the extent of the change. It is important to note here that sensors for taking such data are, at this time, either inaccurate or quite expensive. An analysis element armed with a model of the column flotation unit will be capable of determining the changes to the feed concentration, just as an analysis element was able to determine changes in the cart-pole environment. Third, a learning element is needed to adjust the control strategy once it has been determined that the column flotation unit has been altered, and the current strategy is outdated.

The adaptive process control system described in this chapter is general in its approach. Certainly, the individual elements are problem dependant, but the overall approach is not. Thus, adapting the approach to different problem environments such as the column flotation unit will require some effort, but the basic parts are all in place. The modular nature of the control system will help in its implementation to various problem environments.

4.7 Future Efforts

Future efforts on the adaptive process control system will focus on exploiting the strength of the system to help overcome one of its weaknesses. Next to the flexibility of the system, its main strength is that most of the computations can be done in parallel. While the control element is manipulating the problem environment, the analysis and learning elements are simultaneously monitoring, evaluating, and adapting the control strategy. The major weakness of the system is its dependence on a computational model of the problem environment. Certainly, in some regions of operation, one computational model will outperform other models. The key is to use the right model at the right time. Since the analysis and learning elements can perform their tasks in parallel with the control element, there is no reason why they cannot be allotted their own processors for monitoring and adapting the control strategy. Thus, there is no reason why multiple analysis and learning elements cannot be written and implemented; each of which can access different models of the problem environment. In this way, the shortcomings of a particular model can be circumvented.

Efforts are ongoing to provide the adaptive process control system with multiple analysis and learning capabilities. This amounts to adding branches to the basic system architecture shown in Figure 4.3. When this addition is completed, the adaptive process control system will be able to explore additional regions of the control space. In its present form the control scheme is limited by the scope of the control region for which a particular computer model of the problem environment is accurate.

Of course, the other major effort on the system is focused on implementing the adaptive process controller to industrial systems. As mentioned in the previous section, the system is currently being implemented on column flotation units. There are also plans to develop an adaptive control system for flying helicopters and for managing mineral beneficiation processes other than flotation such as grinding. Successful implementation into industrial settings will add credence to the adaptive process control system, and will increase the interest in the synergism of artificial intelligence based techniques such as the link established here between GAs and FLCs.

References

[Agar91] Agar, G. E., Huls, B. J., and Hyma, D. B. (Eds.) *Column '91*. The Canadian Institute of Mining and Metallurgy and Petroleum, 1991.

[Bart83] Barto, A. G., Sutton, R. S., and Anderson, C. W. "Neuronlike adaptive elements that can solve difficult learning control problems." *IEEE Transactions on Systems, Man, and Cybernetics*, SMC-13(5), 834-846, 1983.

[Davi85a] Davis, L. "Job shop scheduling with genetic algorithms." *Proceedings of an International Conference on Genetic Algorithms and their Applications*, 136-140, 1985a.

[Davi85b] Davis, L., and Smith, D. "*Adaptive design for layout synthesis.*" Internal report, Dallas, TX: Texas Instruments, 1985b.

[Dejo75] De Jong, K. A. "Analysis of the behaviour of a class of genetic adaptive systems." Ph. D. Thesis, Ann Arbor, MI: University of Michigan, 1975.

[Eick93] Eick, C. F., and Jong, D. "Learning bayesian classification rules through genetic algorithms." *Proceedings of the Second International Conference on Information and Knowledge Management*, November, Washington, DC, 1993, to appear.

[Evan89] Evans, G. W., Karwowski, W., and Wilhelm, M. R. "*Applications of fuzzy set methodologies in industrial engineering*", Amsterdam: Elsevier, 1989.

[Feld93] Feldman, D. S. "Fuzzy network synthesis with genetic algorithms." *Proceedings of the Fifth International Conference on Genetic Algorithms*, 312-317, 1993.

[Finc90] Finch, J. A., and Doby, G. S. *Column flotation*. Toronto: Pergamon Press, 1990.

[Gold86] Goldberg, D. E., and Samtani, M. P. "Engineering optimisation via genetic algorithm." *Proceedings of the Ninth Conference on Electronic Computation*, 471-482, 1986.

[Gold89] Goldberg, D. E. "*Genetic algorithms in search, optimisation, and machine learning.*" Reading, MA: Addison-Wesley, 1989.

[Herb89] Herbst, J. A., and Rajamani, K. "Models for the dynamic optimisation of mineral processing plant performance." In K. V. S. Sastry and M. C. Fuerstenau (Eds.) *Challenges in mineral processing*. SME, 709-739, 1989.

[Holl75] Holland, J. H. "*Adaptation in natural and artificial systems*. Ann Arbor, MI: The University of Michigan Press, 1975.

[Holl71] Hollstien, R. B. "Artificial genetic adaptation in computer control systems." Ph. D. Thesis, Ann Arbor, MI: University of Michigan, 1971.

[Homa93] Homaifar, A., and McCormick, E. "Simultaneous design of membership functions and rule sets for fuzzy controllers using GAs." *IEEE Transactions on Fuzzy Systems*, 1993, to appear.

[Karr89] Karr, C. L. "Analysis and optimisation of an air-injected hydrocyclone." Ph. D. Thesis, Tuscaloosa, AL: The University of Alabama, 1989.

[Karr91a] Karr, C. L., Stanley, D. A., and Scheiner, B. J. "*A genetic algorithm applied to least squares curve fitting*" (Report of Investigations number 9339). Washington, DC: U.S. Department of the Interior, Bureau of Mines, 1991a.

[Karr91b] Karr, C. L. "Genetic algorithms for fuzzy controllers", *AI Expert*, **6** (2), 26-31, 1991b.

[Karr92] Karr, C. L. "Applications of advanced computing to mineral processing." Presentation at the International Mining Congress' MineExpo, November, Las Vegas, NV, 1992.

[Lutt93] Luttrell, G. H., Adel, G. T., and Yoon, R. H. "*Modelling of column flotation*." SME Annual Meeting, Preprint number 87-130, 1987.

[Proc78] Procyk, T. J., and Mamdani, E. H. "A linguistic self-organising process controller." *Automatica*, **15**, 15-30, 1978.

[Sast78] Sastry, K.V.S. "*Beneficiation of mineral fines, problems and research needs*." Report of a workshop held at Sterling Forest, New York, 1978.

[Sast88] Sastry, K. V. S., and Lofftus, K. "Mathematical modelling and computer simulation of column flotation." *Column Flotation '88*, SME, 57-63, 1988.

[Suge85] Sugeno, M., ed. "*Industrial applications of fuzzy control*." Amsterdam: Elsevier, 1985.

[Surm93] Surmann, H., Kanstein, A., and Goser, K. "Self-organising and genetic algorithms for an automatic design of fuzzy control and decision systems." *Proceedings of EUFIT'93*, 1993, to appear.

[Sun93] Sun, C., and Jang, J. "Using genetic algorithms in structuring a fuzzy rulebase." *Proceedings of the Fifth International Conference on Genetic Algorithms*, 655, 1993.

[Thri91] Thrift, P. "Fuzzy logic synthesis with genetic algorithms." *Proceedings of the Fourth International Conference on Genetic Algorithms*, 509-513, 1991.

[Vale91] Valenzuela-Rendon, M. "The fuzzy classifier system: A classifier system for continuously varying variables." *Proceedings of the Fourth International Conference on Genetic Algorithms*, 346-353, 1991.

[Wass89] Wasserman, P. D. "*Neural computing: theory and practice*." New York: Van Nostrund Reinhold, 1989.

[Wate89] Waterman, D. A. "*A guide to expert systems*." Reading, MA: Addison-Wesley, 1989.

[Wu91] Wu, K., and Nair, S. S. "Self-organising strategies for fuzzy control." *Proceedings of the North American Fuzzy Information Processing Society 1991 Workshop*, 296-300, 1991.

[Ynch88] Ynchausti, R. A., Herbst, J. A., and Hales, L. B. "Unique problems and opportunities associated with automation of column flotation cells." *Column Flotation '88*, SME, 27-33, 1988.

[Zade65] Zadeh, L. A. "Fuzzy sets", *Information and Control*, **8**, 338-351, 1965.

5

Neural Network Weight Selection Using Genetic Algorithms

David J. Montana

5.1 Introduction

Neural networks are a computational paradigm modelled on the human brain that has become popular in recent years for several reasons. First, despite their simple structure, they provide very general computational capabilities [Horn89]. Second, they can be manufactured directly in VLSI hardware and hence provide the potential for relatively inexpensive massive parallelism [Mead89]. Most importantly, they can adapt themselves to different tasks, i.e. learn, solely by selection of numerical "weights". How to select these weights is a key issue in the use of neural networks. The usual approach is to derive a special-purpose weight selection algorithm for each neural network architecture. Here, we discuss a different approach.

Genetic algorithms are a class of search algorithms modelled on the process of natural evolution. They have been shown in practice to be very effective at function optimisation, efficiently searching large and complex (multimodal, discontinuous, etc.) spaces to find nearly global optima. The search space associated with a neural network weight selection problem is just such a space. In this chapter, we investigate the utilisation of genetic algorithms for neural network weight selection.

Intelligent Hybrid Systems, Edited by S. Goonatilake and S. Khebbal.

5.2 Overview of the Technologies Involved

5.2.1 Neural Networks

Neural networks generally consist of five components

1. A directed graph known as the network topology whose nodes represent the processing elements and whose arcs represent the connections,
2. A state variable associated with each node,
3. A real-valued weight associated with each connection,
4. A real-valued bias associated with each node,
5. A transfer function f for each node such that the state of the node is $f(\Sigma \omega_i x_i - \beta)$, where β is the bias of the node, ω_i are the weights on the incoming connections, x_i are the states of the nodes on the other end of these connections.

To execute a neural network, one must initialise it by setting the states of a subset of its nodes. In particular, all the input nodes, i.e. those nodes which have no incoming connections, must have their states set at initialisation. Then, at successive time ticks (or continuously in the case of analog-networks), each node updates its state as computed by its transfer function based on the states of the other nodes feeding into this one. After enough time ticks that the network has reached steady-state (i.e., the node states have stopped changing significantly), the states of a chosen subset of the nodes are the network's output. In particular, the state of any output node, i.e. a node with no outgoing connections, is part of the network output (or else superfluous).

A feedforward neural network is one whose topology has no closed paths. To execute a feedforward network, the states of the input nodes are set. After n time ticks, where n is the maximum path length from an input node to an output node, all the nodes have assumed their steady-state values. Note that the computation from each of these n times ticks can be performed in parallel, and hence feedforward neural networks execute very quickly when fully parallel.

As mentioned above, many networks are able to adapt to the data, i.e. learn. In most cases, this requires a separate stage called "training". During training, the network topology and/or the weights and/or the biases and/or the transfer functions are selected based on some training data. In many approaches, the topology and transfer functions are held fixed, and the space of possible networks is spanned by all possible values of the weights and biases. It is this problem of weight selection which we discuss in this chapter.

5.2.2 Genetic Algorithms

Genetic algorithms are algorithms for optimisation and machine learning based loosely on several features of biological evolution [Holl75]. They require five components [Davi87]:

1. A way of encoding solutions to the problem on chromosomes.

2. An evaluation function which returns a rating for each chromosome given to it.
3. A way of initialising the population of chromosomes.
4. Operators that may be applied to parents when they reproduce to alter their genetic composition. Standard operators are mutation and crossover (see Figure 5.1).
5. Parameter settings for the algorithm, the operators, and so forth.

(19.3, 0.05, -1.2, 345, 2.0, 7.7, 68.0) —mutation→ (19.3, 0.05, -1.2, 345, 2.0, 8.2, 68.0)

(19.3, 0.05, -1.2, 345, 2.0, 7.7, 68.0)
(17.6, 0.05, -3.5, 250, 3.0, 7.7, 70.0) —crossover→ (17.6, 0.05, -1.2, 345, 3.0, 7.7, 68.0)

Figure 5.1 — The mutation and crossover operators.

Given these five components, a genetic algorithm operates according to the following steps:

1. Initialise the population using the initialization procedure, and evaluate each member of the initial population.
2. Reproduce until a stopping criterion is met. Reproduction consists of iterations of the following steps:
   a. Choose one or more parents to reproduce. Selection is stochastic, but the individuals with the highest evaluations are favoured in the selection.
   b. Choose a genetic operator and apply it to the parents.
   c. Evaluate the children and accumulate them into a generation. After accumulating enough individuals, insert them into the population, replacing the worst current members of the population.

When the components of the genetic algorithm are chosen appropriately, the reproduction process will continually improve the population, converging finally on solutions close to a global optimum. Genetic algorithms can efficiently search large and complex (i.e., possessing many local optima) spaces to find nearly global optima.

A key concept for genetic algorithms is that of schemata, or building blocks. A schema is a subset of the fields of a chromosome set to particular values, with the other fields left to vary. As originally observed in [Holl75], the power of genetic algorithms lies in their ability to implicitly evaluate large numbers of schemata simultaneously and to combine smaller schemata into larger schemata.

5.3 Introduction to the Application Domain

The fixed-features supervised pattern classification problem is the following. Consider a k-dimensional feature space and M separate classes. Different instances of a particular class generally have different values for the features and hence correspond to different points in feature space. Our only knowledge of how the instances of a particular class are distributed in feature space comes from a finite number of exemplars which are known to be of this class and whose feature vectors are known. The set of exemplars from all classes is called the training set. Then, the problem is: given a training set, select the most likely class for any feature vector not in the training set.

Training consists of selecting a network which minimises some error criterion (often taken to be the sum of the squares of the errors on the training data) which predicts how well the network will perform on test data (i.e., data not in the training set). Hence, training is an optimisation problem, requiring search through the space of possible networks for an optimal network. When the topology and transfer functions are fixed, this is optimisation over $\Re^n$, where n is the number of weights and biases left to vary.

We now discuss two different neural network architectures and their approach to pattern classification.

5.3.1 Sigmoidal Feedforward Networks

The most common neural network is a fixed-topology feedforward sigmoidal network. A sigmoidal network is one such that the transfer functions of all its nodes are sigmoids of the form $f: x \mapsto \frac{1}{1+e^{-x}}$. Such a network is picture in Figure 5.2.

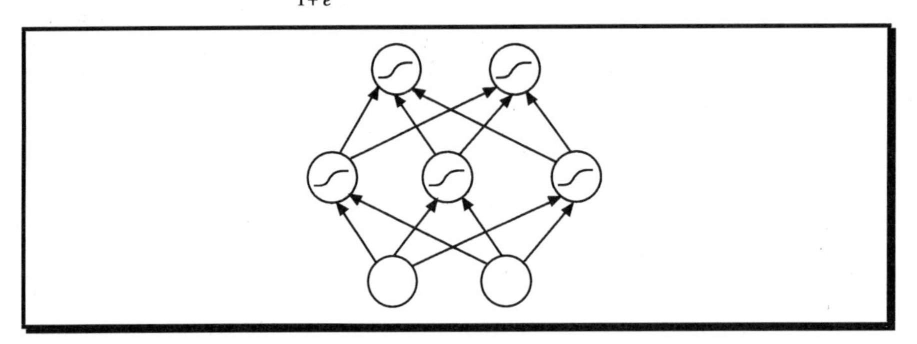

Figure 5.2 — A feedforward sigmoidal network.

For a pattern classification problem with k features and M classes, a feedforward network must have k input nodes, M output nodes, and an arbitrary number and configuration of hidden nodes, where hidden nodes are those which have both incoming and outgoing connections. When the states of the input nodes are set to particular feature values and the states are propagated through the network to the output nodes, the values of the output nodes are the likelihoods for each class. The evaluation function optimised during training is often the sum over the training exemplars of the output errors squared. These networks are usually trained using the backpropagation algorithm, which is a variant on gradient search and is described in [Rume86]. Below, we examine an alternate training algorithm.

5.3.2 Weighted Probabilistic Neural Networks (WPNN)

WPNN [Mont92] is a pattern classification algorithm which is an extension of Probabilistic Neural Networks (PNN) [Spec90] and which works as follows. Let the exemplars from class i be the k-vectors $\vec{x}^i_j$ for $j = 1,\ldots,N_i$ Then, the likelihood function for class i is defined to be

$$L_i(\vec{x}) = \frac{1}{N_i(2\pi)^{k/2}(\det\Sigma)^{1/2}} \sum_{j=1}^{N_i} e^{-(\vec{x}-\vec{x}^i_j)^T\Sigma^{-1}(\vec{x}-\vec{x}^i_j)} \quad (5.1)$$

where the covariance matrix Σ is a positive-definite kxk symmetric matrix whose entries are the free parameters. The conditional probability for class i is

$$P_i(\vec{x}) = L_i(\vec{x}) / \sum_{j=1}^{M} L_j(\vec{x}) \quad (5.2)$$

Note that the class likelihood functions are sums of identical (generally) anisotropic Gaussians centred at the exemplars.

Training WPNN consists of selecting the entries of Σ. This is equivalent to selecting a particular distance metric in feature space. Hence, our work is similar to that of [Atke91] and [Kell91], which are other nearest-neighbour-like algorithms which use modified metrics. Like [Atke91] and [Kell91], we have so far restricted ourselves to diagonal covariances, and we can hence think of the inverse of each of the diagonal entries of Σ as a weight on the corresponding feature. We therefore refer to these entries as "feature weights". Selection of feature weights is done by a genetic algorithm as described in section 5.6.

The neural network implementation of the WPNN algorithm is as follows. There are k nodes in the input layer. There are M nodes in the output layer, each with a normal (i.e., mean 0 and variance 1) Gaussian transfer function and a zero bias. There are kN nodes in the single hidden layer, where N is the total number of exemplars, each with a linear transfer function. The $(ik + j)^{th}$ hidden node (i) is connected to the jth input node with weight 1.0, (ii) has bias equal to the jth feature of the ith example, and (iii) is connected to the lth output node, where l is the class of the ith exemplar, with weight f_j, where f_j is the selected feature weight for the jth feature. Such a network is pictured in Figure 5.3.

5.4 The Rationale for Hybrid Processing

There are a few reasons why it can be beneficial to use genetic algorithms for training neural networks. With regard solely to the problem of weight (and bias) selection for networks with fixed topologies and transfer functions (which is all that we address in this chapter), there is the issue of local optima. As the number of exemplars and the complexity of the network topology increase, the complexity of the search space also increases insofar as the error function obtains more and more local minima spread out over a larger portion of the space. Gradient-based techniques often have problems getting past the local optima to find global (or at least nearly global) optima.

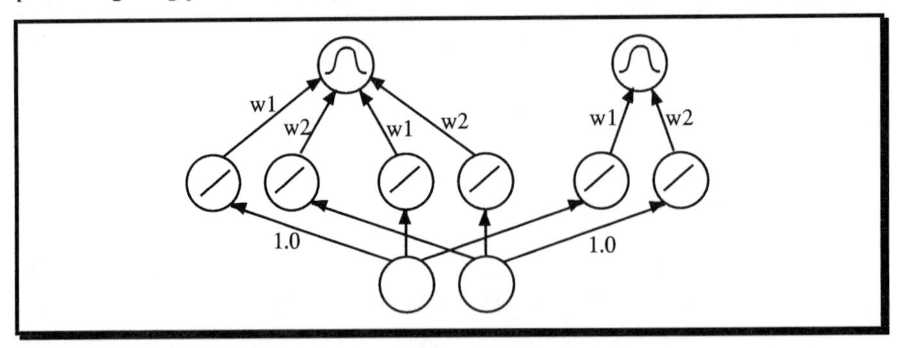

Figure 5.3 — A WPNN network corresponding to a pattern classification problem with two features, two classes, and three exemplars, two of class 1 and one of class 2.

However, genetic algorithms are particularly good at efficiently searching large and complex spaces to find nearly global optima. As the complexity of the search space increases, genetic algorithms present an increasingly attractive alternative to gradient-based techniques such as backpropagation. Even better, genetic algorithms are an excellent complement to gradient-based techniques such as backpropagation for complex searches. A genetic algorithm is run first to get to the "right hill" (i.e., a part of the space in a neighbourhood of a nearly global optimum) and the gradient-based technique climbs the hill to its peak [Mont89, Bele90].

A second advantage of genetic algorithms is their generality. With only minor changes to the algorithm, genetic algorithms can be used to train all different varieties of networks. They can select weights for recurrent networks, i.e. networks whose topologies have closed paths. They can train networks with different transfer functions than sigmoids, such as the Gaussians used by WPNN or even step transfer functions [Koza91, Coll91, Born91], which are discontinuous and hence not trainable by gradient techniques. Furthermore, genetic algorithms can optimise not just weights and biases but any combination of weights, biases, topology, and transfer functions. Possibly the most important type of generalisation allowed by genetic algorithms is the use of an arbitrary evaluation function. This allows potentially the use of discontinuous evaluation functions based on the principle of Minimum Description Length (MDL) [Riss89] to make the trained network generalise better. This ability to use arbitrary evaluation function becomes increasingly important as we attempt to expand neural networks beyond pattern classification and function approximation and towards domains such as control logic for autonomous vehicles or early visual processing, where performance is not measured as a simple sum of squares of the error.

A third reason to study genetic algorithms for learning neural networks is that this is an important method used in nature. While it is not clear that evolutionary learning methods are used in individual organisms (although some such as [Edel87] would claim so), it is relatively certain that evolutionary learning at a species level is a major method of "wet" neural network training. For example, in parts of the brain such as the early visual processing component, the neural network weights (in addition to the topology and transfer functions) are essentially "hard-wired" genetically and hence were "learned" via the process of natural selection.

5.5 Survey of Related Hybrid Systems

Genetic algorithms have been used together with neural networks in a variety of ways, each with their own body of literature. One such way is to use genetic algorithms for the pre-processing of neural network data, usually using the genetic algorithm to perform feature selection [Chan91, Deck93]. A second use of genetic algorithms with neural networks is to use the genetic algorithm to select the neural network topology [Harp89, Scha90]. A third such application is to employ genetic algorithms for selecting the weights of a neural network [Mont89, Mont91]. Some approaches optimise both the weights and the topology simultaneously [Koza91, Grua92, Coll91, Born91]. A fourth application is to use genetic algorithms to learn the neural network learning algorithm [Harp89, Bele90, Scha90]. We do not attempt to give a survey of the whole field. Entire papers have been devoted to this endeavour, and interested readers are referred to [Scha92] as a good overview. In this chapter, we concentrate solely on the problem of weight selection.

There are two basic differences between the different approaches to using genetic algorithms for weight selection. The first difference is that the weight selection is performed on different architectures. Examples of the different network architectures to which genetic weight selection algorithms have been applied are: feedforward sigmoidal [Mont89, Whit90, Bele90], WPNN [Mont92], cascade correlation [Karu92, Pott92], recurrent sigmoidal [Wiel91] recurrent linear threshold [Torr91], feedforward with step functions [Koza91], and feedback with step function [Coll91, Born91]. The second difference, and the one on which this chapter concentrates, is the difference in the genetic algorithms. However, we do not discuss these differences in this section; instead, in section 5.6.1 we will compare and contrast each of the five components of our genetic algorithms with other genetic algorithms after we discuss each component.

5.6 Case Studies

In this section, we describe how we have applied genetic algorithms to the problem of weight selection for the two different neural network architectures discussed in section 5.3. section 5.6.1 details the five components of the genetic algorithm used for each architecture and compares and contrasts them with other approaches. Section 5.6.2 gives some experimental results.

5.6.1 The Genetic Algorithms

5.6.1.1 Representation

WPNN: The genetic algorithm we used with WPNN employs a "logarithmic" representation. A chromosome is a fixed-length string of integers with values between 1 and K, with each location in the string corresponding to a particular feature weight. The map from an integer n in the chromosome to its corresponding feature weight values is $n \mapsto B^{n-ko}$. Here, K, k_0 and B are parameters we choose depending on the desired range and resolution for the feature weight values. We use a logarithmic representation because it provides large dynamic range with proportional resolution (and hence proportional representation in a randomly selected population) at all scales.

Feedforward Sigmoidal: The genetic algorithm we used with feedforward sigmoidal neural networks does not use any encoding scheme for the networks but rather uses the networks themselves as the chromosomes. This allows us to define genetic operators such as Mutate-Nodes and Crossover-Nodes (see below), which utilise the structure of the network when generating children. Note that when we use the weight-based genetic operators such as Biased-Mutate-Weights and Crossover-Weights (see below) rather than node-based operators, the representation we use is equivalent to using a real-valued fixed-length string representation.

Other Approaches: Some other genetic algorithms for weight selection use fixed-length binary string representations [Bele90, Wiel91, Whit89]. Their philosophy, which is common in the field, is that using a binary representation maximises the number of schemata and hence the amount of implicit parallelism in the search. While this philosophy is valid when we have no *a priori* knowledge of what the good schemata are, for the current problem we know that the bits which represent a single weight are inextricably linked and hence are much more likely than random bits to form a good schema. Forcing the genetic algorithm to learn what we already know makes the search much less efficient. [Whit90] provides experimental evidence of this theoretical argument.

We have taken this concept of treating as an indivisible unit elements which we know *a priori* to form such a unit one step further by using the operators Mutate-Nodes and Crossover-Nodes (see below). These operators treat all the incoming weights to a node as an indivisible unit, which makes sense because it is only the values of these weights relative to each other which matter. The experimental results discussed in section 5.6.2 validate this reasoning.

When the weights are optimised simultaneously with the topology, then the number of weights is not fixed and the representations are generally much less standard. For example, [Koza91] represents a neural network as a tree of variable size and shape. Similarly, [Grua92] represents a network using a constrained syntactic structure.

5.6.1.2 Evaluation Function

WPNN: To evaluate the performance of a particular set of features weights on the training set, we use a technique called "leaving-one-out", which is a special form of cross-validation [Ston74]. One exemplar at a time is withheld from the training set and classified using WPNN with those feature weights and the reduced training set. The evaluation is the sum of the errors squared on the individual exemplars. Formally, for the exemplar $\vec{x}_j^i$, let $\tilde{L}_q(\vec{x}_j^i)$ for $q = 1,\ldots,M$ denote the class likelihoods obtained upon withholding this exemplar and applying equation (5.1), and let $\tilde{P}_q(\vec{x}_j^i)$ be the probabilities obtained from these likelihoods via equation (5.2). Then, we define the performance as

$$E = \sum_{i=1}^{M} \sum_{j=1}^{Ni} ((1-\tilde{P}_i(\vec{x}_j^i))^2 + \sum_{q \neq i} (\tilde{P}_q(\vec{x}_j^i))^2) \tag{5.3}$$

We have incorporated two heuristics to quickly identify covariances which are clearly bad and give them a value of ∞ the worst possible score. This greatly speeds up the optimisation process because many of the generated feature weights can be eliminated this way. The first heuristic identifies covariances which are too "small" based on the condition that, for some exemplar $\vec{x}_j^i$ and all q=1,...M, $\tilde{L}_q(\vec{x}_j^i) = 0$ to within the precision of IEEE double-precision floating-point format. In this case, the probabilities $\tilde{P}_q(\vec{x}_j^i)$ are not well-defined. The second heuristic identifies covariances which are too "big" in the sense that too many exemplars contribute significantly to the likelihood functions. Empirical observations and theoretical arguments show that PNN (and WPNN) work best when only a small fraction of the exemplars contribute significantly. Hence, we reject a particular Σ if, for any exemplar $\vec{x}_j^i$,

$$\sum_{\vec{x} \neq \vec{x}_j^i} e^{-(\vec{x}-\vec{x}_j^i)^T \Sigma^{-1} (\vec{x}-\vec{x}_j^i)} > (\sum_{i=1}^{M} N_i) / P \tag{5.4}$$

Here, P is a parameter which we chose for our experiments to equal four.

Feedforward Sigmoidal: We used the same evaluation function as is standard for use with backpropagation, the sum of the squares of the errors on the training set. We used the same evaluation function to allow us to experimentally compare the genetic algorithm to backpropagation.

Other Approaches: While the evaluation function for the problem of pattern classification is similar across different systems, the evaluation function can in general be as varied as the problems to which the neural networks are applied. One such example is the application of neural networks to control using reinforcement learning [Whit91]. Here, the evaluation function is a score which summarises the feedback from the environment.

5.6.1.3 Initialisation Procedure

WPNN: The entries for each individual are integers chosen randomly with uniform distribution over the range 1, ..., K, where K is defined above.

Feedforward Sigmoidal: The weights and biases of each network in the initial population are real numbers randomly chosen from the two-sided exponential distribution $e^{-|x|}$. This distribution is an attempt to reflect the actual distribution of weights in typical neural networks; the majority of weights are between –1 and 1, but the weights can be of arbitrary magnitude.

Other Approaches: Most approaches use some variety of random seeding of the initial population.

5.6.1.4 Genetic Operators

WPNN: For training WPNN, we used the two standard genetic operators: mutation and crossover. The mutation operator is standard, randomly selecting a fraction of the alleles (i.e., locations in the string) and replacing their values with values randomly selected from the (finite) set of possible values. The crossover operator is uniform crossover, i.e. it selects from which parent to take the value for a particular allele independent of its choices for other alleles. This is in contrast to the point-based crossover operators (most commonly, 1-point and 2-point crossover), which except at a small number of cut points will select the values of neighbouring alleles from the same parent. Point-based crossover can provide an advantage when the low-order schemata tend to be formed from neighbouring alleles, as is the case when real numbers are encoded using a binary representation. However, since we are using a real-valued representation and since there is no particular ordering to the feature weights, uniform crossover is the better choice.

Feedforward Sigmoidal: For training feedforward sigmoidal networks, we experimented with eight different operators: three varieties of mutation (Unbiased-Mutate-Weights, Biased-Mutate-Weights, and Mutate-Nodes), three varieties of crossover (Crossover-Weights, Crossover-Nodes, and Crossover-Features), and two special-purpose operators (Mutate-Weakest-Nodes and Backprop). While some proved more effective than others in our limited experiments (see section 5.6.2), they all illustrate important concepts and all hold potential for the future. We now discuss each of them.

Unbiased-Mutate-Weights: This operator replaces the weights of randomly selected connections in the parent network with values chosen randomly from the same probability distribution used for initialisation.

Biased-Mutate-Weights: This operator does the same as Unbiased-Mutate-Weights except that, instead of replacing the old values with the randomly selected values, it adds the randomly selected values to the old values. So, the probability distribution for the new weight values is a two-sided exponential centred at the original value. The reasoning is that any network selected as a parent (especially as the run progresses) is significantly better than a randomly selected network, and so its individual weights are usually better than randomly selected ones. Hence, a mutation which uses the current weights as starting points will outperform a mutation which ignores the current weights. (This concept is similar to Davis' Creep operator [Coom87].) The experiments (see below) bears this out.

Mutate-Nodes: This operation randomly selects nodes from the set of all hidden and output nodes and performs a biased mutation on all the incoming weights for the selected nodes. As mentioned above, we know *a priori* that the set of incoming weights for a particular node forms a low-order schemata, or building block. A mutation which disrupts as few of these schemata as possible by concentrating its changes in a small number of them should outperform a mutation which disrupts many of these schemata. The experiments (see below) confirms this reasoning. (Note: While the genetic algorithm literature discusses extensively the problem of schema disruption due to crossover, there is little discussion of schema disruption due to mutation. It is perhaps the widespread use of binary representations together with maximally disruptive mutation operators which have cause the genetic algorithm community to devalue mutation.)

Crossover-Weights: This operator performs uniform crossover (described above) on the weights and biases of the two parent networks. Note that if we were to encode the weights in a string so that the incoming weights for any node were all adjacent, point-based crossover would probably be better than uniform crossover because it would disrupt less schemata. However, since we know where these schemata are, an even better approach is the following.

Crossover-Nodes: This operator performs uniform crossover with the basic units exchanged being the set of all incoming weights of a particular node. This preserves what we know *a priori* to be some of the low-order schemata. As discussed in section 5.6.2, our experiments were inconclusive about whether Crossover-Nodes was better than Crossover-Weights because the dominant operator was mutation.

Crossover-Features: This operator attempts to compensate for the problem of "competing conventions," a term first used in [Whit90] for a problem discussed earlier in [Mont89]. This problem arises because hidden nodes which are connected to the same set of other nodes are interchangeable, i.e. if two such nodes exchange all their weights then the output of the network will still always be the same. However, interchangeable nodes generally play different roles in a network. If two parent networks have assigned nodes from a set of interchangeable nodes to different roles, then the two types of crossover discussed above will not be effective.

Mutate-Weakest-Nodes: This operator is a version of node-based mutation which differs from Mutate-Nodes in the following ways. First, instead of randomly selecting which nodes to mutate, it mutates those nodes which contribute the least to the network. The contribution of a node is defined as the total error of the network with the node deleted minus the total error of the network with the node included. (We have devised an efficient method for calculating these contributions.) Second, instead of mutating just the incoming weights, it also mutates the outgoing weights. Third, if the contribution of the nodes is negative, it performs an unbiased mutation rather than a biased mutation. The intuition behind this operator is that performing mutation on nodes that are contributing the least to the network (and especially on those that are contributing negatively) will yield bigger gains on the average. As shown in section 5.6.2, Mutate-Weakest-Nodes does in fact yield large payoffs when at the beginning of a genetic algorithm run, before the population has started converging to good solutions and hence while there are still many nodes making negative contributions. However, it becomes ineffective after the population has converged and all the nodes are making positive contributions because the size of a node's positive contribution is a factor of both the role it plays as well as how well it plays it. Note that since this operator will never change nodes which are already performing well, it should not be the only source of diversity in the population.

Hill-climb: This operator creates a child by taking a single step in the direction of the gradient of the parent of a variable step size. (see [Mont89] for more details.) This operator is detrimental at the beginning of a run because it yields only incremental improvements to the parents at the same time that it acts to greatly reduce diversity. However, it is beneficial at the end of a run in order to help climb the final hill (see section 5.6.2).

Other Approaches: Some genetic algorithms use an additional operator, called the reproduction operator, which just copies a selected parent without change into the next generation. Because we use a steady-state genetic algorithm (see the next section), such an operator is superfluous.

Those genetic algorithms which have representations based on syntactically constrained structures [Koza91, Grua92] need special mutation and crossover operators which respect these syntactic constraints and which are similar in some respects to our node-based operators.

5.6.1.5 Parameter Values

WPNN: Some of the key parameter values are as follows:

Population-Size: We used a population size of 1600. We needed such a large population size because roughly 90 of the initial population evaluated to ∞ (i.e., did not pass the heuristic screener). Hence, only around 160 members of the initial population had any information at all, and this was about the minimum needed to provide sufficient diversity for reproduction.

Generation-Size: This parameter is the number of new individuals generated and old individuals discarded from the population during one iteration of the reproduction process. There are two big issues in selecting the value of Generation-Size. The first issue is whether to choose it to be small relative to Population-Size or to choose it equal to or slightly less than Population-Size. The former is known as the "steady-state" approach, while the latter is "generational replacement" [Sysw89]. Assuming that the evaluation function does not change with time (which is the case with our problem), the only argument in favour of the generational replacement approach is that it converges slower than the steady-state approach and hence is more likely to find a nearly global optimum. However, controlling convergence rate by proper selection of Parent-Scalar and Population-Size is seemingly a much more efficient way to protect against premature convergence than throwing away good individuals. The second issue is whether to choose Generation-Size to be one, or a small number greater than one. As discussed in [Mont91], this depends on whether the evaluations are being performed by a single CPU or by multiple CPUs. When there is a single CPU performing evaluations, the only effect of a generation size greater than one is that it imposes a delay between when good individuals are found to be good and when they can be used as parents. Hence, the optimal Generation-Size when using a single CPU is one. However, when using multiple CPUs, it is usually necessary to have Generation-Size greater than one in order to utilise the potential parallelism. Because we did our experiments with a single CPU, all the experiments except those evaluating full generational replacement used a Generation-Size of one.

Parent-Scalar: This is the parameter which determines the parent selection probabilities, i.e. the probability for each member of the population that it will be selected as a parent the next time one is needed for reproduction. We use (as provided by OOGA) an exponential distribution of parent selection probabilities in which these probabilities depend only on relative ranking within the population and the probability of selecting the $(n+1)^{st}$ best individual divided by the probability of selecting the nth best is the same for all n. This ratio is the parameter Parent-Scalar. The value of Parent-Scalar (in addition to Population-Size) controls the convergence rate of the genetic algorithm; making Parent-Scalar smaller (i.e., closer to zero) makes it converge faster, while making Parent-Scalar large (i.e., closer to one) makes it converge slower. We were able to find good solutions relatively quickly with Parent-Scalar set to 0.9.

Feedforward Sigmoidal:

Population-Size: We used a population size of 50. Using such a small population meant that the space was very coarsely sampled by the initial population and forced the run to converge quickly. After convergence, the genetic algorithm essentially became a hill climber using mutations to drive its ascent. It is not clear whether this is the desired behaviour for the genetic algorithm, but we probably should have at least experimented with larger populations and slower convergences.

Generation-Size: We used a generation size of one for the reasons given above.

Parent-Scalar: We used a value for Parent-Scalar which varied linearly from 0.93 to 0.89 as the run progressed. In this manner, we were able to do more exploration at the beginning of a run and more exploitation at the end of a run.

Other Approaches: Population sizes vary widely, but our two genetic algorithms with population sizes of 1600 and 50 are close to the extremes. [Whit91] describes experiments with a population size of 1, although this is correctly referred to as stochastic hillclimbing rather than a true genetic algorithm.

Many genetic algorithms use full generational replacement rather than steady-state replacement.

There are a variety of approaches to parent selection, including tournament selection and fitness-proportional selection, and there is not yet any clearly superior approach.

5.6.2 Experimental Results

5.6.2.1 WPNN

Because WPNN was a new and untested algorithm, our experiments with WPNN centred on the overall performance rather than on the training algorithm. To test the performance of WPNN, we constructed a sequence of four datasets designed to illustrate both the advantages and shortcomings of WPNN. Dataset 1 is a training set we generated during an effort to classify simulated sonar signals. It has ten features, five classes, and 516 total exemplars. Dataset 2 is the same as Dataset 1 except that we supplemented the ten features of Dataset 1 with five additional features, which were random numbers uniformly distributed between zero and one (and hence contained no information relevant to classification), thus giving a total of 15 features. Dataset 3 is the same as Dataset 2 except with ten (rather than five) irrelevant features added and hence a total of 20 features. Like Dataset 3, Dataset 4 has 20 features. It is obtained from Dataset 3 as follows. Pair each of the true features with one of the irrelevant features. Call the feature values of the ith pair f_i and g_i. Then, replace these feature values with the values $0.5(f_i+g_i)$ and $0.5(f_i-g_i+1)$, thus mixing up the relevant features with the irrelevant features via linear combinations. The results of the experiments are shown in Figure 5.4.

Algorithm \ Dataset	1	2	3	4
Backprop	11 (69)	16 (51)	20 (27)	13 (64)
PNN	9	94	109	29
WPNN	10	11	11	25
CART	14	17	18	53

Figure 5.4 — Performance on the four datasets of backprop, CART, PNN and WPNN (parenthesised quantities are training set errors).

Because WPNN fully met its theoretical expectations, the training set must have found good sets of weights. It was able to do this by evaluating relatively few points: about 300 per feature weight (i.e., 3000 for Dataset 1, 4500 for Dataset 2 and 6000 for Datasets 3 and 4) with about half of these eliminated quickly via the two heuristics.

5.6.2.2 Feedforward Sigmoidal

We have run a series of experiments discussed in detail in [Mont89]. We summarise the results here. First, comparing the three general-purpose mutations resulted in a clear ordering: (1) Mutate-Nodes, (2) Biased-Mutate-Weights, and 3) Unbiased-Mutate-Weights (see Figure 5.5). Second, a comparison of the three crossovers produced no clear winner (see Figure 5.6).

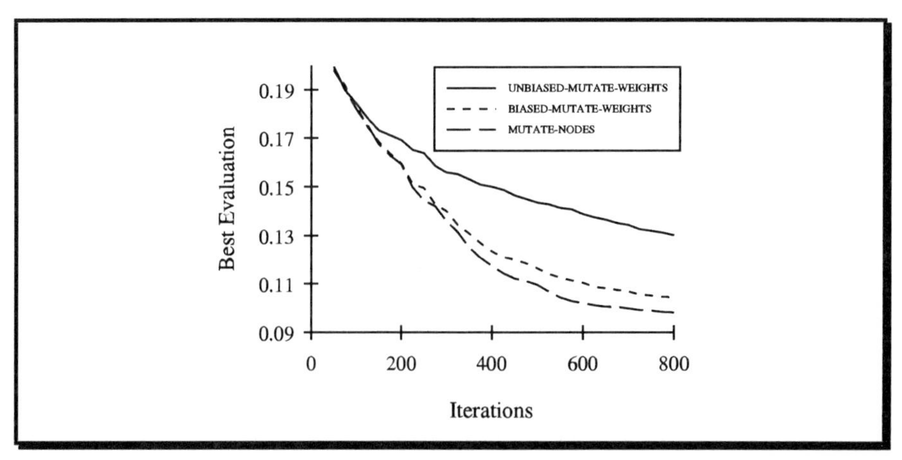

Figure 5.5 — Comparison of mutations.

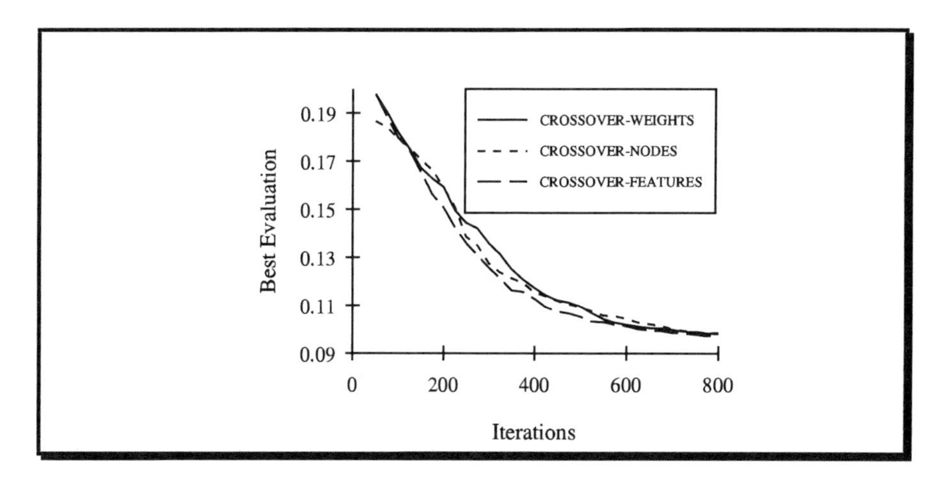

Figure 5.6 — Comparison of crossovers.

Third, when Mutate-Weakest-Node was added to a mutation and crossover operator, it improved performance only at the beginning; after a certain small amount of time, it decreased performance (see Figure 5.7). Fourth, a hill-climbing mode with just a Hill-climb operator and a population of size two gave temporarily better performance than Mutate-Nodes and Crossover-Nodes but would quickly get stuck at a local minimum. Finally, a genetic training algorithm with operators Mutate-Nodes and Crossover-Nodes outperformed backpropagation on our problem (see Figure 5.8).

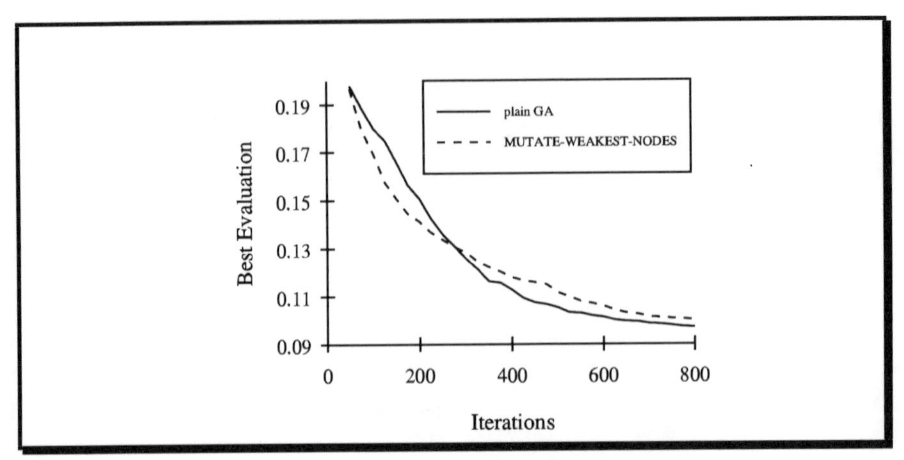

Figure 5.7 — Effect of Mutate-Weakest-Nodes.

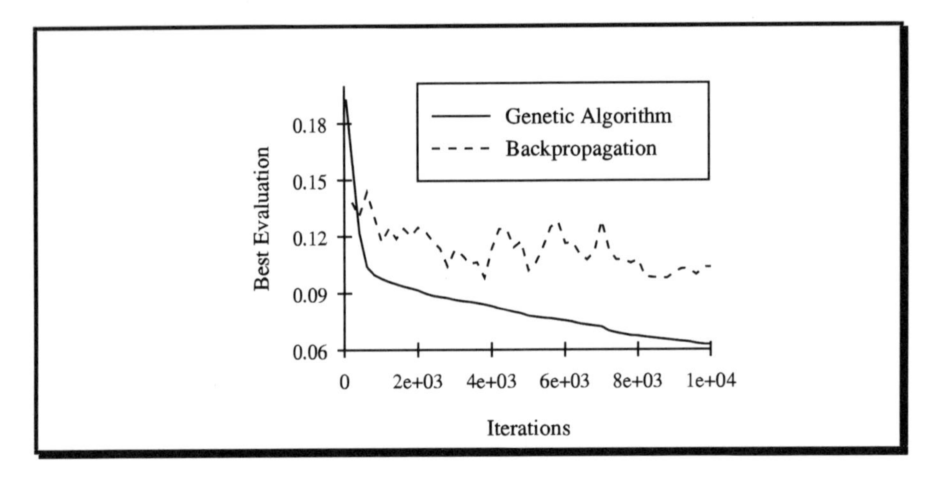

Figure 5.8 — Comparison of genetic algorithm with backpropagation.

5.6.2.3 Others

A variety of experimental evidence from other work supports some of the main conclusions of our work: first, genetic algorithms can find nearly globally optimal weight values and are especially useful when used in conjunction with a local hill climber [Bele90, Deck93], and second, genetic algorithms can train network architectures for which other training algorithms do not exist [Torr91, Koza91].

5.7 Future Directions

The application of genetic algorithms to neural network weight selection, and more generally neural network training, is a rapidly growing area of research with a few different trends. The first trend is towards more rigorous experimentation on "standard" datasets. In early work, researchers often used their own datasets on which gradient-based techniques did not work well (and hence which required a different type of training algorithm). However, it is now time to compare genetic algorithms against gradient-based techniques on well-known and well-characterised datasets on which gradient-based techniques have achieved success.

The second trend is towards applying genetic algorithms to new neural network architectures. In particular, whole new architectures, such as WPNN, are being developed because it is known that genetic algorithms will be able to train them.

The third, and probably the most important, trend is towards the use of more complex evaluations functions. The most interesting potential consequence of this is the acceleration of neural network development in areas other than pattern recognition and function approximation. While neural networks have provided small improvements over previous algorithms for these two tasks, neural networks will not reach their potential for making computers "smarter" until they can perform tasks such as early visual processing and control of autonomous vehicles. Without use of genetic algorithms, researchers have had to hand-design neural networks to perform such tasks, which cannot be evaluated as the sum of the errors squared. Genetic algorithms provide a potential method to learn neural networks for such tasks and hence to simplify the development of such networks and to make them more robust.

References

The work on training sigmoidal networks was done with Dave Davis using an early version of his OOGA code [Davi91].

[Atke91] Atkeson, C. G., "Using locally weighted regression for robot learning," *Proceedings of the 1991 IEEE Conference on Robotics and Automation*, pp. 958-963, 1991.

[Bele90] Belew, R. K. , McInerney, J., and Schraudolph, N., "Evolving Networks: Using the Genetic Algorithm with Connectionist Learning," *Artificial Life II*, pp. 511-547, 1992.

[Born91] Bornholdt, S. and Graudenz, D. *General Asymmetric Neural Networks and Structure Design by Genetic Algorithms*, Deutsches Electronen-Synchrotron, 1991.

[Chan91] Chang, E. J. and Lippmann, R. P., "Using Genetic Algorithms to Improve Pattern Classification Performance," *Advances in Neural Information Processing 3*, pp. 797-803, 1991.

[Coll91] Collins, R. and Jefferson, D. "An Artificial Neural Network Representation for Artificial Organisms," *Parallel Problem Solving from Nature,* 1991.

[Coom87]Coombs, S. and Davis, L., "Genetic Algorithms and Communication Link Speed Design: Constraints and Operators," *Proceedings of the Second International Conference on Genetic Algorithms and Their Applications,* 1987.

[Davi87] Davis, L. (ed.), "*Genetic Algorithms and Simulated Annealing,*" Pitman, London, 1987.

[Davi91] Davis, L., "*Handbook of Genetic Algorithms,*" von Nostrand Reinhold, New York, 1991.

[Deck93] Decker, D, and Hintz, J. C., "A Genetic Algorithm and Neural Network Hybrid Classification Scheme," *Proceedings of AIAA Computers in Aerospace 9 Conference*, 473-482, 1993.

[Edel87] Edelman, G. M., "*Neural Darwinism,*" Basic Books, New York, 1987.

[Gold88] Goldberg, D., "*Genetic Algorithms in Search, Optimisation and Machine Learning,*" Addison-Wesley, Redwood City, CA, 1988.

[Grua92] Gruau, F., "Genetic Synthesis of Boolean Neural Networks with a Cell Rewriting Development Process," *Proceedings of the IEEE Workshop on Combinations of Genetic Algorithms and Neural Networks,* 1992.

[Harp89] Harp, S. A., Samad, T. and Guha, A., "Towards the Genetic Synthesis of Neural Networks," *Proceedings of the Third International Conference on Genetic Algorithms,* pp. 360-369, 1989.

[Holl75] Holland, J. H., "*Adaptation in Natural and Artificial Systems,*" University of Michigan Press, 1975.

[Horn89] Hornik, K., Stinchombe, M. and White, H., "Multilayer Feedforward Networks are Universal Approximators," *Neural Networks,* vol. 2, pp. 359-366, 1989.

[Karu92] Karunanithi, N., Das, R. and Whitley, D., "Genetic Cascade Learning for Neural Networks," *Proceedings of the IEEE Workshop on Combinations of Genetic Algorithms and Neural Networks,* 1992.

[Kell91] Kelly, J. D. Jr., and Davis, L, "Hybridizing the genetic algorithm and the k nearest neighbours classification algorithm," *Proceedings of the Fourth International Conference on Genetic Algorithms*, pp. 377-383, 1991.

[Koza91] Koza, J. R., and Rice, J. P., "Genetic Generation of Both the Weights and Architecture for a Neural Network," *Proceedings of the IEEE Joint Conference on Neural Networks,* pp. 397-404, 1991.

[Mead89] Mead, C. A., "*Analog VLSI and Neural Systems,*" Addison-Wesley, Reading, MA, 1989.

[Mont89] Montana, D. J., and Davis, L., "Training Feedforward Neural Networks Using Genetic Algorithms," *Proceedings of the International Joint Conference on Artificial Intelligence,* pp. 762-767, 1989.

[Mont91] Montana, D. J., "Automated Parameter Tuning for Interpretation of Synthetic Images," in [Davi91], pp. 282-311.

[Mont92] Montana, D. J., "A Weighted Probabilistic Neural Network," *Advances in Neural Information Processing Systems 4,* pp. 1110-1117, 1992.

[Pott92] Potter, M.A., "A Genetic Cascade-Correlation Learning Algorithm," *Proceedings of the IEEE Workshop on Combinations of Genetic Algorithms and Neural Networks,* 1992.

[Riss89] Rissanen, J., "Stochastic Complexity in Statistical Inquiry", *World Scientific,* N.J., 1989.

[Rume86] Rumelhart, D. E., Hinton, G.E. and Williams, R. J., "Learning Internal Representations by Error Propagation," in *Parallel Distributed Processing, vol. 1,* MIT Press, Cambridge, MA, 1986.

[Scha90] Schaffer, J. D., Caruana, R. A. and Eshelman, L. J. "Using Genetic Search to Exploit the Emerging Behaviour of Neural Networks," in S. Forrest (ed.) *Emergent Computation,* pp. 102-112, 1990.

[Scha92] Schaffer, J. D., Whitley, D. and Eshelman, L. J., "Combinations of Genetic Algorithms and Neural Networks: A Survey of the State of the Art," *Proceedings of the IEEE Workshop on Combinations of Genetic Algorithms and Neural Networks,* 1992.

[Spec90] Specht, D. F., "Probabilistic neural networks," *Neural Networks*, vol. 3, no. 1, pp. 109-118, 1990.

[Ston74] Stone, M. "Cross-validatory choice and assessment of statistical predictions," *Journal of the Royal Statistical Society*, vol. 36, pp. 111-147, 1974.

[Sysw89] Syswerda, G., "Uniform Crossover in Genetic Algorithms," *Proc. Third International Conference on Genetic Algorithms,* pp. 2-9, 1989.

[Torr91] Torreele, J., "Temporal Processing with Recurrent Networks: An Evolutionary Approach," *Proc. Fourth International Conference on Genetic Algorithms,* pp. 555-561, 1991.

[Whit89] Whitley, D., "Applying Genetic Algorithms to Neural Network Learning, *Proceedings of the Seventh Conference of the Society of Artificial Intelligence and Simulation of Behaviour,* pp. 137-144, 1989.

[Whit90] Whitley, D., Starkweather, T. and Bogart, C., "Genetic Algorithms and Neural Networks: Optimising Connections and Connectivity," *Parallel Computing,* vol. 14, pp. 347-361.

[Whit91] Whitley, D., Dominic, S. and Das, R., "Genetic Reinforcement Learning with Multilayer Neural Networks," *Proceedings of the Fourth International Conference on Genetic Algorithms,* pp. 562-569, 1991.

[Wiel91] Wieland, A. P., "Evolving Neural Network Controllers for Unstable Systems," *IEEE Joint Conference on Neural Networks,* pp. 667-673, 1991.

Part Two

Intercommunicating Hybrids

6

A Unified Approach For Engineering Design

David J. Powell, Michael M. Skolnick and Shi Shing Tong

6.1 Introduction

Engineer's productivity and their ability to produce optimal designs is severely limited by the lack of a unified approach to design optimisation. Engineers have developed thousands of computer programs to simulate the performance of a product, but lack the means to most effectively use them in the iterative design process. Even though the experienced engineer has a lot of knowledge of the underlying, real world component that the computer code simulates, no single, constrained optimisation approach exists to exploit and supplement this knowledge and automatically iterate the simulation codes to an optimal design. Design optimisation is difficult because of the complexity of the parameter spaces and the time constraints imposed on the optimisation process. The parameter spaces modelled by the codes are high-dimensional, non-linear and can have both continuous and discontinuous regions. The time allotted to an engineer for a particular design is limited. As a result, the engineer is less concerned with obtaining the optimal design but rather the best design possible before the deadline. An efficient technique must focus on interim performance and obtain the greatest optimisation gain in as few runs of the code as possible. The existing optimisation techniques of expert systems, numerical optimisation and genetic algorithms cannot individually solve the engineer's problem.

A hybrid approach offers a solution to the engineer's problem. This approach uses a combination of expert systems, numerical optimisation and genetic algorithms.

Intelligent Hybrid Systems, Edited by S. Goonatilake and S. Khebbal. ©1995 John Wiley & Sons Ltd.

This approach has the individual advantages of each separate optimisation technique but offsets their disadvantages. Hence, it is called a "unified" approach to design optimisation. This "unified" approach is very general and allows the engineer to concentrate on the design problem without having to worry about the selection and fine tuning of optimisation algorithms.

The "unified" approach has been validated against a test suite of engineering codes, used in the preliminary design of the turbine for General Electric's largest commercial engine and in the design of General Electric's DC-motors. The test suite is proposed as a standard test suite for the genetic algorithm community. The use of the "unified" approach to optimisation within General Electric has improved the productivity of engineers by a factor of ten and has developed a turbine which surpasses the efficiency of turbines designed with other optimisation techniques.

6.2 Survey of Existing Optimisation Techniques

Several optimisation techniques have been used for design automation. These techniques are categorised as: expert systems, numerical optimisation and genetic algorithms. The major advantages and disadvantages of each technique are presented. The intent is to show that an advantage of one technique is commonly a disadvantage of another. The unified approach to design automation takes advantage of this and will integrate the above techniques to maximise their advantages and offset their disadvantages.

6.2.1 Expert Systems

Expert systems codify domain specific knowledge in the form of IF THEN rules. These rules are manipulated by forward and backward inference techniques to determine solutions to design problems. Expert systems have developed optimal designs in many domains such as the designs for refrigerator fans [Tong86], VLSI circuits [Jabr87], and bridges [Adel86]. The main advantages of using expert systems are: solution efficiency, use of engineers' experience, and solution explanation. The disadvantages of expert systems are knowledge acquisition, rule completeness, adaptability to change, and domain dependency. Knowledge acquisition is a difficult process because of the mismatch between the way the engineer expresses knowledge and the format required by the rule representation in the computer. Knowledge acquisition is also hindered by the inability of humans to express all of their knowledge and errors made in the transfer process [Zhou87].

6.2.2 Numerical Optimisation

Numerical optimisation has had tremendous development over the past 20 years. Non-linear constrained optimisation, sequential linear programming, sequential quadratic programming, modified method of feasible directions and the generalised reduced gradient method have emerged as the best numerical optimisation techniques

[Gabr89, Vand87, Vand89]. They have been successfully applied to optimisation problems in a large number of domains, e.g. chemical process design, aerodynamic optimisation, non-linear control systems, mechanical component design and structural design [Vand85]. These optimisation techniques use gradient approximations to calculate search directions. The advantages of numerical optimisation are its mathematical underpinning, general applicability to engineering designs and its wide application base. The disadvantages of numerical optimisation are: inability to exploit domain knowledge, extreme sensitivity to both problem formulation [Gero88] and algorithm selection, large amount of required computational effort, and the assumptions of both design variable independence and continuity within the parameter space.

6.2.3 Genetic Algorithms

Genetic algorithms, (GA), are a class of general purpose search strategies that strike a near optimal balance between exploration and exploitation of a parameter space [Holl75]. They are based on the heuristic assumptions that the best solutions will be found in regions of the parameter space containing relatively high proportions of good solutions, and, that these regions can be explored by the genetic operators of selection, crossover and mutation. The advantages of genetic algorithms are: it searches from a set of designs and not from a single design, it is not derivative based, it works with discrete and continuous parameters, and it both explores and exploits the parameter space [Gold89]. The major disadvantage of a genetic algorithm is the excessive number of runs of the design code required for convergence.

6.3 Description of Unified Optimisation Approach

The “unified” approach gets the greatest gain in as few runs of the code as possible by using explicit, domain knowledge about the underlying parameter space. This knowledge focuses the search in areas that the engineer knows will be the most beneficial. However, the engineers’ knowledge is typically incomplete. The “unified” approach supplements this incomplete knowledge with “implicit” knowledge supplementation search techniques.

6.4 Knowledge Directed Search

The engineer provides the explicit knowledge to the “unified” approach in the form of rules. An intuitive menu based interface lets the engineer add knowledge to the system by hiding the underlying rule syntax. This eliminates the need for a separate knowledge engineer to extract the information from the design engineer. These rules are manipulated by the knowledge directed search techniques of direct search and numerical optimisation. The content of a sample rule is shown in Figure 6.1. This figure shows the four major components of a rule: the goal, applicability conditions, rule actions and rule weight.

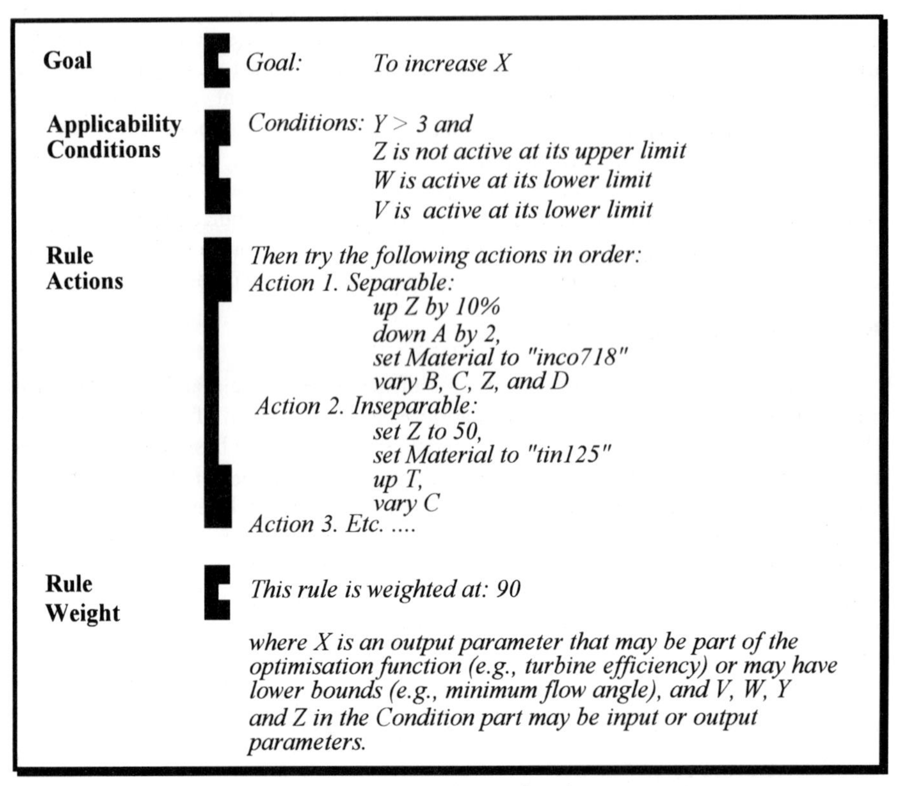

Figure 6.1 — A sample rule.

The goal indicates an output of a design code and the direction that it should be moved. For instance, the goal of Figure 6.1 is to increase X. This goal indicates that the output parameter X needs to be increased or maximised.

The applicable conditions under which this rule is valid consists of 0 or more expressions. If no expression is specified then the rule is always applicable. However, the engineer can supply some additional information that can narrow the range of applicability of the rule. This additional information can be through "LISP" and symbolic constraint expressions. LISP expressions allow the engineer to exactly specify parameter space conditions. For example, Figure 6.1 depicts the conditions that parameter Y must be greater than 3. Frequently, applicable conditions are based on the fact that a parameter is either on, inside or outside a lower or upper bound. These conditions are expressed in the form of symbolic constraints. For instance, Figure 6.1 has the symbolic constraints that parameter Z is not at its upper limit and parameters W and V are at their lower limits. This is called a symbolic constraint since the rule only specifies that a parameter is at a lower bound or an upper bound but not the actual numerical value of the bound.

If the goal of the rule is the goal of the current optimisation and furthermore, if the applicability conditions are satisfied then the rule actions specify which input parameters to change and how to change them. The rule actions are made up of action operators and action types. The allowable action operators are: up/down, up-by/down-by, up-by-percent/down-by-percent, set and vary. The up and down operators are specified when the engineer knows the direction to move a parameter but not the amount of the change. The amount will be determined by the current optimisation technique. For example, if numerical optimisation is used, then the amount will be determined by possibly a golden section method. The up-by and down-by operators are used when the engineer knows the direction of change and the relative amount that the parameter should be changed by. This relative amount is specified with a "LISP" expression which when evaluated determines the amount of relative change. For instance, the action "down A by 2" in Figure 6.1 specifies that A should have a relative decrease specified by the LISP expression which evaluates to 2. The up-by-percent and down-by-percent operators are similar to the up-by and down-by operators except that the LISP expression evaluates the percentage to change the parameter by. For example, the action "up Z by 10%" would increase the value of Z by 10%. The set operator is used to set a parameter to an exact value. A LISP expression is evaluated and the parameter is set equal to the value determined from the expression's evaluation. The last operator is the vary operator. The engineer knows that a parameter is important but does not know which direction to vary it. The optimisation technique will calculate gradient information to determine the direction of change.

The rule actions may specify many parameters to vary but the determination of whether some or any of these get varied is initially determined by the rule type of "separable" or "inseparable". Inseparable means that all of the parameters specified to be varied are both allowable design variables for this design and can be moved in the specified direction. For example, for a particular design action-1 of Figure 6.1, the action "down A by 2" is not possible if A is already at its lower limit. The action is "inhibited". In this case, since the rule is inseparable then the entire action-1 is not tried. This action type is important when it only makes sense to try an action if all of the specified parameters can be simultaneously varied. A rule type of separable means that if some but not all individual actions of the rule action are "inhibited" then the non "inhibited" parameters make up the rule action.

The last part of a rule is the rule weight. This weight is used to determine which rule to try first when more than one rule is applicable. The weight allows the engineer to rank the rules based on importance.

6.5 Manipulation of Explicit Knowledge

The "unified" approach uses the knowledge provided in the form of rules to focus on the key input parameters to vary. Instead of varying all of the design variables, those design variables which are most likely to have an impact are varied first. This reduces the number of design variables and the wasting of computer time in determining gradient approximations for parameters which have little or no effect. When a rule is applied, if the rule actions contain the operators vary, up or down then the "unified" approach uses either a direct search technique or a numerical optimisation technique to determine the amount that each parameter should be changed by. The direct search technique will vary each parameter individually by a increment that gets dynamically doubled or halved based on the success in improving the optimisation value. This technique is not concerned with continuous halving of a parameter increment until a tight convergence has been obtained. Rather, it focuses on getting close to a optimum value quickly for a individual parameter and then changing to another design variable. The direct search approach is not based on gradient calculations and can be quite effective in discontinuous or gradient insensitive regions.

The numerical optimisation technique does not vary a single parameter at a time but rather attempts to determine a single, composite search direction for all of the parameters with an action operator of vary, up or down. This search direction is determined by finite difference techniques and is pursued until specified convergence criteria are met. The ability to specify multiple operators for the same parameters within a rule gives the numerical optimisation technique the ability to avoid local optimum, constraint boundaries and gradient, insensitive regions. For example, the "unified" approach would use the action, (up-by A 10%) (vary A), to increase A by 10% and to then calculate gradients for A from this new point. If the jump to the new value of A and its subsequent movement by the numerical optimisation technique does not increase the optimisation value within a limited number of iterations then the previous design state is restored.

6.6 Knowledge Supplementation

Parameter spaces are infinite and constantly changing with various constraints placed on a design. Most engineers have limited experience in a given domain. Their knowledge is used to prune the design space but other optimisation techniques are needed to supplement incomplete "explicit" knowledge. The "unified" approach uses numerical optimisation and the genetic algorithm for this purpose. Numerical optimisation was used in knowledge directed optimisation with only design variables which appeared in rules. In this phase, numerical optimisation starts from the best design obtained by the knowledge directed search but uses all of the allowable design variables. The intent is to quickly climb to the local optimum. Exploration is sacrificed for exploitation to get the greatest gain in optimisation value in as few runs of the code.

Genetic algorithms play a crucial but purposely limited role within the "unified" approach. The genetic algorithm is entered only after the "more efficient" exploitation techniques have been used. The genetic algorithm provides a robust fail-safe technique to avoid the traps that limited the progress of the exploitation techniques. These traps include one or more of the following: a local optima, a gradient insensitive region, constraint boundaries, discontinuities or non independent input parameters.

Implicit knowledge is gained about the parameter space during its exploration and exploitation by direct search and numerical optimisation techniques. Genetic algorithms provide a means to exploit the implicit knowledge by using the states visited to *seed* the initial genetic algorithm population. This seeding provides "fertile" schemas for propagation and combination within the genetic algorithm. These schemas would eventually be discovered but at the expense of thousands of runs of the design code. To further enhance the efficiency of the genetic algorithm, a database of all previous designs is maintained. This database is used so that a genetic algorithm does not rerun a design code for a design that it has already run.

The role of the genetic algorithm is limited. It is entered only after the more efficient exploration techniques of numerical optimisation have been exhausted. Furthermore, the genetic algorithm is only run until an improvement in the objective function is found. Once a better design is found then the "unified" approach pursues this better design with the numerical optimisation techniques (hill climbing). When the numerical techniques again arrive at a impasse then the genetic algorithm is reseeded with the best designs discovered by the numerical optimisation and the past best designs from the last genetic algorithm population.

6.7 Interdigitation

Interdigitation involves the use of multiple optimisation techniques. The specification of each technique, the internal parameters for the technique, the run order of the techniques and the integration of the techniques is called an optimisation plan. A default optimisation plan has been developed that is very robust at obtaining an optimal design for a wide variety of parameter spaces. This default plan frees the engineer from having to know anything about optimisation or having to tune the optimisation technique to the engineering problem. The default optimisation plan is shown in Figure 6.2. This plan uses knowledge directed search followed by knowledge supplementation search techniques. After running through the optimisation plan, a transition is made back to the knowledge directed search techniques and the plan is rerun. The intent is to get the greatest gain in optimisation in as few runs as possible but if time permits then the parameter space will continue to be explored. The transitions from the genetic algorithm back to the other numerical and direct search techniques indirectly creates a genetic algorithm hill climb operator. The best design from the genetic algorithm is used as a new starting design for the next run through the optimisation plan. Whenever the genetic algorithm is entered on subsequent iterations through the optimisation plan then the genetic algorithms population is reseeded. The reseeding involves the replacement of the worse members of the population with the better designs found, if any, from the exploitation techniques.

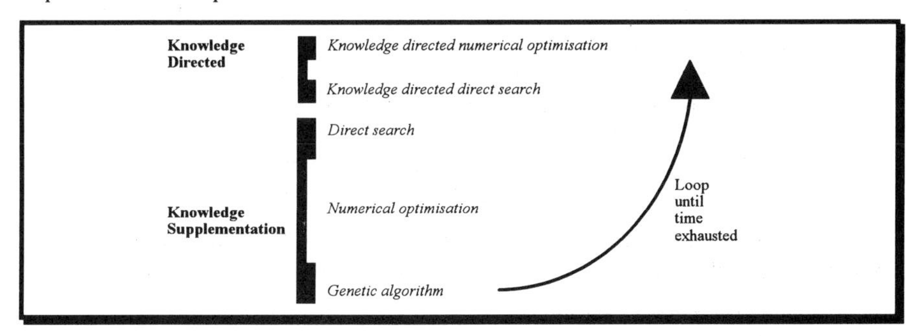

Figure 6.2 — A default optimisation plan.

The default plan is robust but the engineer can modify the optimisation plan or specify a optimisation plan directly tailored for a particular parameter space. For example, the user may decide that the problem is directly suited for numerical optimisation and may specify an optimisation plan that consists of the numerical optimisation technique, exterior penalty method, followed by the modified method of feasible directions.

6.8 Results

The robustness and validity of the "unified" approach is verified by its performance on first, a test suite of difficult engineering problems and secondly, on the design of a commercial airline turbine.

6.8.1 Test Suite of Engineering Problems

Six engineering problems were selected as adequate representations of different types of parameter spaces that a single optimisation approach must be able to handle. These problems are listed in Table 6.1. These problems were originally proposed by Sandgren [Sand77] but problems 2, 3, 4 and 6 have been dropped in numerical optimisation studies because "of their multiple optima and computer time required for a solution" [Gabr88]. It is exactly for this reason that they have been included in this study. Real world engineering problems are very complex and the time required for a single run of a program ranges from seconds to a couple of hours for a finite element analysis. In Sandgren's study and subsequent optimisation studies no single optimisation technique has solved all of these problems. In fact, the best any of the 25 optimisation algorithms used by Sandgren could do was to solve two of these problems. The last column of Table 6.1 relates the test suites numbering scheme with the numbering scheme used by Sandgren. For instance, problem 1 of the test suite is problem 3 of Sandgren's study.

Some of the features represented by the design problems of Table 6.1 are:

1. Variation of design variables, N, from 3 on problem 2 to 16 on problem 4.
2. Inclusion of inequality constraints, J, and equality constraints, K.
3. Initial starting points which are feasible (problems 1,2,3,6) and unfeasible (problems 4, 5).
4. Continuous and discontinuous parameter spaces (problem 3).
5. Large and small feasible regions. For example, problem 4 has a very small feasible region where just finding a feasible solution is difficult.
6. Well scaled and poorly scaled problems. For instance, problem 4 has a variation in scale among the design variables from .1E-5 to 1E+3.

Test results can vary widely depending on the testing conditions. Some of the characteristics of the controlled optimisation on the problems in Table 6.1 are:

1. No violation of constraints was allowed for a feasible solution.
2. All of the engineering problems were run in double precision. No

modifications to the code were made to obtain better results.

3. All the optimisation was done on a Sun 4/330.
4. The numerical optimisation algorithms used single precision and forward difference for all gradient calculations.
5. The elitist strategy was used for the genetic algorithm along with a standard crossover rate of 60% and a mutation rate of .1%.
6. The same optimisation settings were used by the unified approach for all the test problems. No settings were changed to enhance the performance of the unified approach for a individual problem.
7. No rules were available from design engineers for these engineering problems so only the knowledge supplementation features of the "unified" approach were tested.

Problem	N	J	K	Feasible Starting Point	f(x0)	f(x*)	Active J	San
1. Process design (MAX)	5	6	0	Y	3.0373	3.066 published 3.066 MMFD 3.055 SQP 3.066 unified	2	3
2. Chemical Reactor Design (MIN)	3	9	0	Y	-0.8756	-4.244 published (-415114) Sun -1.3 MMFD -1.3 SQP -3.19 unified	0	9*
3. Gear ratio selection (MIN)	5	4	0	Y	0.2802	0.2679 published 0.2737 MMFD 0.2737 SQP 0.2678 unified	0	13*
4. Five stage membrane separation (MIN)	16	19	0	No	284.739	174.786 published 294.24 MMFD 181 unified	16 (3 violated) 15 (0 violated)	22*
5. Geometric program-ming (MAX)	7	4	0	No	2205.86	1809.76 published 1822 MMFD 1820 unified	2 (1 violated)	23
6. Lathe (MIN)	10	14	1	Y	2931	-1614.938 published +1300 MMFD +1300 SQP -996.2 unified	3 (0 violated)	29*

Table 6.1 — Test set of engineering problems

The test results of Table 6.1 validate the robustness of the "unified" approach.

The column titled f(x*) shows the optimal value achieved either from published results [Sand77, Gabr88], the numerical optimisation technique of modified method of feasible directions, MMFD, [Vand85], the numerical optimisation technique of sequential quadratic programming, SQP, [Vand85] or the "unified" approach. Since the "unified" approach uses MMFD and SQP then the results of the "unified" approach will always be at least as good as the values shown for SQP and MMFD. On problems 1 and 5 the numerical optimisation technique was sufficient to obtain the published optimum. However, the "unified" approach needed the combination of genetic algorithms and numerical optimisation to achieve the published optimum on problems 3, and 4 and to come very close to the published optimum on problems 2, and 6. On problems 2, 3, 4, and 6 the numerical optimisation technique made progress but by itself it could not get to the optimum. For instance on problem 2 the numerical optimisation could only get to –1.3 but with the combination of genetic algorithms and numerical optimisation the optimisation could get to –3.17. Since the engineer does not know which optimisation algorithm is best for the engineering design problem, then the "unified" approach offers a robust, general purpose strategy for obtaining an optimal design.

On problems 2 and 6, the final solution from the "unified" approach varied by a significant amount from the published optimum. The reason for this variance is that the "unified" approach uses single precision and the published results are based on using double precision for the optimisation technique. The sensitivity of problem 2 to the precision of its inputs is easily shown by analysing the variation in the problems output when run at double and single precision. The result of running problem 2 in double precision at the optimum input vector of [7828.7954, 188.81406, 113.81406] was –4.15114 which is significantly different from the published result of –4.24. When problem 2 is run at single precision for the optimum input vector then the result is –4.28 but constraint #3 is now significantly violated. Its value is now –3.82909 instead of 0.4706. The main result from these two problems is that the "unified" approach did significantly better than using a single optimisation technique of MMFD or SQP.

6.8.2 Aircraft Engine Turbine Preliminary Design

A modern high-bypass engine turbine consists of multiple stages of stationary and rotating blade rows inside a cylindrical duct. In the so-called preliminary design stage all of the critical dimensions (e.g., the inner and outer wall) and flow characteristics are determined without defining the detailed blade shape. For turbine design there are approximately 110 input parameters and 20 relevant output parameters per stage. The code used to model the turbine is approximately 10,000 lines of FORTRAN code, uses single precision and takes approximately 30 seconds to run on a Sun 3/260.

Table 6.2 shows the results for a two stage turbine design. The same optimisation plan was used for the turbine design as for the test suite. However, unlike the test suite, knowledge was available. An experienced engineer defined 16 rules which involved eight different design variables. The use of the unified approach outperformed numerical optimisation by 0.7%, a very significant number for turbine efficiency. The final turbine was an unconventional design with some of the parameter distributions opposite to what was done traditionally. Analysis of the optimisation history shows that although the use of genetic algorithm resulted in only a small gain in efficiency, it does appear to have pushed the optimisation process away from being trapped in constraint boundaries so that the local hill climbing process could continue. In addition, the use of the "unified" approach found one more active output constraint and four more active side constraints than numerical optimisation.

Problem	N	J	K	Feasible Starting Point	f(x0)	f(x*)		Active J	San
Two stage Turbine (MAX)	20	18	0	Y	93.45	95.1	SQP	1	8
						95.87	unified	2	12

Table 6.2 — Optimisation results for a two stage aircraft engine turbine design.

6.9 Conclusions and Future work

No single optimisation technique exploits both domain knowledge and works for all parameter spaces. A "unified" approach of expert systems, numerical optimisation and genetic algorithms provides an answer to the engineer's problems. This approach runs the engineers simulation codes unmodified and allows the engineer to concentrate on the design problem instead of the optimisation algorithm. The genetic algorithm provides a fail safe technique which enables parameter space traps to be avoided. To make the genetic algorithm an efficient optimisation technique, it is used after exploitation techniques; it is seeded with good designs; and it uses a database to avoid rerunning the same design twice.

A test suite was used to validate the "unified" approach. This test suite consists of engineering problems and involves constrained optimisation. This test suite is recommended as a standard test suite both for the genetic algorithm and the design automation communities. Finally, the "unified" approach shows a significant benefit over individual numerical optimisation techniques when applied to the design of a two stage turbine at General Electric .

Future work on the unified approach is focused on automatically learning domain specific rules and to better control the genetic algorithm.

6.9.1 Learning Domain Specific Rules

The amount by which the expert system can increase the efficiency of numerical optimisation and genetic algorithms is dependent on the amount and quality of the knowledge available about the parameter space. If the domain specific knowledge is incomplete or inaccurate then domain independent knowledge supplementation techniques will discover better designs. One of the major disadvantages of not using domain knowledge with numerical and genetic algorithms was the extra computational effort required by more evaluations of the simulation code. Techniques of machine learning can be used to analyse the improved designs achieved by knowledge supplementation techniques to develop new rules which express the parameters to change and the conditions when they are applicable. This additional knowledge can now be used in future designs. The system will learn more about the parameter space with each design. This will make the optimisation of subsequent designs more efficient, the rule base more complete and provide more insight for the engineer of the parameter space in the form of symbolic rules.

6.9.2 Control over genetic algorithm

Currently, the genetic algorithm is run until a better design is found and then numerical optimisation techniques are used for exploitation. The genetic algorithm must find not only a better niche but also a design with a better optimisation value. The performance of the genetic algorithm might be greatly enhanced if only a design from a different niche had to be found. This design could then be exploited by a numerical optimisation technique.

Acknowledgements

The authors would like to thank Garret Vanderplaats for providing his ADS numerical optimisation package and John Grefenstette for providing his GENESIS genetic algorithm [Gref84].

References

[Adel86] Adeli, H., "Artificial intelligence in structural engineering", *Engineering Analysis*, vol. 3, Nov. 1986.

[Gabr88] Gabriele, G., "The Generalised Reduced Gradient Method for Engineering Optimisation", Chapter 2.5, *Engineering Design*, Elsevier Science Publishing Company, 1988.

[Gabr89] Gabriele, G., "Design Optimisation Course Notes", Course at Renselaer Polytechnic Institute, Troy, NY, 1989.

[Gero88] Gero, J., "Development of a knowledge-based system for structural optimisation", *Structural Optimisation*, Kluwer Academic Publishers, 1988.

[Gold89] Goldberg, D. E., *Genetic Algorithms in Search, Optimisation and Learning*, Addison-Wesley Publishing Company, 1989.

[Gref84] Grefenstette, J. *Genesis Software Package*, 1984.

[Holl75] Holland, J., "Adaptation in Natural and Artificial Systems", Univ. of Michigan Press, 1975.

[Jabr87] Jabri, M., "Implementation of a knowledge base for interpreting and driving integrated circuit floor planning algorithms", *AI in Engineering*, 1987, vol.2.

[Sand77] Sandgren, E., "The utility of non-linear programming algorithms", 1977, PhD thesis, University of Purdue.

[Tong86] Tong, S.S., "Coupling Artificial Intelligence and Numerical Computation for Engineering Design," AIAA-86-0242, *AIAA 24th Aerospace Sciences Meeting*, Reno, Nevada, January 6-9, 1986.

[Vand85] Vanderplaats, G., "ADS - A FORTRAN Program for Automated Design Synthesis", May 1985.

[Vand87] Vanderplaats, G., "Lecture Notes on Computer Aided Optimisation in Engineering Design", Union College 1 week course, July 27-31, 1987.

[Vand89] Vanderplaats. G., Personal conversation, July 1989.

[Zhou87] Zhou, H., "CSM: a genetic classifier system with memory for learning by analogy", PhD thesis, Vanderbilt University, 1987.

7

A Hybrid System for Data Mining

Randy Kerber, Brian Livezey, Evangelos Simoudis

7.1 Introduction

Database mining is the process of finding patterns and relations in large databases. The extracted information can be (1) organised by a data analyst into a model, (2) used to refine an existing model, or (3) provide a summary of the target database. The resulting model is subsequently employed for predicting the values of user-defined attributes based on new data. A number of database mining techniques have been developed and used, one at a time, in domains that range from space exploration to financial analysis. However, a single database mining technique has not proven appropriate for every domain and data set. Instead, several techniques may need to be integrated into hybrid systems and used co-operatively during a particular database mining operation. In this paper we present the Recon database mining framework that includes artificial intelligence and conventional data analysis techniques which can be used independently or co-operatively to extract information from databases with structured data. Recon has been applied in the domains of financial analysis, manufacturing process analysis, and analysis of demographic information.

The role of database mining for model creation is discussed in the next section. Recon's architecture is described in section 7.3. The database mining techniques used by Recon are described in section 7.4. An example of using Recon in the domain of stock portfolio selection is described in section 7.5. Related work is discussed in section 7.6. Our conclusions are presented in section 7.7.

Intelligent Hybrid Systems, Edited by S. Goonatilake and S. Khebbal.

7.2 Database Mining and Model Creation

The models generated through a database mining operation are generally statistical, neural, or symbolic. A typical statistical model is generated through the use of a regression method, (e.g., linear regression, non-linear regression). Each such model is represented as a system of equations. A neural model is represented as an architecture of weighted, linked nodes each of which incorporates a thresholding function. Associated with a particular architecture, (e.g. feedforward networks) is a learning function (e.g. the Delta rule). Symbolic models are represented either as sets of IF ... THEN ... rules, or as decision trees. They are usually generated through symbolic inductive algorithms [Kerb91, Quin86]. It also has to be noted that there exist techniques for extracting certain types of rules from trained neural networks [Gall88].

We view model generation as a process of incremental refinement performed by interweaving two operations: *pattern validation* and *data exploration*. Analysts often hypothesise value-prediction patterns. For example, a securities analyst may hypothesise that "if a public corporation reports high cash flow for three consecutive quarters, then the return on investment of its security during the next quarter will be greater than 5%." Before being incorporated into a model, each such pattern must be tested against the database to be mined, called the *target database*. For example, the previously mentioned pattern may be tested against a target database, call it *tdb1*, containing quarterly technical data of securities from the Fortune 1000 companies between 1985 and 1992. The database mining system performs this testing as part of the pattern validation operation. The testing could lead the analyst to accept the pattern and incorporate it into the model being developed, refine the pattern, or even reject it. For example, after testing the above pattern the analyst may modify its right hand side, changing the return on investment to 3%. Alternatively, the analyst may want to modify the contents of the target database, e.g., add to *tdb1* the performance of the Fortune 1000 company securities between 1980 and 1984.

In addition to patterns hypothesised by the analyst, the target database may support additional important patterns that can only be identified by intelligently exploring its contents. For example, through data exploration in *tdb1* the database mining system forms a pattern stating that "if a company's earnings per share growth is greater than 50% during six consecutive quarters, then the return on investment of its stock during the quarter following this period will be greater than 5%." Data exploration results in the generation of several patterns only few of which may be valuable. The analyst selects the subset of the best extracted patterns, possibly edits them, and incorporates them into the model being developed.

In addition to interweaving pattern discovery and data exploration, a database mining system must include several techniques for performing each operation due to the diverse characteristics of each target database, and the domain-specific goals of each value-prediction task. For example, a supervised symbolic inductive learning algorithm may be appropriate for developing models that can rank a set of companies based on the performance of their securities. A neural network using the backpropagation learning algorithm may be more appropriate for creating models for trading a set of selected securities based on the characteristics of the stock market.

Finally, two or more of the techniques integrated in a database mining system may need to be used co-operatively on the target database. For example, the contents of the *tdb1* database may first be clustered (a data exploration technique), then the contents of each cluster visualised (yet another data exploration technique), and finally, selected clusters be provided to a supervised rule induction component. Therefore, database mining systems are hybrid because of the two basic operations they must perform and the variety of techniques they must include to perform each operation.

The appropriate database mining operation and technique that must be used while developing a model is selected by the analyst. Each such selection is based on: (1) the amount and type of knowledge that is available before the database mining process begins, and (2) the characteristics of the target database which include (a) the target database's completeness (i.e., missing values), (b) the characteristics of the data (e.g., distributions), (c) the degree to which the attribute values are representative of the data that will be presented to the model during prediction, and (d) the amount of noise in the data.

7.3 Recon's Architecture

Recon consists of: (1) a graphical user interface, (2) six database mining components, each of which is implementing a different technique, (3) a server component, (4) a knowledge repository, and (5) interfaces to relational database management systems. Two of the database mining components (deductive database processor and case-based reasoner) are used for performing pattern validation. The remaining four (inductive learning, conceptual clustering, data visualisation, and statistical analysis) are used for performing data exploration. Other techniques/components, e.g., neural networks, can also be incorporated in the system on an as-needed basis. Recon's architecture is shown in Figure 7.1.

Recon's architecture is based on a distributed client/server model. Each mining component is a client to the Recon server that accesses and maintains the target database, and the knowledge repository. The server accesses existing heterogeneous databases via commercial-off-the-shelf database and network connectivity software, e.g., SQL Connect [1], OMNISQL[2], etc.

A database mining technique (1) may only be able to operate on data of specific types, e.g., neural inductive systems, statistical analysis systems, and data visualisation systems can only work on numeric data, and (2) has a degree of tolerance to missing values and noise in a data set. Recon assumes that each of its database mining components is responsible for transforming the target database to the necessary formats (e.g., transforming nominal values to numeric), filling missing values if necessary, and locally maintaining the transformed data. The mined knowledge, and all user-provided domain knowledge are stored in Recon's *knowledge repository*. In this way, the user-defined knowledge, which consists of high-level concepts, called *derived attributes*, and

---

[1] SQL Connect is a registered trademark of the Oracle Corporation.
[2] OMNISQL is a registered trademark of Sybase Inc.

interrelations among attributes and among values in the target database, can be shared among several models and analysts.

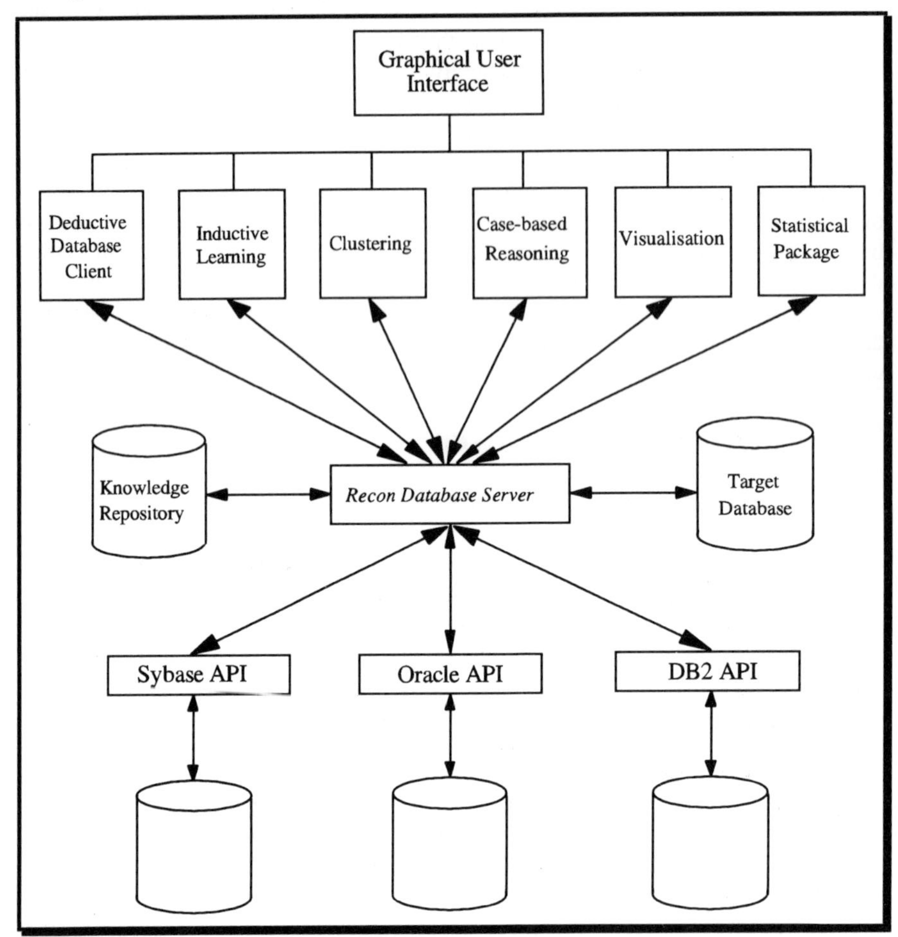

Figure 7.1 — The architecture of the Recon database mining system.

7.3.1 Creating the Target Database

The analyst creates, and later refines, the target database by interacting with the Recon server. In particular, the analyst first connects to a database. The Recon server automatically extracts the database's schema and displays its contents to the analyst. The displayed information includes a description of the database's relations (tables), called *base relations*, complete with the attributes of each relation, the type of each attribute's values, the number of different values each attribute takes, the minimum and maximum values of each attribute, etc. From the schema the analyst selects the base relations, attributes, etc., that will comprise the target database. The Recon server performs the appropriate join operations thus forming the target database.

7.4 Database Mining Techniques

In this section we describe each database mining technique/component integrated in the Recon system, explain how each technique is used during database mining, and justify why these particular techniques were selected.

We begin with the techniques used for pattern validation: deductive database processing and case-based reasoning. We continue with the techniques used for data exploration: supervised inductive learning, conceptual clustering, data visualisation, and statistical analysis.

7.4.1 Deductive Database Processing

Recon's deductive database processing component consists of (1) the client module that is used for (a) defining *concepts*, (b) formulating queries, (c) representing patterns that must be validated, and (d) refining existing concepts and patterns, and (2) the *Recon server* that is used for (a) maintaining the target database and Recon's knowledge repository, (b) distributing knowledge from the knowledge repository to the other database mining components, (c) processing queries issued by the client module, and (d) validating the patterns formed by the analyst.

7.4.1.1 The Deductive Database Client Module

In Recon, a concept is defined in terms of its name, a set of attributes, and one or more functions that provide alternative ways of establishing its values. An attribute may either belong to one of the target database's base relations, or be part of another, previously-defined, concept. Each function accepts as inputs the values of a subset of the included attributes. For example, assume that the target database contains annual fundamental data about the Fortune 1000 companies between 1982 and 1992 and wants to create a new concept called *strong security*. The analyst then selects the base relation called *market-year* from the target database. Of the eight attributes this relation includes, the analyst selects the following four to include in the new derived relation: *ticker symbol*, *week*, *liabilities*, *book value*. The analyst then finally defines the following logical function:

$$\textit{strong-security} = \textit{liabilities} < 10\% * \textit{book value}$$

The analyst also uses the client module of the Recon's deductive database processor to form queries that may include either base relations or user-defined concepts, and IF... THEN ... patterns. The antecedents and consequents of these patterns consist of defined concepts. For example, a financial analyst may want to validate the following pattern against a target database with corporate technical and fundamental data of the Fortune 1000 companies between 1982 and 1992: "IF a security is *strong* at time t, THEN its *return on investment* two quarters later will be greater than 5%." Assume that *return on investment* is a previously-defined concept.

Concepts and patterns discovered by Recon's other database mining components can be refined by the analyst using the client module, and then tested through Recon's server. For instance, the analyst can test patterns that have been generated by Recon's inductive learning component.

7.4.1.2 The Recon Server

Once defined, each concept is sent to Recon's server which automatically incorporates it in the target database's schema, and stores it in the knowledge repository. A database containing such user-defined knowledge is called a *deductive database*. The values of concepts are not stored in the target database. They are computed dynamically via deductive inferences in response to an analyst's query. In particular, after a query is formulated by the analyst, it is sent to the Recon server which expands it automatically until it consists solely of base relations and computable functions. If several alternative value-derivation functions are associated with a concept, then the resulting queries tend to be very complex and computationally expensive to compute. Recon's server transforms each expanded query into a set of optimised Structured Query Language (SQL) expressions. The expanded query is then posed to the target database and the answers returned to the user.

The Recon server can provide justifications to query responses. The justifications detail the exact chain of reasoning that led to the result in question. In this way, the analyst can better understand the results and determine whether the definitions of particular concepts need to be modified/refined, or whether the data in the target database is anomalous.

In addition to queries, every stated pattern is sent to the Recon server for validation. The server first expands the concepts included in the pattern until they consist solely of base relations and computable functions, and then automatically creates two queries that it poses to the target database. The first query returns the set of records that support the pattern. The second query returns the set of records that do not match the pattern's right-hand side.

The server supports interaction with Recon's other database mining components, providing them with stored patterns and concepts. For example, the analyst may communicate to the rule induction component the definition of the concepts *strong security*, and *return on investment*, as well as the previously-stated hypothesis that relates the two concepts. The inductive component can incorporate these concepts in rules it discovers. Furthermore, it can eliminate, generalise, or specialise the communicated rules. In a similar manner, the server can provide analyst-defined knowledge to Recon's conceptual clustering component. In particular, a hierarchy of analyst-defined concepts can become the "scaffold" on which Recon's conceptual clustering component can add more concepts.

7.4.2 Case-based Reasoning

Case-based reasoning (CBR) is a technique that uses the characteristics of, and the actions taken during previous problem-solving experiences to reason about new situations. It is a particularly appropriate technique when the target database is small and, consequently, statistically significant rules cannot be induced. CBR is used in two ways. First, to define new concepts, represented as sets of related patterns/characteristics, in terms of similarity to a prototypical instance. By contrast, deductive database processing is used to define concepts in terms of a set of conditions. Second, during the pattern validation operation, the data analyst uses previously defined concepts to establish sets with similar characteristics.

A case-based reasoning system consists of two major components: a *case memory* and a *problem-solver*. The case memory contains descriptions of previously solved and unsolved problems, termed *cases*. The case memory in Recon is the target database. For example, a case may include prototypical descriptions of a set of strong securities and the average return on investment of each such security over a ten-year period. One description may include features of pharmaceutical companies with low debt, while another include the features of publicly traded, medium-size, high-technology companies.

In general, the problem-solver of a CBR system consists of two components: a *case retriever* and a *case reasoner*. The case retriever identifies from the case memory the cases that are most appropriate to a particular problem description, and presents them to the case reasoner that adapts them as necessary to solve the new problem. In database mining applications, the problem-solver includes only a case retriever.

The simplest retrieval method, called *nearest neighbour comparison* [Duda73], compares the stated characteristics of a new problem to the corresponding features of each case stored in the case memory, and returns the cases that most closely match these characteristics. Matching may be exact or fuzzy. For example, the analyst may specify that only a subset of the stated characteristics of a new situation need match before a case is retrieved. Furthermore, two characteristics being compared may have similar but not identical values. This type of fuzzy matching cannot be performed using deductive database processing techniques. Fuzzy matching provides the ability to mine information that represents near misses of concepts stated by the analyst. It allows the analyst to ask for the instances (records) that are the best examples of a concept, ordered according to their degree of similarity to the prototype. The retrieved cases are manually examined by the analyst through Recon's user interface.

Nearest neighbour comparison is computationally expensive because each time it checks exhaustively every case in the target database. In order to avoid performing such an exhaustive comparison without compromising the accuracy of the case retriever, the contents of the target database can be indexed. Recon's CBR component can automatically index the target database using inductive, statistical, or knowledge-based techniques.

7.4.3 Supervised Inductive Learning

While pattern validation allows the analyst to express knowledge and test it against the target database, supervised inductive learning, or supervised induction, is used to automatically explore the target database to extract knowledge. Supervised induction refers to the process of automatically creating a classification model from a set of instances (examples), called the *training set*. The training set may either be a sample of the target database or the entire database. The instances in the training set must belong to a *small* set of classes, also known as *concepts*, that have been predefined by the analyst. For example, assume that the target database contains historical data about the quarterly performance of the securities listed in the New York Stock Exchange over the past seven years, including the quarterly return of investment of each security. Further assume that the analyst defines two classes/concepts: "high return on investment," for securities with quarterly return on investment that is greater than 5%, and "low return on investment," for the rest of the securities. The induced model consists of patterns, essentially generalisations over the instances, that are useful for distinguishing the classes. Once a model is induced it can be used to automatically predict the class of other unclassified instances. Symbolic methods create models that are represented either as *decision trees*, or as IF ... THEN ... *rules*. The former are generated using algorithms such as ID3, and CART [Brei88]. The latter are generated by algorithms such as FOIL [Quin90].

Recon's supervised inductive learning component is based on the OTIS system. OTIS is a symbolic induction algorithm that represents a classification model as a collection of rules. It was selected for inclusion into Recon over other such algorithms for three reasons. First, because of its ability to produce high quality models even when the data in the training set is noisy and incomplete. Second, because it represents the classification model using rules, the same representation used by Recon's deductive database processor. Finally, because of its ability to accept knowledge established during the pattern validation operation. Such domain knowledge can expedite the rule-induction task while simultaneously improving the quality of the induced model. Domain knowledge and the training set are provided to the supervised induction component by the Recon server.

Since it is rarely possible to find a single, perfect discriminating pattern that matches all examples in one category and no examples in any other category, Recon's supervised induction component produces a collection of patterns, each of which is assigned a strength factor to indicate how well it predicts each of the classes. For example, using a training set with quarterly, technical and fundamental data of 1,500 securities traded in the New York Stock Exchange over a seven-year period the supervised induction component derived 90 patterns, two of which are shown below:

1. IF $64.5 < \frac{\text{sales}}{\text{debt}} < 66.5$ THEN *return-on-investment* < 0.2 strength $= 0.8$

2. IF $0.056 < \frac{\text{net-worth}}{\text{size}} < 0.064$ THEN *return-on-investment* > 0.2 strength $= 0.77$

Symbolic supervised induction systems, such as Recon's, offer several advantages over statistical model-creation methods, and neural induction methods. In particular, the induced symbolic patterns can be based upon local phenomena while many statistical measures check only for conditions that hold across an entire population with well understood distribution. For example, an analyst might want to know if one attribute is useful for predicting another in a population of 10,000 instances. If, in general, the attribute is not predictive, but for a certain range of 100 values it is very predictive, a statistical correlation test will almost certainly indicate that the attributes are completely independent because the subset of the data that is predictive is such a small percentage of the entire population. Recon's supervised induction component is likely to find such a pattern since its evaluation functions would rate such a pattern very highly.

A significant advantage of symbolic inductive learning systems, especially ones that represent the created models in rule form, over connectionist systems is that the constructed classifiers communicate to the analyst symbolically what has been extracted from a data set. The explicit representation of the classification patterns has two advantages. First, through Recon's user interface, the analyst has the ability to examine the constructed patterns and (1) eliminate the ones he feels are useless or (2) examine closely others that provide unexpected results. Second, as was mentioned in section 7.4.1, the analyst can test induced patterns using the deductive database processing component, and manually refine them as appropriate in an effort to increase their predictiveness. Thus, the analyst is able to fine-tune a set of patterns rather than being required to put blind faith in a black-box classifier that provides little explanation, feedback, or insight.

7.4.4 Clustering

Clustering is used in Recon to partition the target database into subsets, the clusters, with the members of each cluster sharing a number of interesting properties. The results of a clustering operation are used in one of two ways. First, as an input to other data exploration methods, e.g., supervised inductive learning. A cluster is a smaller and more manageable data set to the supervised inductive learning component. Second, for summarising the contents of the target database by considering the characteristics of each created cluster rather than those of each record in the database.

Clusters can be created either statistically, or using neural and symbolic unsupervised induction methods. The various methods are distinguished by (1) the type of attribute values they allow the records in the target database to take, e.g., numeric, nominal, structured objects, (2) the way they represent each cluster, and (3) the way they organise the set of clusters, i.e., hierarchically or into flat lists.

Statistical and neural methods can only cluster records with numeric-valued attributes. Statistical methods represent a cluster as a collection of instances. It is difficult to decide how to assign a new example to existing clusters since one must define a way for measuring the distance between a new instance and the instances already in the cluster. It is also difficult to predict the attributes of members of a cluster. One can identify the attribute's value by applying a statistical procedure to the entire data set. Neural clustering methods, e.g., feature maps [Koho82], represent a cluster as a prototype with which they associate a subset of the instances in the data set being clustered. Both statistical and neural methods organise clusters into flat lists. Symbolic clustering methods, e.g., [Mich83, Lebo87, Fish88], operate primarily on instances with nominal values. They consider all the attributes that characterise each instance and use artificial intelligence-based search methods to establish the subset of these attributes that will describe each created cluster.

Recon uses a symbolic clustering component that is based on the COBWEB algorithm [Fish88]. Two other clustering algorithms have been analysed considered for inclusion into Recon: UNIMEM [Lebo87], and AUTOCLASS [Chee88]. The results of this analysts are described in [Kerb93]. COBWEB represents each cluster as a probability distribution over the space of possible attribute values. The produced clusters are organised hierarchically. COBWEB has been extended to deal with both numeric- and nominal-valued attributes, and utilise domain-specific knowledge about the attributes. Once the target database is clustered, the analyst can examine the created clusters to establish the ones that useful or interesting using either Recon's deductive database processor or its visualisation component.

Clustering differs from Recon's other database mining techniques in that the objective is generally far less precise. Because of this lack of a precise objective, clustering algorithms will often not cluster the instances along dimensions the analyst finds useful. Major contributing factors are a sensitivity to redundant and irrelevant features and because all attributes are usually given equal importance. This problem is alleviated in Recon by permitting the analyst to direct the clustering component to ignore a subset of the attributes that describe each instance. In addition, the analyst can assign a weight factor to each attribute; increasing the weight of an attribute increases the likelihood that the algorithm will cluster according to that attribute. The analyst can use Recon's statistical component to aid in determining weights by investigating which attributes are predictive of other attributes.

7.4.5 Data Visualisation

Recon's visualisation component provides analysts with visual summaries of data from the target database and Recon's other analysis components. Features that are difficult to detect by scanning rows and columns of numbers often become obvious when viewed graphically. Recon's visualisation component is based on the *Rave* system [Simo94]. *Rave* employs interactive visualisation techniques that allow the analyst to quickly and easily change the type of information displayed, as well as the particular visualisation method used (e.g., change from a histogram display to a scatter plot display). It uses both well-known visualisation templates, e.g., scatter plots, and domain-specific visualisation templates developed by Lockheed. *Rave* interacts with commercially available visualisation systems, e.g., AVS, PV-Wave.

Visualisations are particularly useful for noticing phenomena that hold for a relatively small subset of the data, and thus are "drowned out" by the rest of the data when statistical tests are used since these tests generally check for global features. In addition, visualisations will often be more effective than statistics for distributions of data that do not conform to the well-behaved, mathematically-correct distributions statisticians tend to rely on.

The advantage of using visualisation is that the analyst does not have to know what type of phenomenon he is looking for in order to notice something unusual or interesting. For example, with statistical tests the analyst must ask rather specific questions, such as "does the data fit this condition?" Often, the analyst wants to discover something unusual or interesting about a set of instances or an attribute. However, he must ask very directed questions, such as "is the distribution skewed?" or "is this set of values consistent with the Poisson distribution?" No general statistical test can answer the question "is there anything unusual about this set of instances?"; there are only tests for determining if the data is unusual in a particular way. Visualisation compensates for this: humans tend to notice phenomena in visually displayed data precisely *because* they are unusual.

The most severe limitation of visualisation is its inability to display information on more than a few attributes simultaneously. The relationship between three attributes can be displayed effectively using visualisations such as three-dimensional scatter plots. At least two additional dimensions can be displayed by superimposing one visualisation on top of another, e.g., a mesh plot on top of a three-dimensional bar plot, and adding colour to indicate intensity. Finally, if one of the variables being displayed is "time", then animation can be used to display this additional dimension. In order to extract information from data of high dimensionality, the other data exploration components included in Recon will need to first be used to provide analysts with clues as to which aspects of the data might be most promising to examine visually. Therefore, in many cases it is best to collaboratively use supervised induction, clustering, statistics, and visualisation.

7.4.6 Statistical Analysis

In Recon, statistical analysis is used to (1) summarise data in the target database and from Recon's other database mining components, (2) answer specific questions about the statistical validity of the analyst's hypotheses, and (3) support the analyst in using statistical measures for non-directed data exploration in order to establish or refine concept definitions.

Statistical data summarisation uses measures such as the mean, standard deviation and degree of correlation to address many of the same objectives as visualisation: to give the analyst an understanding of the properties and characteristics of the target database's contents to facilitate further analysis. For example, analysts often want to obtain a statistical summary either of the instances returned by a particular query in the deductive database processor, or of the members of a cluster that has been created by Recon's clustering component. In addition, analysts use statistical measures to define new attributes and concepts in conjunction with Recon's deductive database processing component. These can subsequently be used by the supervised inductive learning, clustering and visualisation components.

The analyst can use Recon's statistical analysis component to ask specific questions about the statistical validity of hypotheses that are generated. For example, when examining data about a set of securities, the analyst might notice that a particular security has price/earnings ratio that is 11% greater than the securities of all other Fortune 1000 companies in the same industry sector. The analyst would want to use the statistical component to determine whether such a difference is meaningful, i.e., statistically significant, or if it is within the expected degree of variation.

Finally, the analyst can use Recon's statistical analysis component to discover relationships in a data set. For example, determining which attributes are independent and which are strongly correlated can help the analyst decide which properties of instances to use when refining concept definitions or constructing queries in the deductive database processing component. Furthermore, the statistical analysis component might indicate that two numeric attributes are slightly correlated, which will lead the analyst to inspect the analysis results visually via a scatter plot.

7.5 Using Recon for Stock Portfolio Selection

We now provide a simple example of how to use Recon to identify a set of securities to be included in an investment portfolio based on historical fundamental and technical data about securities. This is a very appropriate domain for database mining for two reasons. First, because the number of available securities being traded in the various exchanges, both American and international, is very large. Identifying appropriate securities for the goals of a particular portfolio is based on the close examination of the performance of these securities. Without the use of database mining techniques, analysts can only closely examine small amounts of such data. Second, analysts are able to state criteria for identifying securities that can potentially meet a set of investment goals. However, they cannot identify all the necessary criteria. Furthermore, even after a set of securities is identified, large volumes of data relating to these securities still has to be examined in order to fine-tune the stated performance criteria, as well as identify others not previously considered by the analyst. For this example, we describe how Recon's deductive database processor, its supervised symbolic inductive learning, and visualisation components are being used collaboratively.

We assume that the analyst has access to a database with monthly data of securities. The analyst begins by using the Recon server to create the target database. Assume that the target database contains data about 1500 securities over a period of seven years. Such data may be purchased from vendors such as Valueline Inc. The analyst is trying to create a portfolio of 150-200 securities which he or she will subsequently trade. After connecting to the Valueline database, the server automatically extracts the database's schema and displays the information it contains in Recon's Query Matrix. Part of the extracted information is shown in Figure 7.2.

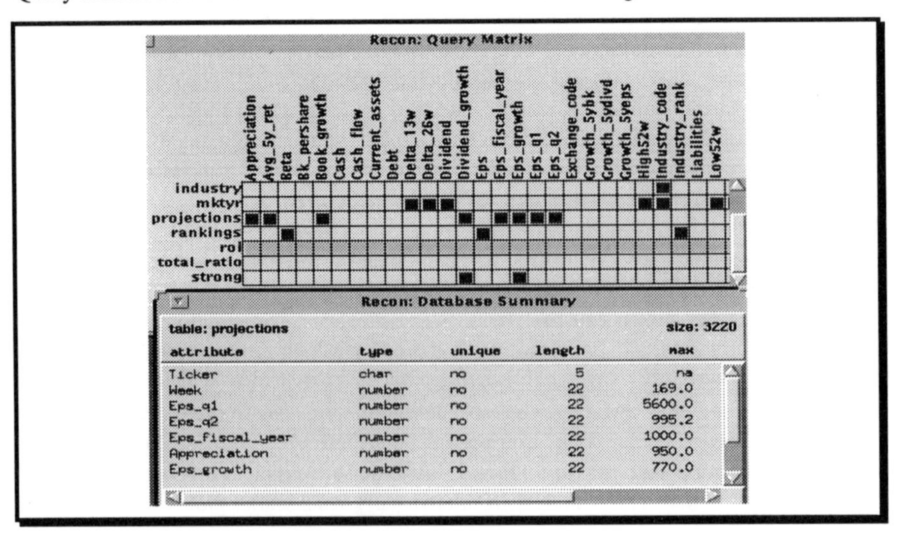

Figure 7.2 — Information about the schema of the assessed database.

The Query Matrix presents the database's tables along the vertical axis and attributes along the horizontal axis. The squares in the matrix indicate which attributes appear in which tables.

The analyst proceeds to use Recon's visualisation component to display information about the securities included in the database. The visualised data allows the analyst to extract initial information about the contents of the database that can either be used in the creation of the target database or later be expressed using Recon's deductive database client component. For example, Figure 7.3 shows two scatter plots produced by Recon's visualisation component.

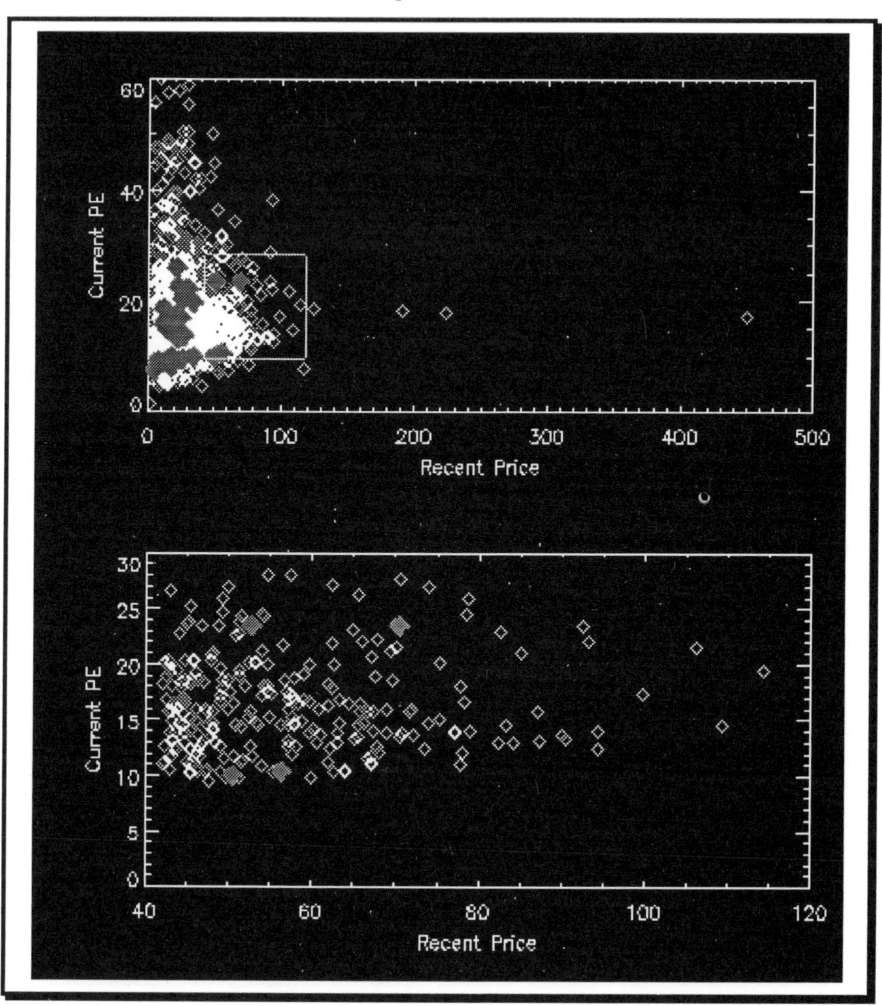

Figure 7.3 — Sample output from Recon's visualisation component.

The top scatter plot displays the relationship between the current price/earning ratio and each security's recent price for a set of securities. The stocks of interest are displayed in a solid grey colour. The bottom scatter plot provides a detail of the area indicated by a rectangle on the top scatter plot.

Next, the analysts use the deductive database client module to interactively create and test their knowledge about indicators that can discriminate high-return securities. The first step in this process, called *screening*, is to quickly identify whether the target database contains data of interest. Recon's graphical, spreadsheet-like interface permits the user to formulate complex queries quickly and easily. In our example, we assume that the analyst is interested in stocks with high earnings-per-share growth and high dividend growth, where "high" in both instances means greater than 50%. The analyst defines the concept of *strong security* as one whose *earnings-per-share-growth* and *dividend-growth* is greater than 50%. The definition of this concept is shown in Figure 7.4.

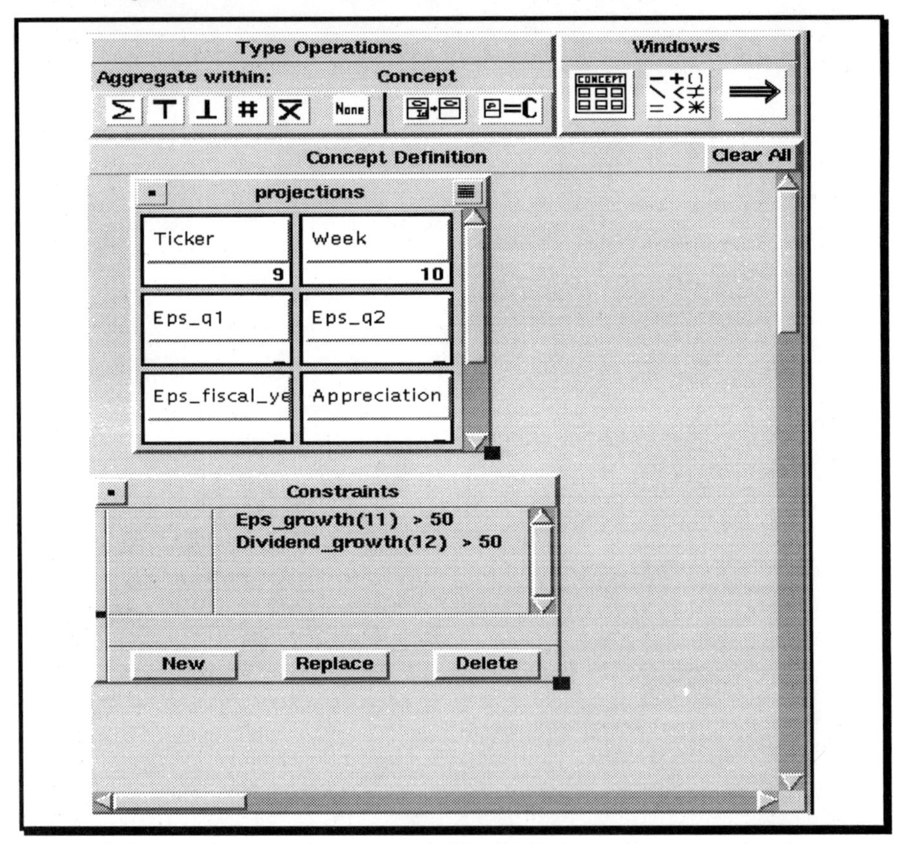

Figure 7.4 — The definition of a concept using Recon's deductive database client module.

Next, the analyst uses the deductive database processor client module to formulate a pattern (hypothesis) and test its validity on the target database. The hypothesis states that "IF a security is *strong* at time t, THEN its *return on investment* two quarters later (or 26 weeks in Recon) will be greater than 5%." In addition to selecting the concepts that will be included in this pattern, the analyst must also specify the bindings between the attributes of the selected concepts. Therefore, the analyst must specify the temporal relation between the values of the *week* attribute of the *strong-security* concept to the corresponding attribute of the *return-on-investment* concept. Figure 7.5 shows the definition of this hypothesis.

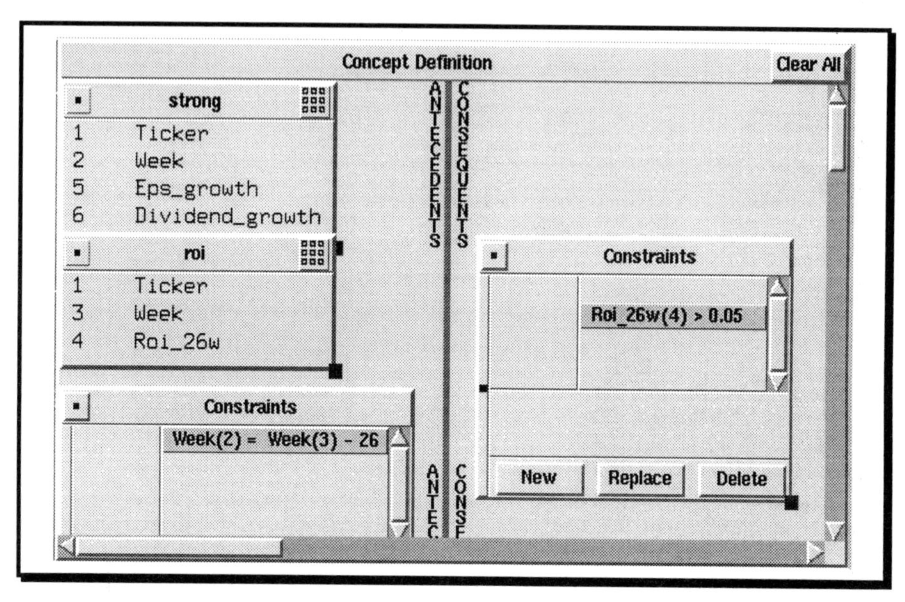

Figure 7.5 — A hypothesis stated in Recon's deductive processing client module.

Once the analyst instructs Recon to validate the hypothesis, the system returns two sets of answers: the records (securities) that support the hypothesis and those that refute it. A sample of the returned answers is shown in Figure 7.6. After inspecting this evidence, the analyst decides that he has discovered a useful indicator for high return on investment, and incorporates the pattern into the model being developed.

While the analyst interactively creates indicators and validates patterns, Recon's supervised inductive learning component, that has been previously invoked by the analyst, automatically discovers other relevant patterns from the target database. Before presenting them to the analyst, the inductive learning component sorts these patterns based on their predictive ability which is determined by the percentage of the instances from each class in the target database that the pattern matches.

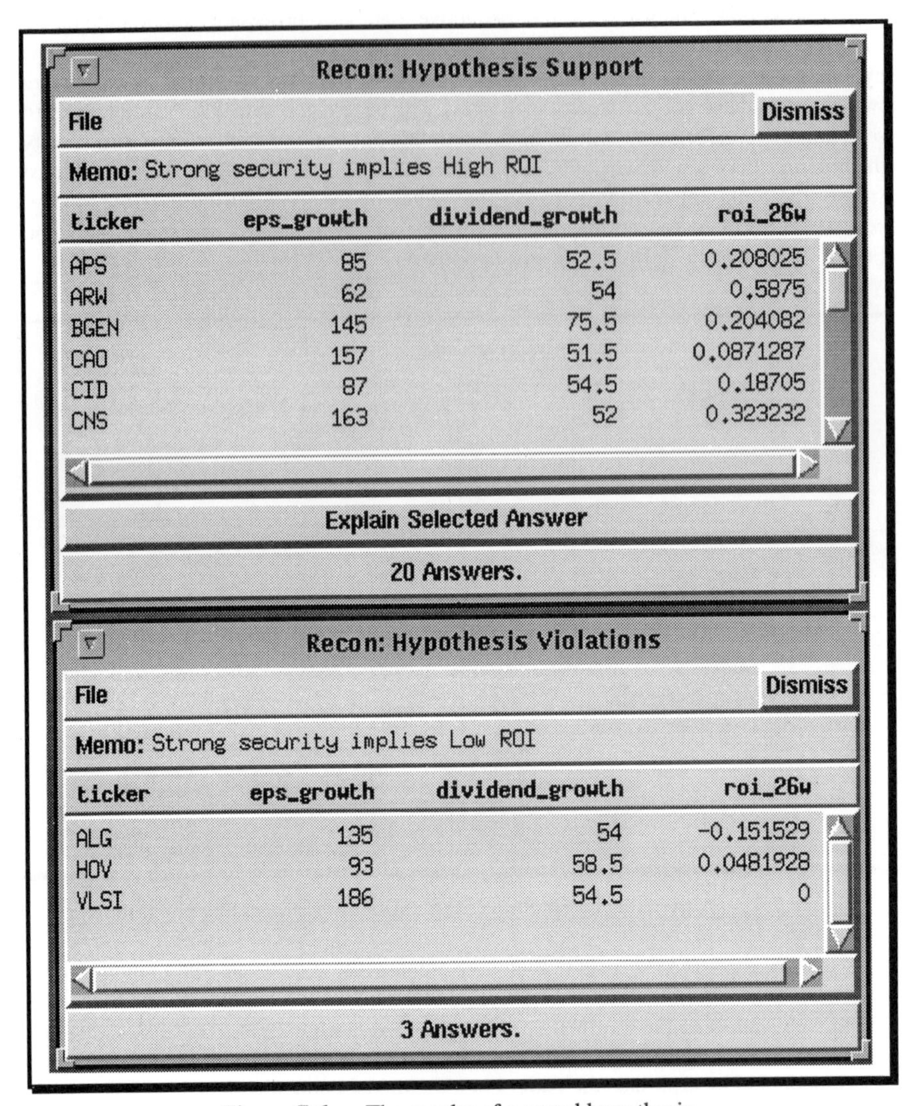

Figure 7.6 — The results of a posed hypothesis.

Figure 7.7 shows a subset of the patterns that have been created by the inductive learning component on a randomly selected subset of the target database.

The analyst tests the discovered patterns by applying them to another data set, called the *test set*, that was not originally used to discover the patterns. With each security to be classified the inductive learning component returns to the analyst its prediction, along with the confidence level associated with the prediction, as well as the actual return on investment. This information is shown in Figure 7.8.

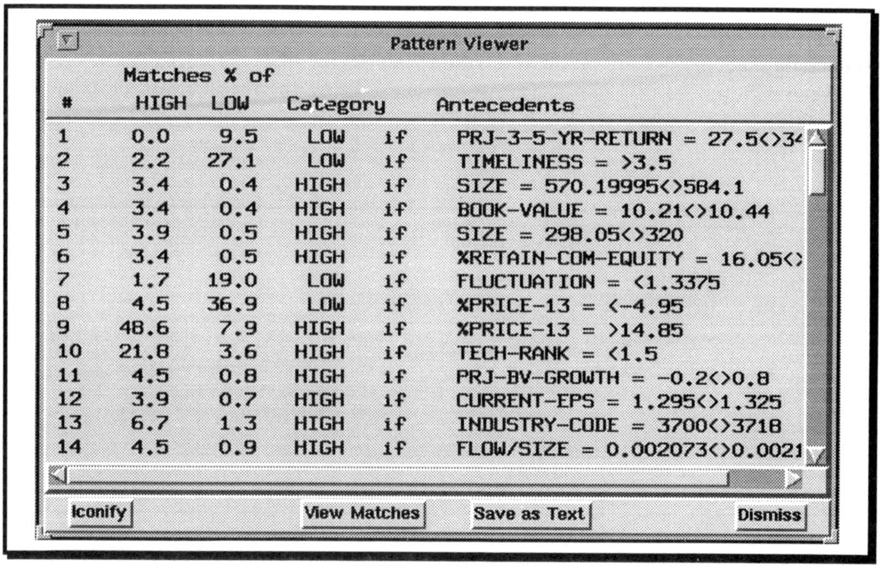

Pattern Viewer

#	Matches % of HIGH	LOW	Category		Antecedents
1	0.0	9.5	LOW	if	PRJ-3-5-YR-RETURN = 27.5<>3
2	2.2	27.1	LOW	if	TIMELINESS = >3.5
3	3.4	0.4	HIGH	if	SIZE = 570.19995<>584.1
4	3.4	0.4	HIGH	if	BOOK-VALUE = 10.21<>10.44
5	3.9	0.5	HIGH	if	SIZE = 298.05<>320
6	3.4	0.5	HIGH	if	%RETAIN-COM-EQUITY = 16.05<
7	1.7	19.0	LOW	if	FLUCTUATION = <1.3375
8	4.5	36.9	LOW	if	%PRICE-13 = <-4.95
9	48.6	7.9	HIGH	if	%PRICE-13 = >14.85
10	21.8	3.6	HIGH	if	TECH-RANK = <1.5
11	4.5	0.8	HIGH	if	PRJ-BV-GROWTH = -0.2<>0.8
12	3.9	0.7	HIGH	if	CURRENT-EPS = 1.295<>1.325
13	6.7	1.3	HIGH	if	INDUSTRY-CODE = 3700<>3718
14	4.5	0.9	HIGH	if	FLOW/SIZE = 0.002073<>0.0021

Iconify | View Matches | Save as Text | Dismiss

Figure 7.7 — Patterns discovered by Recon's supervised inductive learning component.

Result Viewer

Correct: 126 Wrong: 24 Error Rate: 16.00%

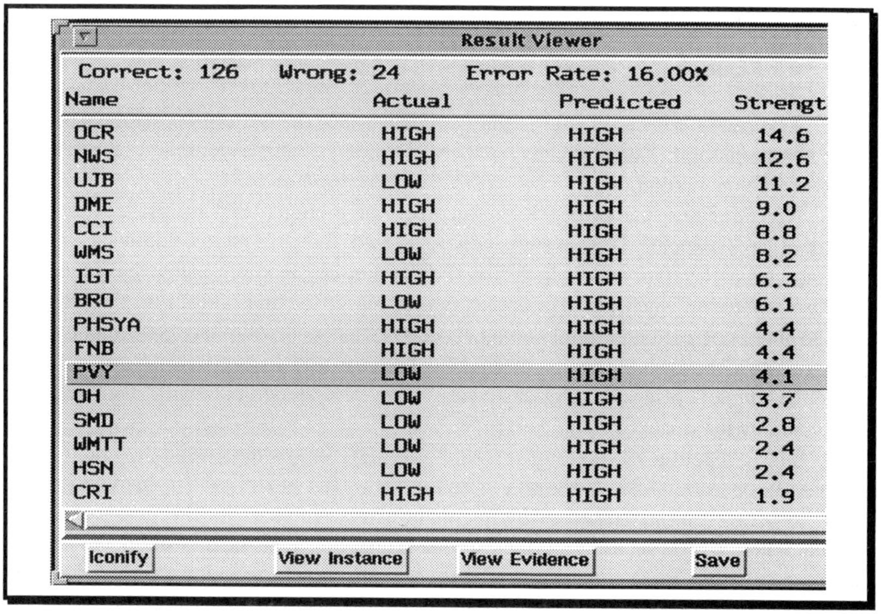

Name	Actual	Predicted	Strengt
OCR	HIGH	HIGH	14.6
NWS	HIGH	HIGH	12.6
UJB	LOW	HIGH	11.2
DME	HIGH	HIGH	9.0
CCI	HIGH	HIGH	8.8
WMS	LOW	HIGH	8.2
IGT	HIGH	HIGH	6.3
BRO	LOW	HIGH	6.1
PHSYA	HIGH	HIGH	4.4
FNB	HIGH	HIGH	4.4
PVY	LOW	HIGH	4.1
OH	LOW	HIGH	3.7
SMD	LOW	HIGH	2.8
WMTT	LOW	HIGH	2.4
HSN	LOW	HIGH	2.4
CRI	HIGH	HIGH	1.9

Iconify | View Instance | View Evidence | Save

Figure 7.8 — The test results returned by the inductive learning component.

The analyst can select individual results from this list and request the rationale for the prediction. Of particular interest are those cases where the inductive learning component made the wrong prediction with a high degree of confidence. The analyst is presented with the set of patterns that were used to make the prediction. For example, after selecting the stock with ticker symbol PVY, the analyst is shown the patterns in Figure 7.9.

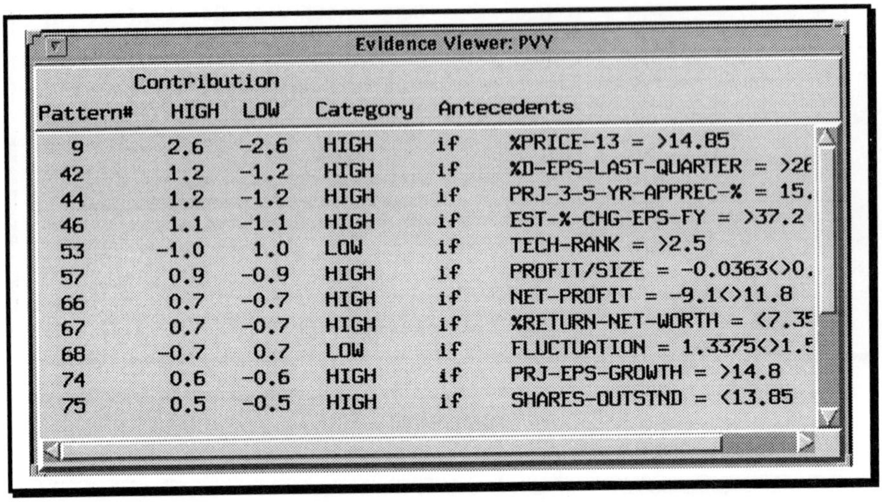

Evidence Viewer: PVY

Pattern#	Contribution HIGH	LOW	Category		Antecedents
9	2.6	-2.6	HIGH	if	%PRICE-13 = >14.85
42	1.2	-1.2	HIGH	if	%D-EPS-LAST-QUARTER = >2
44	1.2	-1.2	HIGH	if	PRJ-3-5-YR-APPREC-% = 15.
46	1.1	-1.1	HIGH	if	EST-%-CHG-EPS-FY = >37.2
53	-1.0	1.0	LOW	if	TECH-RANK = >2.5
57	0.9	-0.9	HIGH	if	PROFIT/SIZE = -0.0363<>0.
66	0.7	-0.7	HIGH	if	NET-PROFIT = -9.1<>11.8
67	0.7	-0.7	HIGH	if	%RETURN-NET-WORTH = <7.3
68	-0.7	0.7	LOW	if	FLUCTUATION = 1.3375<>1.
74	0.6	-0.6	HIGH	if	PRJ-EPS-GROWTH = >14.8
75	0.5	-0.5	HIGH	if	SHARES-OUTSTND = <13.85

Figure 7.9 — Patterns contributing to a particular classification.

Associated with each rule is the confidence that the pattern contributed to the particular prediction. The individual pattern confidence levels determine the overall confidence in the prediction.

By examining the evidence for incorrect predictions, the analyst can identify rules that are suspect. This examination may lead the analyst to eliminate certain attributes from the data set or to refine the definitions of others using the deductive database processor. Such iterative application of top-down and bottom-up data mining leads to more correct and complete models and higher prediction accuracy. Finally, the analyst selects the desired set of discovered patterns and incorporates them into the predictive model being developed where they are integrated with the patterns discovered by the deductive database processor.

The collection of patterns created through the pattern validation and data exploration operations forms the rule-based model. This model is then applied on the set of securities that are targeted by the analyst to be included in the portfolio being created. The top 200 of the securities whose return on investment is predicted by the model to be "high", i.e., greater than 5% over a six month period are included in the portfolio.

7.6 Related Work

A growing number of systems that perform database mining using artificial intelligence techniques have been developed over the past three years. In this section we describe the ones that have been most widely reported in literature and compare them to Recon. We evaluate these systems from three dimensions: (1) the type of database mining operations they support, (2) the degree of integration among their components, and (3) how each component/method is selected to perform a database mining operation. We will compare the following systems: IMACS [Brac93], INLEN [Kauf91], LEAD [Lu92], KDW [Shap91a], MLT [Uszy92], and ALEX [Wink93]. Other similar systems at varying degrees of maturity also exist. The reader should refer to [Shap89, Shap91b, Shap93] for a more complete list.

Our general observation of the cited systems is that none of them supports both the pattern validation and data exploration operations we associate with data mining. Existing systems support either the former or the latter. Furthermore, with the exception of the IMACS system, the rest of the frameworks support only the data exploration operation. IMACS supports pattern validation through techniques that can be compared to those of Recon's deductive database processor. Knowledge is expressed using the CLASSIC [Borg89] knowledge representation system with which IMACS is tightly coupled. Recon's deductive database processing component represents hypotheses and concepts internally in Prolog. Because Recon can interact with commercial database management systems the knowledge representation capabilities of its deductive database processor is only limited by the languages of these systems, e.g., SQL. IMACS does not interact with databases. For this reason it can take advantage of the full representation capabilities of CLASSIC.

From the data exploration perspective, Recon is most closely related to the KDW system in three respects. First, both systems use the same four methods (supervised induction, clustering, visualisation, and statistical analysis) to perform data exploration. Second, both Recon and KDW allow knowledge extracted by one component to be utilised by the other integrated components. Third, the component/techniques used by each system are loosely integrated. The systems INLEN, and MLT are also similar to Recon because they integrate a number of inductive learning methods to perform the data exploration operation. However, the components in these systems are more tightly integrated than the components in Recon and KDW. The high degree of component integration is also evident by the fact each of these two systems uses a custom-developed knowledge representation that allows the components to share knowledge. The systems ALEX and LEAD also use a common knowledge representation that allows the learning components they include to share extracted knowledge. However, these systems use a smaller set of techniques to perform data exploration.

Each of the cited systems MLT, ALEX, and LEAD includes a method selection component that is used to automatically select the most appropriate induction technique to apply on a particular data set. The rest of the systems, including Recon, rely on the user to perform this selection. The selection of the appropriate technique, e.g., neural induction versus decision trees for performing data exploration, is based on a mapping between the various data exploration methods, and the characteristics of the data and on how these characteristics relate to a set of analysis goals. These mappings are domain-specific and extremely hard to develop due to the complexities of the real-world data.

7.7 Conclusions

We have successfully applied Recon in: (1) financial domains to develop prediction models from technical and fundamental data about securities and commodities, (2) manufacturing domains to extract information that allowed for the correction of manufacturing processes, (3) demographic data to identify potential customers in product marketing campaigns. Our work to date has allowed to reach the following conclusions:

1. Effective database mining is performed through the interweaving of pattern validation and data exploration operations. Furthermore, database mining cannot be performed using a single technique. The selection of which technique to use is often influenced not only by the analyst's understanding of the data being target, but also of the processing performed by the techniques being used, and the desired representation of the resulting model.

2. Two or more techniques may need to be used co-operatively to achieve the desired result. Recon provides an analyst with techniques for both pattern validation and data exploration. These techniques have been integrated in a way that enables them to be used co-operatively, when the need arises.

Current work on Recon includes the improvement of the search techniques used by its artificial intelligence components, refinement of the user interface, improvement of the server's functionality, and the application of the system in additional domains.

References

[Borg89] Borgida, A., McGuiness, D., and Resnick, L., Classic: "A structural data model for objects." In *Proceedings 1989 ACM SIGMOD International Conference of Management of Data*, ACM, 1989.

[Brac93] Brachman, R., Selfridge, P., Terveen, L., *et al.* "Integrated support for data archaeology." In *Proceedings 1993 AAAI Workshop on Knowledge Discovery in Databases*, pp 197-211. AAAI, 1993.

[Brei88] Breiman, L. *et al. Classification and Regression Trees,* Wansworth International, 1988.

[Chee88] Cheeseman, P., Kelly, J., Matthew, S., Stutz, J., Taylor, W., and Freeman, D., "Autoclass: A bayesian classification system." In *Proceedings of the Fifth International Conference on Machine Learning*, pp 54-64. Morgan Kaufmann, 1988.

[Duda73] Duda, R.O. and Hart, P.E., *Pattern Classification and Scene Analysis,* John Wiley and Sons, 1973.

[Fish88] Fisher, D. and Langley, P., "Conceptual clustering and its relation to numerical taxonomy". In William Gale, editor, *Artificial Intelligence and Statistics*, pp 77-116 Addison-Wesley,1988.

[Gall88] Gallant, S., "Connectionist expert systems." *Communications of the ACM*, 31:153-168, 1988.

[Kauf91] Kaufman, K., Michalski, R., and Kerschberg, L., "Mining for Knowledge in databases: Goals and general description of the INLEN system." In *Proceedings of the 1991 AAAI Workshop on Knowledge Discovery in Databases*, pp 35-51. AAAI, 1991.

[Kerb91] Kerber, R., "Learning classification rules from examples." In *Proceedings 1991 AAAI Workshop on Knowledge Discovery in Databases.* AAAI 1991

[Kerb93] Kerber, R., "Comparison of the UNIMEM, COBWEB, and AUTOCLASS clustering algorithms." Technical report, Lockhead Artificial Intelligence Center, 1993.

[Koho82] Kohonen, T., "Self-organised formation of topologically correct feature maps." *Biological Cybernetics*, 43:59-69, 1982.

[Lebo87] Lebowitz, Michael, "Experiments with incremental concept formation: Unimem." *Machine Learning*, 2:103-138, 1987.

[Lu92] Lu, S.Y., "Learn to use machine learning in real world applications." In *Proceedings of the National Communications Forum*, 1992.

[Mich83] Michalski, R. and Stepp, R., "Automated construction of classifications: Conceptual clustering versus numerical taxonomy." *IEEE Transactions on Pattern Analysis and Machine Intelligence*, 5:396-409, 1983.

[Quin86] Quinlan, J.R., "Induction of decision trees", *Machine Learning*, 1:81-106,1986

[Quin90] Quinlan, J.R. "Learning logical definitions from relations." *Machine Learning*, 5:239-266, 1990

[Shap89] Piatetsky-Shapiro, G., *Proceedings of the 1989 AAAI Workshop on Knowledge Discovery in Databases*, AAAI, 1989.

[Shap91a] Piatetsky-Shapiro, G., and Matheus, C., "Knowledge Discovery Workbench: An exploratory environment for discovery in business databases." In *Proceedings of the 1991 AAAI Workshop on Knowledge Discovery in Databases*, pp 11-24. AAAI, 1991.

[Shap91b]Piatetsky-Shapiro, G., *Proceedings of the 1991 AAAI Workshop on Knowledge Discovery in Databases*, AAAI, 1991.

[Shap93] Piatetsky-Shapiro, G., *Proceedings of the 1993 AAAI Workshop on Knowledge Discovery in Databases*, AAAI, 1993.

[Simo94] Simoudis, E., Klumpar, D., and Anderson, K., "Rapid visualisation environment: Rave." In *Proceedings of the 9th Goddard Conference on Space Applications of Artificial Intelligence*, May 1994.

[Uszy92] Uszynski, Marc, "Machine learning toolbox" Technical report, *European Economic Community*, Esprit II, 1992.

[Wink93] Winkelbauer, L. and Berka, P. "New algorithms for ALEX: Expanding an integrated learning environment." In *Proceedings of the 1993 Workshop on Integrated Learning Architecture's*, 1993.

8

Using Fuzzy Pre-processing with Neural Networks for Chemical Process Diagnostic Problems

Casimir C. Klimasauskas

8.1 Introduction

The increasing use of computers in industry has led to the collection of unprecedented quantities of data. The computerised Data Acquisition Systems which control entire chemical plants collect megabytes of data each day, most of which is discarded in a matter of hours. Most large pharmaceutical companies collect, on average, 300 megabytes of data daily (100 gigabytes of data annually) on physician prescription patterns. Similarly, in the finance sector most securities houses collect very large amounts of data from their trading floors.

Unfortunately, data does not necessarily lead to information. *Measuring* the world does not necessarily improve our *understanding* of it. This is the challenge that neural networks, fuzzy logic, and other "intelligent" technologies face. They all hold tremendous promise in addressing the problem of converting data into forms from which decisions can be made. In many situations, one single approach alone is not sufficient. A combination of techniques produces a more powerful solution than is possible with just one technique in isolation.

Intelligent Hybrid Systems, Edited by S. Goonatilake and S. Khebbal. ©1995 John Wiley & Sons Ltd.

This chapter describes the analysis of data from a batch-process chemical plant. The degree of non-linearity in the process coupled with the number of process variables make the problem particularly challenging. The solution required an appropriate coupling of fuzzy pre-processing of the data for input to a neural network model.

8.2 The Application Domain

Though this chapter discusses a problem from the chemical process industry, the application of this technique is quite broad. Substantial benefits have been seen in problems as diverse as market timing [Wong94], bankruptcy prediction [Smit93], and insurance risk prediction.

The specific problem arose during the creation of polyethylene film. A simplified summary of the process is shown in Figure 8.1 [Hydr93]. The basic process consists of taking ethylene, often supplied directly from a nearby petrochemical refinery, and converting it to polyethylene through a catalytic process. The output of the process is granular polyethylene. The catalytic process is highly non-linear and subject to a variety of un-measurable or hard to measure disturbance variables. One of the major problems in the process is the creation of long-chain molecules. This is a known problem, and several solutions are available during the melt and extrude phase to address this. If during the melt and extrude phase the long-chain molecules are not sufficiently dispersed through the film, the surface chemistry of the film may change sufficiently, subsequently preventing proper coating. A typical customer acceptance test is that the film must uniformly accept a customer's coating.

Unfortunately, the precise conditions under which either the catalytic process or the melt and extrude process fail are not typically well known. The primary reason is that pilot plants are typically highly inconsistent, and when scaled up by a factor of 10-5000 times, to a fully functioning plant, the characteristics of the process change. This is largely because certain aspects of the process scale as the square of the scale-up factor, and others as the cube. As such, it is not possible to generalise operating experience on a pilot plant to the fully scaled up plant. Performing extensive testing on a fully functional plant is often far too costly in the raw materials consumed. Furthermore, the operating characteristics change with season, raw material impurities, and slight variations in the catalyst. The net result is that the operating limits imposed on the various process variables represent the experience of the process engineer, and often contain un-explored operating regions where the process may fail or degrade below acceptable standards. The limits set by the process engineer are typically designed to provide a reasonably consistent product, subject to the control capabilities of the operators and computer systems. These type of issues are addressed in [Ande94].

In this particular application, when the polyethylene film was delivered to the customer, it would not uniformly accept the coatings on approximately one third of all the batches. Reviewing the data logs of both the good and failed batch runs, all of the approximately 50 process variables were within the "acceptable" operating ranges set by the process engineer. Cursory examination showed no obvious anomalies in the product. Standard statistical techniques for linear regression, and k-means nearest neighbour failed to identify the problem.

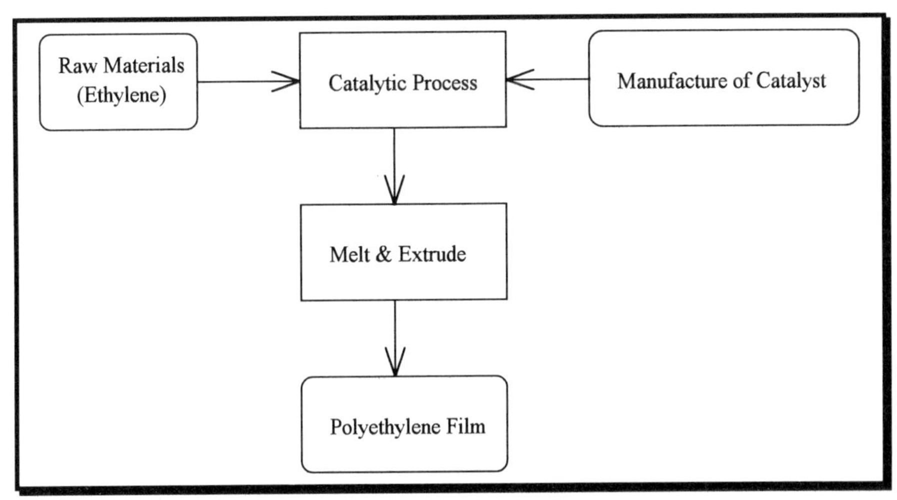

Figure 8.1 — Overview of the basic process of creating polyethylene film.

Typical of most chemical processes is that operating limits are set for each of the controlled as well as measured variables. Each operating variable has a target (setpoint) and a range in which it can vary. The setpoints and ranges are usually set based on the experience of the process engineers. When all of the process variables are within the acceptable ranges, the process is operating within specifications. Within the acceptable operating ranges, the *average* operating point may change over time. This is the result of the interaction between seasonal changes (average outside temperature), raw material impurities, and intentional or un-intentional operator intervention.

In this particular instance, over time, the ratio of bad batches eventually reduced to zero. A combination of intentional (attempts to modify operating parameters within process specifications) and un-intentional (changes in the raw materials) ultimately moved the process into an average operating regime where consistently acceptable product was produced. Unfortunately, the precise actions or circumstances which corrected the problem were never fully understood. This concerned the process engineer. The suspect was a possible high-order interaction between a variety of process variables which shifted the chemical equilibrium away from specific desirable products toward those less desirable. Figure 8.2 shows graphically how the yield of one of the key components of the reaction might look as a function of two process variables. This particular kind of problem is often seen in processes which rely on catalytic effects. This change in yields can be the result of little explored complex interactions between variables.

The challenge is to find a model which can be used to predict when the process will fail to provide the appropriate levels of yields or uniformity of dispersion necessary to achieve the desired end product characteristics.

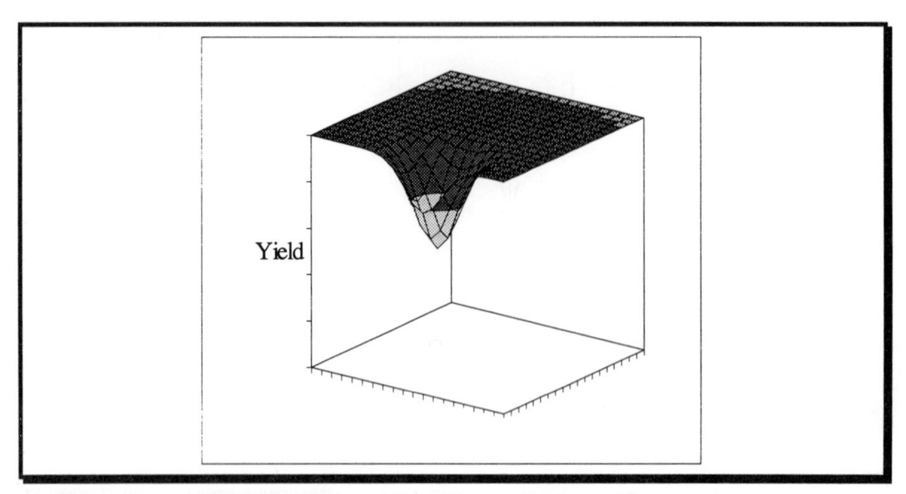

Figure 8.2 — Yield of the desired product as a function of two process variables.

8.3 The Rationale for Hybrid Processing

This particular application posed two problems: First, the plethora of input variables; and second, the high degree of correlation between many of them. Either of these alone is enough to make traditional statistical approaches difficult if not untenable. The first attempt to solve the problem with neural networks was done by appropriately scaling the input data and attempting to train a non-linear feedforward network with a single hidden layer with 30 hidden units. Often, neural networks are able to identify interactions between several input variables, and build a model which can adequately predict an output. However, in this application all our attempts using neural networks failed. We used Scott Fahlman's Cascade Correlation Algorithm [Fahl90] in all simulations. This is an neural network algorithm which "grows" hidden layers and units as demanded by the input data.

The next approach was to further transform each of the input variables in a way which might make it easier for the neural network to develop a predictive relationship. One of the methods selected for this was to map each of the continuous input variables to a series of over-lapping fuzzy sets.

Fuzzy membership sets often have the shape of truncated pyramids as shown in Figure 8.3. The formulas used to transform a value along the x-axis to a degree of membership in the set are shown in Figure 8.4. The algebra to derive these formulas is straightforward.

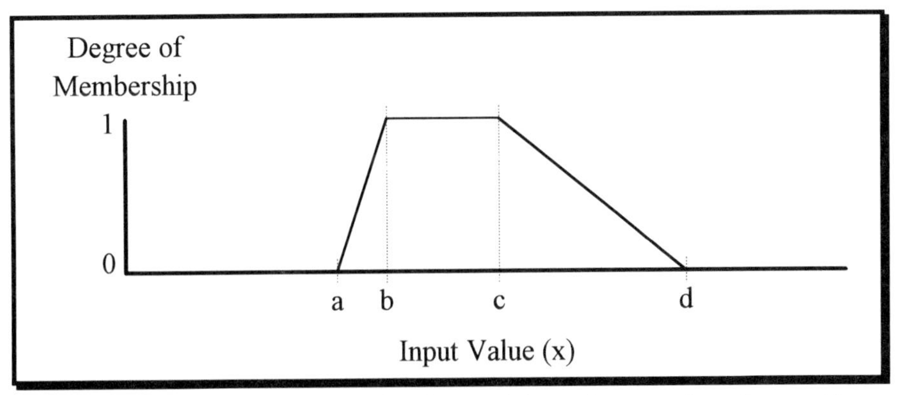

Figure 8.3 — A typical fuzzy membership set. Points a, b, c, and d are the transition points which define the membership set.

$$k_1 = \frac{1}{b-a}$$

$$k_2 = \frac{1}{c-d}$$

$$g = \frac{a * k_1 - d * k_2}{k_1 - k_2}$$

$$h = (g-a) * k_1$$

$$M(x) = \min(1, \max(0, h - [(g-x) * (x \le g ? k_1 : k_2)]))$$

Figure 8.4 — Formulas to compute the degree of membership in the fuzzy set shown in Figure 8.3. k_1, and k_2 are the slopes of the sides of the trapezoid. The two sides of the trapezoid intersect at the point (*g, h*). *M*(*x*) is the degree of membership of the value *x* in this set. As a condition for these formulas to work, the following inequalities must be met: $a<b\le c<d$.

Figure 8.5 shows an example of how these sets overlap in our application. Each continuous input variable was mapped onto four fuzzy ranges. The total network inputs for the neural networks then increased from 50 to 200.

There are two reasons for this approach. First, appropriately transforming each of the inputs into four overlapping fuzzy membership sets provides an isomorphic mapping of the data to properly constructed membership values, and as such, no information is lost [Klim92]. Second, it is easier for a neural network to identify and model high-order interactions when the data is transformed in this way. To understand how this works, consider the problem in Table 8.1. This is a simple two input classification problem in which the output is one in the central convex section and zero around the edges.

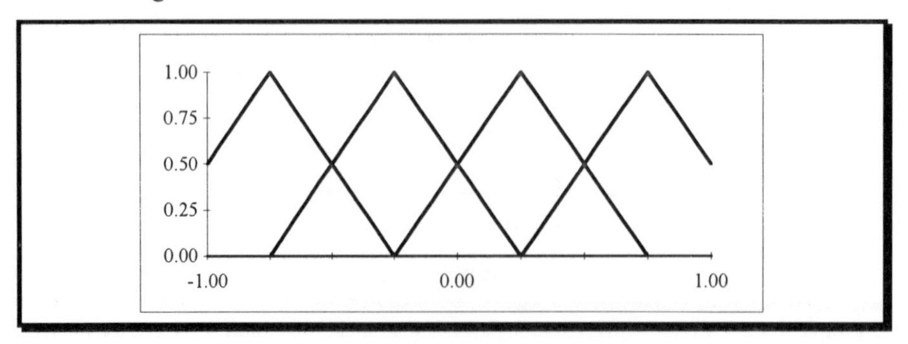

Figure 8.5 — Overlapping fuzzy membership sets used to transform each continuous input variable.[1]

Figure 8.6 graphically shows a typical neural network solution requiring three hidden units. This tends to be the preferred solution, even when the network has more than three hidden units. The other hidden units tend to be ignored in the solution.

X1/X2 Input	-.75	-.25	.25	.75
.75	0	0	0	0
.25	0	1	1	0
-.25	0	1	1	0
-.75	0	0	0	0

Table 8.1 — Simple classification problem used to illustrate the power of fuzzy encoding.[1]

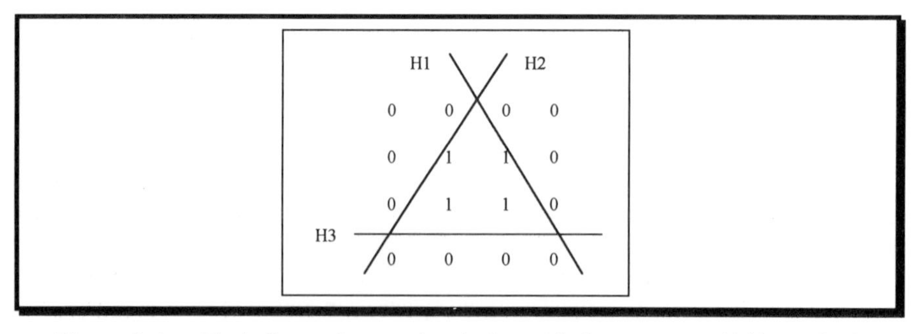

Figure 8.6 — Typical neural network solution with three or more hidden units.[1]

[1] Reproduced by permission of *Advanced Technology for Developers* [Klim92].

When the data is transformed, so that each input is mapped into four fuzzy membership sets using the equations in Figure 8.7, the problem, in the transformed space, is linearly separable.

$$Z_1 = \max(0, 1 - 2abs(X_1 + 0.75))$$
$$Z_2 = \max(0, 1 - 2abs(X_1 + 0.25))$$
$$Z_3 = \max(0, 1 - 2abs(X_1 - 0.25))$$
$$Z_4 = \max(0, 1 - 2abs(X_1 - 0.75))$$
$$Z_5 = \max(0, 1 - 2abs(X_2 + 0.75))$$
$$Z_6 = \max(0, 1 - 2abs(X_2 + 0.25))$$
$$Z_7 = \max(0, 1 - 2abs(X_2 - 0.25))$$
$$Z_8 = \max(0, 1 - 2abs(X_2 - 0.75))$$

Figure 8.7 — Fuzzy transformations of the input variables *X*1, and *X*2. Due to the symmetric nature of the transformations ($b = c$, and $k_1 = k_2$ in Figure 8.4), the formulas were simplified as shown above.

Neighbourhood techniques, such as radial basis functions, can also effectively build classifiers which involve high-levels of interactions between variables. It is possible to build a neighbourhood-based classifier to effectively model a system with multiple dis-joint convex regions associated with each class. However, because neighbourhood classifiers interpolate class membership from individual points in space, they require on the order of k^m data examples uniformly distributed over the input space (where k is the granularity of the input data, typically 2-10, and m is the number of inputs to the classifier). For problems with few input variables, this works well.

Non-linear feedforward networks are able to work effectively with hundreds of inputs and they combine together planar discriminants to form class boundaries. However, even when the class boundaries are distinct, they have difficulty building classifiers where more than two or three variables are interacting to form one or more convex regions.

When the problem involves many variables (more than 20), and entails convex class boundaries based on more than three variables, neither non-linear feedforward networks, nor neighbourhood-based classifiers provide acceptable solutions to the problem. We believe that in these situations the above mentioned fuzzy pre-processing method is very effective.

In our chemical process problem, the combined fuzzy pre-processing and neural network method was able to correctly identify 85% of the unacceptable batches. The neural network used 200 inputs, one output, and evolved five hidden units using the Cascade Correlation algorithm. A threshold (zero) midway between the limits of the output (−1 to +1) was used to convert the network output into a +1 for acceptable or −1 for unacceptable. Analysing the network using sensitivity analysis indicated that the interaction of five process variables were responsible for the unacceptable product. In the transformed space these five process variables were essentially linearly separable.

This technique for transforming data in a way which reduces complex high-order interactions to low-order, or linearly separable problems, has been applied in a series of other domains. Francis Wong has incorporated a methodology for decomposing selected input variables into fuzzy sets for automatically creating market timing systems [Wong94]. In a completely unrelated application, Murray Smith discusses their use in developing bankruptcy prediction models [Smit93]. In this application, decomposing continuous variables into fuzzy membership sets improved the ability of models to generalise, providing a more robust solution. This is contrary to what one may expect from increasing the number of inputs. A possible explanation is that the fuzzy transformation allows portions of the neural network to specialise for distinct regions of the input space (high-order interactions) in ways that do not occur using untransformed data.

8.4 Future Directions

Fuzzy transformation of continuous data has the potential to convert difficult problems into much simpler, even linearly separable ones. One area of on-going research is to quantify the effectiveness of non-linear transformations, such as log or square-root transforms, prior to fuzzy transformation.

A commercially available program, DataSculptor with DataWizard (available from NeuralWare, Inc., Pittsburgh, Pennsylvania, USA), has been developed which implements these algorithms in a computer program. Internal tests on 29 data sets indicated that networks trained using the transformed data performed better in 25 of the 29 instances. In four of the instances, performance was comparable.

Other areas of application of the approach include direct marketing and financial forecasting. Financial forecasting, particularly market timing, is a very noisy problem. Many rules are used by professional traders to decide what position (long, out, short) to take, and when. These rules are of the form, “When RSI (relative strength indicator) is over X, and OBV (on balance volume) is decreasing, then buy”. Converting the output of these technical indicators into fuzzy sets is a natural transformation. Preliminary results suggest that this can improve the performance of market timing systems.

Another area of application is marketing data analysis. Product marketing often uses questionnaires to collect preference data. This data is often in the form of a level of feeling from strongly disagree to strongly agree with a particular statement. We have applied fuzzy encoding of input data to the task of predicting customer satisfaction to a given product. Each preference input is divided into three fuzzy sets: one which encodes degree of disagreement, one approximately neutral, and one for degree of agreement. One reason this may improve performance is that it provides for a more tolerant encoding of preferences reducing the effects of systematic bias.

A key challenge is to develop a methodology for determining when to apply fuzzy transformations to neural network models. Applying them to a variety of problems is one of the ways to collect the data necessary to develop such a methodology.

References

[Ande94] Anderson, Jim, Backx, Ton, Van Loon, Jost, and King, Myke (1994) "Getting the most from Advanced Process Control" in *Chemical Engineering*, March 1994, pp. 78-90.

[Falh90] Fahlman, Scott E., and Lebiere, C. (1990) "The Cascade-Correlation Learning Architecture" in D. S. Touretzky (ed.), *Advanced in Neural Information Processing Systems 2*, Morgan Kaufmann.

[Hydr93] "Petrochemical Handbook 1993" in *Hydrocarbon Processing*, March 1993, pp. 199-204.

[Klim92] Klimasauskas, Casimir C. "Hybrid Fuzzy Encoding for Improved Backpropagation Performance" in *Advanced Technology for Developers*, Volume 1, September 1992. High-Tech Communications, 103 Buckskin Court, Sewickley, PA 15143, USA, pp. 13-16, 1992.

[Smit93] Smith, Murray. *Neural Networks for Statistical Modelling*, Van Nostrand Reinhold, New York, pp. 155-159, 1993.

[Wong94] Wong, Francis, Tan, Clarence, "Hybrid Neural, Genetic and Fuzzy Systems" in *Trading on the Edge: Neural, Genetic and Fuzzy Systems for Chaotic Financial Markets*, Guido J. Deboeck (ed.), John Wiley, New York. pp. 243-261, 1994.

9

A Multi-agent Approach for the Integration of Neural Networks and Expert Systems

Andreas Scherer and Gunter Schlageter

9.1 Introduction

The high complexity and the inherent heterogeneity of real world problems is still one of the major challenges of current *artificial intelligence* (AI) systems. Due to the necessity of using different problem solving techniques, the interest in hybrid systems is rapidly growing as shown by the diverse approaches found within this book. For supporting the integration of intercommunicating hybrids [Goon92], this chapter advocates the use of *distributed* artificial intelligence (DAI) techniques. The main advantages of this approach include the encapsulation of different paradigms, the separation of control and domain knowledge, and the reduction of the complexity of individual problem solvers.

Before examining the field of DAI, we introduce and analyse two intelligent techniques, namely knowledge-based systems and neural networks, that will be integrated in our distributed application. Section 9.2, concludes by reviewing related methods for hybrid problem solving.

In section 9.3, the field of DAI is examined with regard to its main components and issues. This section concludes by discussing the potential for using DAI as an integration method for building hybrid systems. Having established the potential of DAI as an integration method, a case study of a DAI application within the domain of

Intelligent Hybrid Systems, Edited by S. Goonatilake and S. Khebbal. ©1995 John Wiley & Sons Ltd.

macro-economic predication, is presented. This includes an overview of the domain and the predictive tools usually used in this area. After introducing some simple economic relationships, a description of how neural networks can be used in multivariate prediction is presented. Also some experimental results from this study are shown. Finally, a distributed problem solving framework, called PREDICTOR, is described. Within this framework neural networks are used for prediction and are integrated with other problem solving techniques such as knowledge-based systems.

9.2 Analysis of the Technologies

The intelligent problem solving techniques (neural networks and knowledge-based systems) used in our application, have been examined in detail by various researchers (see amongst others: [Becr91, Caud90, Dunk92, Gutk91, Roma90]). Most of these researchers agree that both these approaches have their advantages and disadvantages. In the following section, these advantages and disadvantages are briefly summarised for both knowledge-based systems and neural networks.

9.2.1 Knowledge-based Systems

The theory of knowledge-based systems is well understood and the practical importance is proven by a variety of applications in various domains (For an overview see for example [Jack90]). The following general characteristics of knowledge-based systems are well known:

Advantages

- **shallow knowledge**: the strength of knowledge-based systems comes from the fact, that shallow knowledge of experts can be represented very easily for example via simple IF-THEN-rules.
- **deep models**: If a more comprehensive approach is required, the domain can be described using deeper and more advanced models of the domain.
- **explanation**: It is important for intelligent systems to be able to explain how they arrived at a particular solution. In knowledge-based systems, this information can be obtained by evaluating the trace generated by the inference engine. More advanced systems have an explanation facility which generates explanations on a higher, more problem oriented level.

Problems

- **availability of knowledge**: Although this aspect sounds trivial it is essential for many applications. It may be that the expert is not co-operating in sharing his/her knowledge, there may be conflicting views on what information is right or wrong, or the experts in this field do not generate and use an explicit representation of their knowledge.
- **knowledge representation**: The process of formalising knowledge is limited by the expressive power of the particular knowledge representation in use.

- **development time**: The development of knowledge-based systems for real world problems is a time consuming iterative process of different engineering phases (knowledge acquisition, formal representation of knowledge, coding and validation).
- **meta-knowledge**: It is yet unknown how to integrate knowledge into a knowledge-based system that is able to decide whether the problem to be solved within the bounds of the system's capabilities or whether its solution would go beyond the current problem solving ability of the system.
- **maintenance**: As the complexity of a knowledge base increases one has to take into account serious problems in maintaining the knowledge base (deleting, changing or introducing new pieces of knowledge).
- **validation**: As in other complex software based systems, the validation of large knowledge bases is still a crucial problem.
- **performance**: It is a well known that large knowledge bases lead to dramatically reduced response times. The efficiency of the problem solver is reduced and depending on how time critical an application is, this aspect can be a problem.

The list of shortcomings described here can be partly compensated by the integration of neural networks. Before discussing this point we give a brief overview of neural network characteristics (see also section 9.3.3).

9.2.2 Neural Networks

In comparison to knowledge-based systems the technology of neural networks is still in its infancy but is maturing rapidly (for an overview of connectionist techniques see [Lipp89] and for a description of popular neural network models e.g. backpropagation, see [Rume86]). Recently there has been a rapidly increasing interest in the theoretical foundations of neural networks, as well as their applications. This has lead to the realisation of the following advantages and shortcomings of neural networks.

Advantages

- **learning**: Neural networks are able to learn from examples and therefore can acquire their own knowledge.
- **rapid execution**: A trained network can usually be executed very quickly.
- **non-linearity**: Neural networks are able to handle non-linear problems.
- **handling of noisy and/or incomplete data**: Neural networks are fault tolerant, they handle noisy and incomplete data.
- **generalisation**: Neural networks are able to solve new problems by exhibiting generalisation.

Problems

- **training time**: In most cases the training of neural networks is a very time consuming task.

- **explanation**: Neural networks do not provide the user with adequate explanations at the problem level.
- ***a priori* knowledge**: Knowledge that may already exist about the domain cannot be brought straight into the network.
- **calibration problem**: Using neural networks in practical applications implies adjustment of a variety of parameters which influences the behaviour of the system (topological aspects, learning rate, learning error). In addition, in some cases there is uncertainty whether the final learning goal is achieved. To validate that the network has learnt something useful, sophisticated test methods are necessary.

9.2.3 Related Work in Hybrid Problem Solving

In recent years an increasing number of researchers have been working in the field of hybrid systems in an attempt to find new ways to integrate two or more technologies to tackle complex real world problems.

9.2.3.1 The HANSA Framework

In the HANSA project an object-oriented hybrid framework has been built to allow developers to rapidly configure and use industry standard tools and novel AI shells (such as neural networks, genetic algorithms and expert systems). The results of this European collaborative project are used in business applications within the areas of marketing, banking, insurance and executive information systems (see [Roch93, Kheb93]).

9.2.3.2 Neural Networks and Expert Systems in Process Fault Diagnosis

In [Becr91], a typical example of the integration of neural networks and expert systems for a process fault diagnosis problem is described. A hierarchy of organised networks is used for fault detection and localising errors, mainly on the basis of sensory information. The networks are used to evaluate non-symbolic information rapidly and important in cases where the exact determination of the actual fault is very difficult. The expert system activates the neural diagnosis to generate potential fault candidates, and then analyses the results and either confirms or suggests an alternative. The expert system uses a simulation model of the domain in addition to the sensory information.

9.2.3.3 Integration of Neural Networks into Federative Systems

Federative systems provide a flexible co-operative environment for multi-agent problem solving. In accordance with [Deen88, Schl88] a federative system can be defined by the characteristics of its local systems. The agents in a federative system

- are autonomous,
- initiate co-operation only if they identify a need for co-operative problem solving,
- are loosely coupled and

- take part in co-operation only if they expect to be able to support the global problem solving process.

In FRESCO the *contract net* architecture [Smit80, Klet89] provides the basis for the communication and co-operation between the agents. The key-idea is that at first a complex task is decomposed into sub-tasks by one agent, which is called the manager. The manager checks which of these sub-tasks can be computed by itself and sends the other agents descriptions of the sub-tasks. These agents' answer with their competence evaluation of how well they can solve the sub-task and therefore become bidders for that task. (How competence knowledge is represented, and how a bid is obtained from that, is explained in detail in [Kirn92]). The manager now computes heuristically a task allocation by distributing each task to one of the bidders. In principle, in the federative environment every agent can be a manager and a bidder at the same time.

Figure 9.1, shows how an agent in FRESCO is structured. The *universal dialog interface (UDI)* supports the interaction between the end user and the whole federative system. In the case that the user wants to consult a specific agent, he uses the *local dialog interface* for that local problem solver.

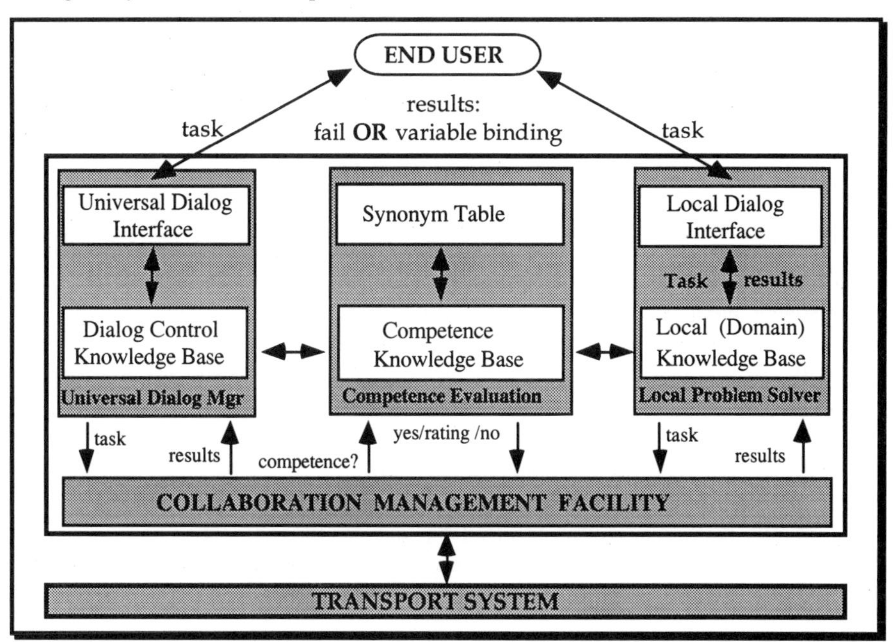

Figure 9.1 — The FRESCO agent model.

The *collaboration management facility* holds the knowledge of how to decompose a complex task, how to find a suitable task allocation and how to synthesise the results. This is a core component with respect to the co-operative problem solving process. The self-model of the agent is part of the *competence evaluation module*. The individual processing components in the co-operative system are represented by the *local problem solver (expert system, neural network, database).* The local problem solver encapsulates standalone implementations of the different processing elements. This approach allows system designers to redesign pre-existing processing techniques to make them co-operative. The overall system co-operation is supported by a transport system (a message passing system) and a global data dictionary. [Dunk92] shows, how neural networks can be embedded into a federative system. In particular, issues concerning the complexity of the integration software are discussed.

9.3 Distributed AI

This section advocates the use of DAI techniques for the integration of different problem solving strategies. After a definition of DAI we give a short overview of the key issues in this research area. At the end of this section, various aspects of the integration characteristics of DAI are discussed.

9.3.1 DAI: A Definition

According to [Bond88], AI research in the past investigated how a single agent can show intelligent behaviour in different areas like *knowledge-based methods, planning, perception, learning,* etc. Due to recent developments in modern information technology, and because of the complexity of real-world problems, there is a growing interest in applying intelligent knowledge processing technology to novel and challenging real-world applications. DAI is a sub-discipline of AI concerned with concurrency and distribution of these AI components.

Advanced information systems are faced with at least three major challenges: the management of a rapidly increasing amount of knowledge, the assistance of collaborating experts and the development of open systems. Due to the different types of reasoning methods used by domain experts there exists a wide spectrum of knowledge representation and evaluation techniques.

Beside the "classic" advantages in managing large domain knowledge [Wied89] we advocate the use of DAI as an integrative paradigm for:

1. extending the complexity barrier,
2. the extendibility of the overall system,
3. maintenance of knowledge representation,
4. performance of problem solving and
5. integration of different reasoning techniques and paradigms by encapsulation.

9.3.2 Issues in DAI

9.3.2.1 Distributed Problem Solving / Multi-agent Systems

DAI discriminates between *distributed problem solving (DPS)* and *multi-agent systems (MAS)* [Bond88]. In DPS the agents are faced with problems that are beyond their individual capabilities, so they must co-operate to achieve their goals. In MAS the agents share an environment so they must act collectively to identify and resolve conflicts, while at the same time taking advantage of the actions of other agents. Consequently, MAS research focuses on co-ordination, providing a lot of promising concepts (e.g. [Smit80, Rose85, Ferb89]). MAS are designed top-down and in general, they do not allow the dynamic addition or removal of agents. MAS use tight coupling where typically, the agents do not decide by themselves whether to co-operate. This seems to be an inherently static aspect of current MAS approaches.

9.3.2.2 Negotiations

Negotiations allow the co-ordination of the agents towards a common global goal. [Conr86] introduced multistage negotiations to deal with incomplete knowledge. Their agents bargain on problems with a restricted global significance and guarantee the overall system converging towards a global consistent solution. [Lâas91] introduced cybernetic aspects to negotiations between co-operative problem solvers. In their system the agents generate goals which are then communicated and allocated to achieve a globally consistent solution. Goals that are not allocated are re-organised and start a new cycle of goal formulation, communication, allocation and achievement. In general negotiation processes increases the autonomy of the single system, but decreases the overall performance because of time consuming communication iterations. System designers have to be aware of this trade-off in developing real world applications.

9.3.2.3 Planning

The basic idea is that agents co-ordinate their behaviour by exchanging plans before they start acting (e.g. see [Bond88]). While multi-agent planning (MP) requires one agent to have all relevant knowledge to generate the global plan, in distributed planning (DP) all agents must act together on one single global plan. To overcome some of the shortcomings [Durf87] suggested, for example, the concept of *partial global plans* (PGPs) in which co-ordination is part of each problem solvers activities and is interleaved with local problem solving. This approach seems to be quite relevant if it is necessary to design multi-agent systems with very flexible co-operation capabilities.

The experience of our work shows that the usage of centralised knowledge can speed up the distributed problem by avoiding expensive communication costs.

9.3.3 DAI as an Integrative Paradigm

In section 9.2 advantages and disadvantages of neural networks and knowledge-based system are discussed. It is becoming evident that both these technologies have intrinsic difficulties in handling complex problems, and cannot be solved by simply using high-end hardware. The nature of real-world problems and also the technologies we want to use for solving them, have reached a degree of complexity and heterogeneity which requires a mix of techniques and innovative methods to integrate them. DAI offers these methods to support co-operative problem solving.

Breaking down a complex system into a set of agents helps to separate functional units within a defined problem solving behaviour. The complexity of a particular unit can be contained and in this way complexity related problems like maintenance, validation, and training time etc., remain on a manageable level.

In addition, the various ways that different problem solvers may interact can be elegantly modelled by defining co-operation strategies on a meta level. This increases the extendibility and flexibility of the overall system. Referring to our discussion on negotiation, the use of decentralised or centralised co-ordination techniques has to be determined taking into account specific characteristics of the problem to be solved.

9.4 PREDICTOR: A Case Study

In this section we demonstrate how co-operative problem solving using different technologies can be performed. Our system PREDICTOR is used to solve prediction tasks in economics. We start with a brief introduction to the domain, explaining the general interest in prediction techniques. Then we discuss currently used techniques in this domain. The most important shortcoming of the existing approaches is their lack of adaptability. Taking this as a motivation to use neural networks for prediction problems we show how they can be used as economic models. The technique used in our experiments is an extension of the windowing techniques used in time series prediction with neural networks. Following this, some experimental results are presented and discussed. Finally, we give an overview of the architecture of PREDICTOR.

9.4.1 An Application Domain: Prediction of Economic Data

The following remarks about the application domain are based on [Fisc89] in which the theoretical foundations of predicting economic data are discussed. To efficiently use the existing economic control instruments such as tax measures, subsidies, and legislation requires advanced modelling, analysis and prediction techniques.

9.4.1.1 "Classical" Prediction Techniques

Single equation models

An *endogenous* variable y is connected via a regression equation with other so called *exogenous* variables x_i. A stochastic term u is used to describe statistical uncertainty. A single equation forms a partial model of the dependencies and relationships between the exogenous variables and a endogenous variables (see formula 9.1). A set of such formulas is used to describe an entire model.

$$y_k = b_0 + b_1 x_{k1} + b_2 x_{k2} + \cdots + b_k x_{km} + u_k,$$
$$k = 1..n,\ b_i = \textit{parameter to be estimated} \quad (9.1)$$

In working with linear regression models a variety of assumptions and limitations of this technique must be taken into account. Some of them refer to the exogenous variables others to the statistical term u. We consider that the most important of these are (see [Kout77], pp. 179ff.):

- **multicollinearity**: this effect is inherently due to the nature of economic models. When a defined number of explanatory variables change in a similar way, it is hard to establish the influence of each regressor on y. The correlation has to be expressed in the model. But in most real world problems there exists a vast amount of such intercorrelating effects.
- **randomness of u**: u is a random variable introducing "error" into the model referring to variables, the form of the model, the measurement of dependent variables, etc. Moreover, it is assumed that each u_i may assume values having a zero mean.
- **homoscedasticity**: the probability distribution of u remains the same over all observations of the x_i.

Of course, there exists a variety of statistical test procedures, which allow further analysis of the nature of the underlying relations. (But the increasing complexity of these tools correlates strongly with the decrease of their practical usage).

Things become even more difficult due to the influence not only of exogenous variables on endogenous variables but also the influence of endogenous variables on exogenous variables. Here, the so-called advanced simultaneous equation models (see [Kout77], p. 331, [Thei71]) have to be used.

Problems of prediction methods

Existing prediction tools has several shortcomings. In regression equation approaches the parameter estimation plays an important role in defining how the assumed relationships between the model variables interact. The assessment of parameters needs a lot of experience and expertise in analysing data and handling statistical tools.

Moreover, the simple regression equations assume linear relationships between the variables, which is a simplification of the nature of economic systems. On the other hand more advanced non-linear methods have several problems in their practical usage. This is the reason why linear approaches have an enormous practical relevance. Because of their adaptive characteristics and their ability to handle non-linearity, neural networks are an interesting alternative to the existing methods.

9.4.2 Forecasting Macro-economic Data with Neural Networks

In this section we describe the forecasting of macro-economic data using a backpropagation network. After introducing our model we describe how neural networks can be used for prediction tasks. A description of the experimental setting is followed by a presentation of our experimental results.

9.4.2.1 Some Domain Specific Dependencies

Based on [Frie73, Keyn36, Walt92] a model of the German economy was developed. For our experiments we extracted some relationships described in this model (see Table 9.1 and Figure 9.2).

IR	interest rates	*P*	price index
I	investment (basis: prices in 1985)	*ER*	savings of private households and retained profits of companies
MS	money supply M3 (basis: prices in 1985)	*SP*	rate of savings; the percentage of ER relative to IN
IM	import	*IN*	income
EX	export	*C*	private consumption
S	public consumption	*NW*	net wages
E	number of persons capable of gainful employment	*t*	time

Table 9.1 — Endogenous and exogenous variables of the domain

$$
\begin{aligned}
E_t &\equiv f_1(I_{t-1}) \\
I_t &\equiv f_2(C_{t-1}, IM_{t-1}, EX_{t-1}, S_{t-1}, IR_{t-1}) \\
P_t &\equiv f_3(NW_{t-1}, C_{t-1}, C_t, I_{t-1}, I_t) \\
\mathrm{S}_t &\equiv f_4(EI_{t-1}, IN_t, IR_{t-1}, IR_t) \\
\mathrm{IR}_t &\equiv f_5(MS_{t-1}, MS_t, I_t, ER_t)
\end{aligned}
$$

Figure 9.2 — Some relationships in our model.

A detailed discussion of the model relationships can be found in the above cited literature. The prediction problem is now defined as follows: find functions f_i such that the historical course of the model variables and the relationships among them are approximated as best as possible. In this case it can be reasonably assumed that f_i also can be used for a prediction step.

9.4.2.2 The Experimental Setting

To forecast the behaviour of a system we want to use domain knowledge about *which* system variables interact. This model is based on theories of how large economies work. The problem remains now to find out *how* the correlation between these variables can be exactly quantified. As we stated in the previous section there are severe problems in finding the right equations to describe the behaviour of the system adequately.

To understand how we use neural networks for this problem we first explain a common technique for time series prediction with neural networks (see [Lape87]). The basic idea of windowing (see Figure 9.3) is to use two windows w′ and w′′ of fixed size. For a given window size a relationship between w'_1 .. w'_n and w''_1 .. w''_m is assumed which is *a priori* unknown. The tupels (w'_i, w''_i) are used as training data. This technique is successfully used in various applications.

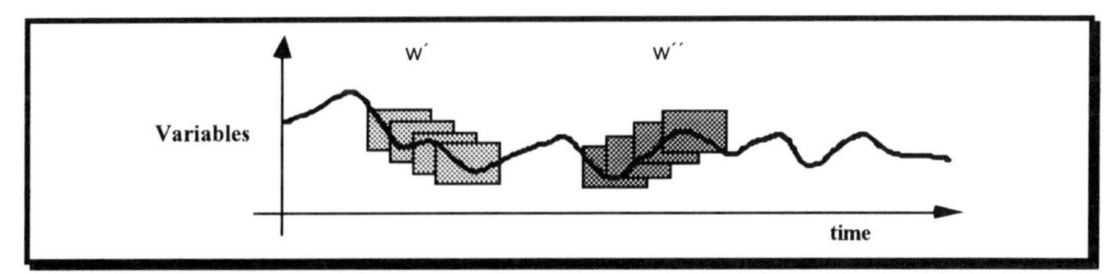

Figure 9.3 — Windowing: detection of correlations in time series.

Figure 9.4 shows a modified time series windowing technique that enables neural networks to learn multivariate dependencies of our model.

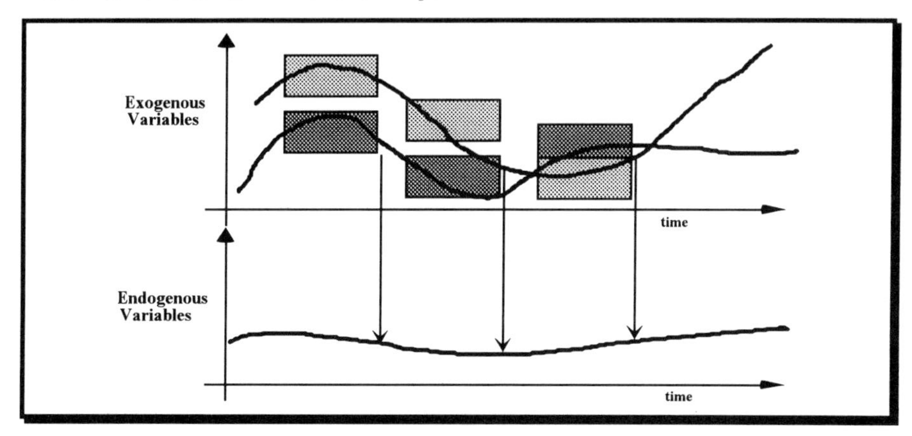

Figure 9.4 — Modified windowing: detection of regularities in correlating variables.

We want to have a function which expresses a macro economic correlation of the form:

$$v_{en} = f(v_{ex(1)}, \ldots, v_{ex(m)}).$$

We assume to have for every exogenous variable $v_{ex(i)}$ a window $w_{ex(i)}$ of a fixed size and for every endogenous variable v_{en} a window w_{en}. We translate the assumed relationship between the exogenous variables and the endogenous variable v_{en} to be predicted in a data set consisting of tupels of the following form: $(w_{1..n,\ ex(1)}, \ldots, w_{1..n,\ ex(m)}, w_{1..m,\ en})$. The task for the neural network is now to learn the relationship between the variables on the basis of their historical values.

In the experiments, the following parameters were varied

Window size

The experiments were based on yearly data. We used three different window sizes for the exogenous variables $n = 8, 11, 14$ years; $m = 1$ year.

Topology

For every prediction problem we developed two topologies. The first with one hidden layer, the second with two hidden layers. The number of neurons in the hidden layers varied in the range from 4 to 12.

Error rate

The error rate ε (see formula 9.2) is the mean square error of actual output o_i and the desired output l_i. We used following error rates: 0.007, 0.004 and 0.001.

$$\varepsilon = \frac{1}{n}\sum_{i=1}^{n}(o_i - l_i) \qquad (9.2)$$

Other parameters

The learning rate λ of the backpropagation learning procedure was set to λ=0.5 and the momentum was set to μ=0.9. The real data was translated into values of the interval [0.3; 0.7] by using a max-min-table.

9.4.2.3 Experimental Results

To train our networks we used data from 1970-1990 describing the historical course of the exogenous and endogenous variables of our model. Combining all experimental parameters (window size, topology, error rate) we had to train 882 different networks using each particular network for a one-year-prediction. The usage of the networks in this way is in our view the most realistic.

A complete overview of the experimental work is documented in [Lech93]. We present now general observations we made in the experiments.

Examples of our prediction results are illustrated in Figures 9.5, 9.6, 9.7. The prediction of investment (see Table 9.2) shows a typical phenomena that could be observed also for other problems. In most cases the results were better for smaller window sizes (8 or 11) with a higher error (0.007). One interpretation of this result is the following. The influence of the different exogenous variables vary over time. Thus, the relationships in early phases of the window have no or a little predictive power. In this view they introduce noise to the learning data and so decrease the quality of the prediction.

Although the results are very promising we mentioned the problems encountered in finding good network configurations (as shown in the experimental setting). An extensive search for the right network parameters is necessary. This can be very time consuming and create a lot of experimental work.

With our experiments we showed that the practical feasibility of backpropagation networks for predication of economic domains. Future work includes the comparison of these results with traditional methods. Using such a convergent validation means it is possible to detect weaknesses of the underlying domain model, because if neither the network approach nor for instance a regression equation approach can prove the assumed relationship it is likely that the underlying model has to be modified.

Window size	Topology	Error	$U(E_t)$
8	type 1	0.007	**0.9528**
		0.004	1.2637
		0.001	1.5502
	type 2	0.007	**0.9728**
		0.004	1.0053
		0.001	1.1173
11	type 1	0.007	1.3275
		0.004	1.3736
		0.001	1.4366
	type 2	0.007	1.3632
		0.004	1.3848
		0.001	1.4987
14	type 1	0.007	1.6632
		0.004	1.1584
		0.001	1.6797
	type 2	0.007	1.6947
		0.004	1.1647
		0.001	1.3083

Table 9.2 — The measured inequality coefficient U for the different resulting curves predicting employment (window: 8, 11, 14 years; topology: type 1 = 1-4-1, type 2 = 1-4-2-1; learning rate).

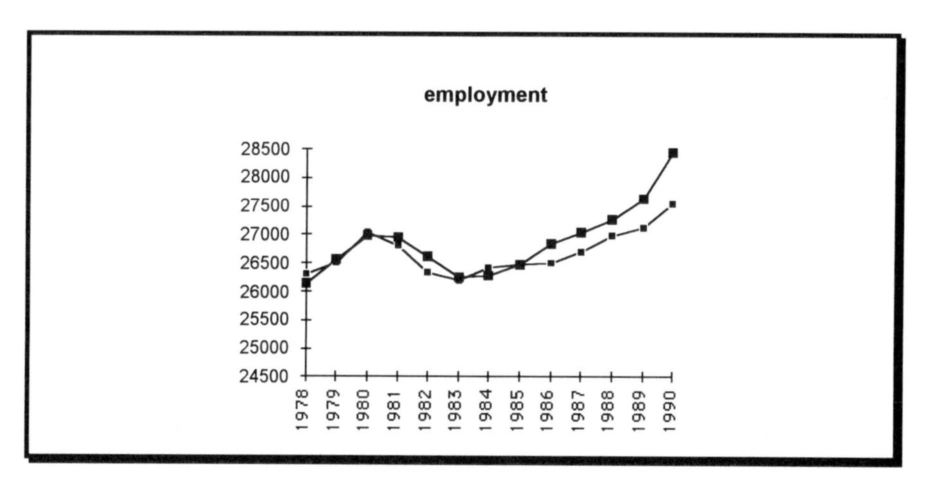

Figure 9.5 — Prediction of employment.

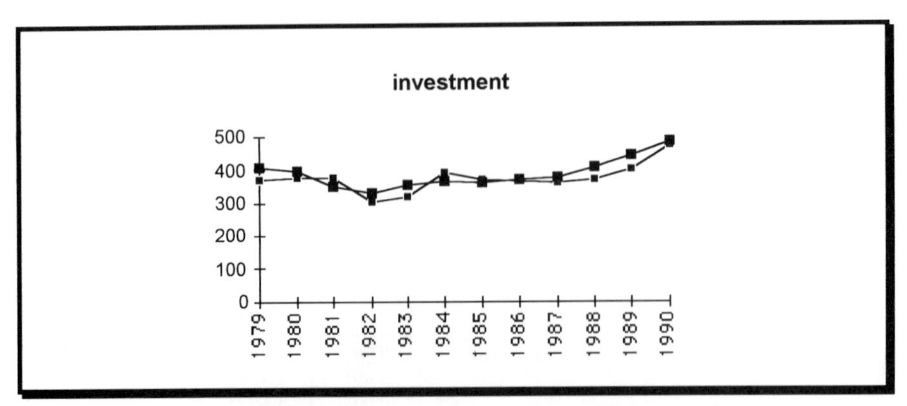

Figure 9.6 — Prediction of investment.

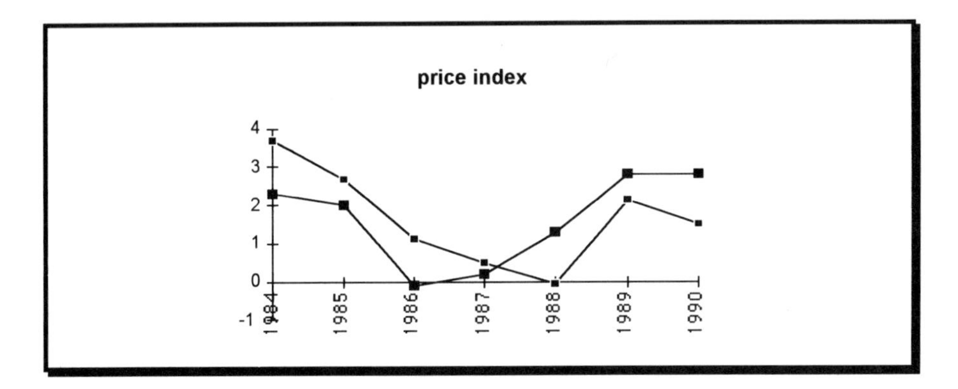

Figure 9.7 — Prediction price index.

The general necessity in prediction to use more than one prediction method leads us to an approach where two or more paradigms can coexist.

9.4.3 A Co-operative Framework

In the co-operative framework called PREDICTOR, neural networks and other problem solving techniques interact in solving problems that are beyond the ability of each individual technique. We use a blackboard architecture for the integration of these heterogeneous problem solvers. After a brief introduction to the basic characteristics of blackboard systems we present the architecture of the PREDICTOR system.

9.4.3.1 Blackboard Systems

In his work on the organisation of problem-solving programs, Newell [Newe66] introduced the idea of blackboard systems in the following way:

Metaphorically we can think of a set of workers, all looking at the same blackboard: each is able to read everything that is on it, and to judge when he has something worthwhile to add to it. This conception is just that of Selfridge's Pandemonium: a set of demons, each independently looking at the total situation and shrieking in proportion to what they see fits their nature.

Currently, blackboard systems are a widely used architecture in DAI applications. For a general introduction see [Jack90]. The basic organisation of a blackboard system can be summarised as follows. The domain is partitioned into different *knowledge sources* (KS). During the problem solving process these KS are controlled by an *scheduler* responsible for the co-ordination of the KS. The *blackboard* is the "memory" that allows the communication between the different knowledge sources.

Knowledge sources are triggered by the appearance of specific objects on the blackboard such as information that is of interest to a specific KS for solving a particular (sub-)task of the problem. These objects are called *knowledge source activation records* (KSAR).

General characteristics of blackboard systems are [Jack90, Zhan91]:

- separation of domain and control knowledge
- parallelisation of the execution of processes
- physical and/or logical distribution of knowledge (thus, inherently heterogeneous)
- open architecture for the integration of new KS.

9.4.3.2 The Architecture of PREDICTOR

PREDICTOR is a blackboard-based framework supporting the hybrid co-operation of different heterogeneous problem solvers. In the sense of [Zhan91] it supports the horizontal co-operation of different experts in predicting economic data.

In Figure 9.8 an overview of the architecture of PREDICTOR is given. Each problem solver has specific knowledge about the domain and the ability to react to messages that are distributed via the communication facility.

To handle the communication aspects in this system, each problem solver has a front end processor which is responsible for managing the co-operative aspects of the problem solver. The communication facility distributes messages (tasks, KSAR's and results) all over the system (scheduler, problem solvers). The blackboard manager provides other nodes with blackboard operations like *search*, *read*, *write* and *update*. The scheduler is responsible for the task analysis, task allocation and task synthesis. Domain dependent knowledge, specific analysis and allocation strategies are stored at a meta level within a knowledge base. This meta knowledge controls the problem solving abilities of the scheduler.

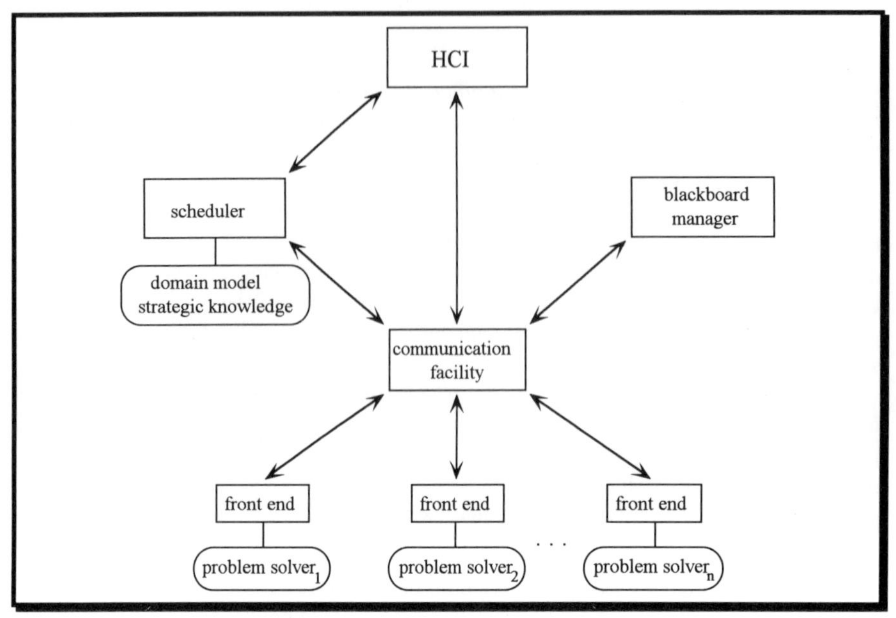

Figure 9.8 — Architecture of PREDICTOR.
(HCI: *human-computer interface*)

The following are some of the important characteristics of the implementation of PREDICTOR :

- the domain knowledge is partitioned in to different knowledge sources (problem solvers)
- the knowledge sources have a condition and an activation component
- problem specification and (partial) results are described on different logical levels within the blackboard (see below)
- the individual modules communicates only via the blackboard
- the problem solvers do not execute each other
- the problem solvers do not exchange data directly
- the problem solving process is inherently parallel
- PREDICTOR is open and extendible.

9.4.4 Logical Levels of the PREDICTOR Blackboard

The blackboard entries can be divided into three different groups. The relationship between these different types of blackboard entries is now described. This will give a better insight into how the communication within PREDICTOR is organised.

BB-level 1. At this level we have entries describing the entire problem or the solution.

- **tasklist([<problem-spec 1>, ... , <problem-spec n>])**: this predicate describes a complex query specified by the user. This Prolog query is generated by the HCI (see Figure 9.8).
  Example: The user is interested in a prediction of the interest rate, the number of employed persons and the public consumption in 1994. So the generated tasklist has following structure:

```
tasklist([    query_for_interest(interest, 1994),
              query_for_employment(empl, 1994),
              query_for_public-cons(public-cons, 1994)]
```

- **resultlist([<problem-sol 1>, ... , <problem-sol n>])**: Finishing the computational process, the resultlist contains the overall solution which is returned to the HCI.
  Example:

```
resultlist([  prediction(INTEREST, 1994, 8.3),
              prediction(EMPL, 1994, 4.8)
              prediction(PUBLIC-CONS, 1994, 2020.6)]
```

BB-level 2. At this level request and partial results are communicated. We have the following entry types:

- **request(variable, year)**: This predicate defines a new sub-problem that has to be solved by one of the agents of PREDICTOR.
  Example:

```
request(INTEREST, 1994)
```

- **ksar(variable, year, id-no, certainty-factor):** This entry is generated by a problem solver within PREDICTOR after it has seen a specific request predicate on the blackboard. The KSAR indicates how well a problem solver, with the identification number id-no, can solve a specific prediction problem for a given year. The quality of the prediction result is expressed via a so called certainty factor which takes a value between 0 and 1. The certainty factor is calculated on the basis of the prediction quality of previous years.
  Example:

```
ksar(INTEREST, 1994, 16, 0.56)
```

  If there are many different KSARs for the same problem, the scheduler usually triggers the agent with the best offer.

- **prediction(variable, year, result):** Via this predicate the agent which was triggered by the scheduler returns its result.
  Example:

```
prediction(INTEREST, 1994, 8.3)
```

BB-level 3. This level of the blackboard allows historical data exchange.

- **fact(variable, year, value)**. Via this predicate historical data can be exchanged between different modules of PREDICTOR.
  Example:

  fact(INTEREST, 1978, 7.4)

The prediction model is defined in the knowledge base of the scheduler. On that basis the task decomposition and synthesis can be computed. The individual problem solvers participate in the co-operative problem solving process only if they can make a contribution to the problem. The agent with the highest competence for a particular sub-problem is triggered by the scheduler. From that moment it is responsible for the sub-task.

9.5 Concluding Remarks

Comparing the advantages of knowledge-based systems and neural networks we come to the conclusion that they have properties and that their integration would help to overcome some of their limitations. For both technologies to coexist, a co-operative framework must be found that allows efficient execution, communication and control of these different problem solving components. In this Chapter we advocate DAI techniques as a suitable hybrid framework. We have presented the co-operative framework PREDICTOR, for predicting economic data.

Our work is also relevant for other domains where predictive models are used, such as:

- **marketing**: prediction of market shares of brands
- **banking**: prediction of currency exchange rates
- **manufacturing**: plant control.

We plan to continue this work further by comparing various prediction techniques and studying combined strategies. We also plan to develop generic tools for integration within PRIDICTOR as new problem solvers. We shall also continue to study advance co-operation strategies.

Acknowledgements

We'd like to thank Dr. Lutz Altenburg, Dipl.Kff. Manuela Grimm and Dr. Nina Vojdani for their comments on previous versions of this paper.

References

[Becr91] Becraft, W.R., Lee, P.L. and Newell, R.B. "Integration of Neural Networks and Expert Systems for Process Fault Diagnosis", *12th Int. Conf. on AI*, Australia, Vol. 2, pp. 832-837, 1991.

[Bond88] Bond, A., Gasser, L. *Readings in Distributed Artificial Intelligence*, Morgan Kaufmann Publishers, San Mateo, CA., 1988.

[Caud90] Caudill, M. "Using Neural Nets: Hybrid Expert Networks", *AI Expert*, pp. 49- 53, Nov. 1990.

[Conr86] Conry, S.E. Meyer, R.A. Lesser, V.R, "Multistage Negotiation in Distributed Planning", in [Bond88], pp. 367.

[Deen88] Deen, S.M. Research Issues in Federated Knowledge-based Systems, British Computer Society Workshop Series: Research and Development in XPS V, *Proceedings of XPS*, Cambridge 1988.

[Dunk92] Dunker, J. Scherer, A. Schlageter, G. "Integrating Neural Networks into a Distributed Knowledge-based System", *12th Int. Conf. on AI*, Expert Systems and Natural Language, France, 1992.

[Durf87] Durfee, E.H. Lesser, V.R. "Using Partial Global Plans to Co-ordinate Distributed Problem Solvers", *IJCAI* 1987, pp. 875.

[Ferb89] Ferber, J. "Eco Problem Solving: How to Solve a Problem by Interactions", *9th DAI-Workshop*, Rosario Texas, 1989, pp. 113.

[Fisc89] Fischer, K. *Theoretische Grundlagen der Konjunkturprognose, Europäische Hochschulschriften*, Peter Lang, 1989.

[Frie73] Friedmann, W. Richter, R. Schlieper, R. "Makroökonomie: *Eine EinführungC*, Springer Verlag, 1973.

[Goon92] Goonatilake, Suran and Khebbal, Sukhdev, "Intelligent Hybrid Systems", *Proceedings of First International Conf on Intelligent Systems, Singapore*, pp. 207-212, Sept 1992.

[Gutk91] Gutknecht, M. Pfeifer, R. Stolze, M. "Co-operative Hybrid Systems", 12th *Int. Conf. on AI*, Australia, Vol. 2, pp. 824-829, 1991.

[Huet82] Huettner, M. "Markt- und Absatzmethoden", Kohlhammer Edition Marketing, 1982.

[Jack90] Jackson, P. *Introduction to Expert Systems*, 2nd Edition, Addison-Wesley, 1990.

[Keyn36] Keynes, J.M. *The General Theory of Employment, Interest and Money*, London, 1936. (see also: *The Collected Writings of John Maynard Keynes*, Vol. XIV, The General Theory and After, Part I and II, Moggridge, D. (ed.), London, 1973.)

[Kheb93] Khebbal, S. Treleavan, P. "An Object-Oriented Hybrid Environment for Integrating Neural Networks and Expert Systems", *Proceedings of First New Zealand International Two-Stream Conference on, Artificial Neural Networks and Expert Systems*, IEEE Computer Society Press, USA, 1993

[Kirn92] Kirn, St. Scherer, A. Schlageter, G. "Problem Solving in Federative Environments: The FRESCO Concept of Cooperative Agents", in *The New Generation Of Information Systems: From Data to Knowledge*; edited by M.P. Papazoglou; J. Zeleznikow, Springer-Verlag, Lecture Notes in AI, 1992.

[Klet89] Klett, G. "Cooperating Expert Systems with Contract Net-Architecture and Their Use in Chemical Applications", FernUniversität Hagen, thesis, Hagen 1989 (in German).

[Kout77] Koutsoyiannis, A. *Theory of Econometrics*, 2nd edition, The Macmillan Press, LTD., 1977.

[Lâas91] Lâasri, B. Lâasri, H., Lesser, V.R. "An Analysis of Negotiations and Its Role For Co-ordinating Cooperative Distributed Problem Solvers", *Proc. of the 11th Conf. on 2nd Generation Expert Systems*, Avignon, France, May 28-31, 1991.

[Lech93] Lechtenberg, D. "Backpropagation Models for economic prediction problems", master thesis (in German), University of Hagen, 1993.

[Lape87] Lapedes, A., Farber, R., "Non-linear Signal Processing using Neural Networks: Prediction and System Modelling", Tech Rep LA-UR-87-2662, Los Alamos National Laboratory, 1987.

[Lipp89] Lippmann, R.P. "Pattern Classification Using Neural Networks", *IEEE Communications Magazine*, 1989.

[Newe66] Newell, A. "Some Problems of Basic Organisation in Problem-Solving Programs", in: Yovits, M.C. Jacobi, G.T. Goldstein, E.D. (eds): *Conference on Self-Organising Systems*, Washington, D.C.: Spartan Books, pp. 393-423, 1966.

[Roch93] Rocha, Paulo, Khebbal, Sukhdev, and Treleavan, Philip, "A Framework for Hybrid Intelligent Systems", in *Proceeding of First New Zealand International Two-Stream Conference on Artificial Neural Networks and Expert Systems*, IEEE Computer Society Press, pp. 206-209, Los Alamitos, USA, November 1993.

[Roma90] Romaniuk, S.G. Hall, L.O. "A Hybrid Connectionist Symbolic Learning System", *Proceedings of 8th National Conf. on Artificial Intelligence*, vol. 2, pp. 783-788, 1990.

[Rose85] Rosenschein, J.S. Genesereth, M.R.: "Deals Among Rational Agents", *IJCAI 85*, pp. 91.

[Rume86] Rumelhart, D.E. Hinton, G.E.; Williams, R.J. "Learning Internal Representations by Error Propagation", *Parallel Distributed Processing*, Rumelhart, D.E. and McClelland, J.L., eds., Chapter 8, Cambridge, MA: MIT Press, 1986.

[Schl88] Schlageter, G. Klett, G. "Federative Knowledge-based Systems", in Cremers, A.B., Geisselhardt, W. *Expert Systems Proceedings, Symposium*, Univ of Duisburg, Germany, pp. 204., 1988.

[Smit80] Smith, R.G. "The Contract Net Protocol: High-Level Communication and Control in a Distributed Problem Solver", *IEEE Transactions on Computers*, vol. C-29, No. 12, December 1980.

[Thei71] Theil, H. *Principles of Econometrics*, John Wiley & Sons, 1971.

[Walt92] Walter, E. "PROKON: Ein Expertensystem zur Konjunkturprognose und sein Einsatz in föderativen wissensbasierten Systemen", FernUniversität Hagen (master thesis), 1992.

[Wied89] Wiederhold, G. *et al.* "Partitioning and Composing of Knowledge", Stanford University, Technical Report, 1989.

[Zhan91] Zhang, C. Bell, D.A. *HECODES: a Framework for Heterogeneous Cooperative Distributed Expert Systems, Data & Knowledge Engineering*, Elsevier Science Publishers (North-Holland), Vol. 6, 1991, pp. 251-273.

Part Three

Polymorphic Hybrids

Part Three

Polymorphic Hydrates

10

Integrating Symbol Processing Systems and Connectionist Networks

Vasant Honavar and Leonard Uhr

10.1 Introduction

The goal of artificial intelligence, broadly defined, is to understand and engineer intelligent systems. This entails building theories and models of embodied minds and brains — both natural as well as artificial. The advent of digital computers and the parallel development of the theory of computation since the 1950s provided a new set of tools with which to approach this problem — through analysis, design, and evaluation of computers and programs that exhibit aspects of intelligent behaviour — such as the ability to recognise and classify patterns; to reason from premises to logical conclusions; and to learn from experience.

The early years of artificial intelligence saw some people writing programs that they executed on serial stored-program computers (e.g. [Newe63, Feig63]); Others (e.g., [Rash60, Mccu43, Self63, Uhr63]) worked on more or less precise specifications of more parallel, brain-like networks of simple processors (reminiscent of today's connectionist networks) for modelling minds/brains; and a few took the middle ground [Uhr73, Holl75, Mins63, Arbi72, Gros82, Klir85].

It is often suggested that two major approaches have emerged – symbolic artificial intelligence (SAI) and artificial neural networks or connectionist networks (CN) and some [Norm86, Schn87] have even suggested that they are fundamentally and

Intelligent Hybrid Systems, Edited by S. Goonatilake and S. Khebbal. ©1995 John Wiley & Sons Ltd.

perhaps irreconcilably different. Others have argued that CN models have little to contribute to our efforts to understand cognitive processes [Fodo88]. A critical examination of the popular conceptions of SAI and CN models suggests that neither of these extreme positions is justified [Bode94, Hona90b, Hona94b, Uhr94a]. Recent attempts at reconciling SAI and CN approaches to modelling cognition and engineering intelligent systems [Hona94c, Sun94b, Levi94, Goon92, Meds94] are strongly suggestive of the potential benefits of exploring computational models that judiciously integrate aspects of both. The rich and interesting space of designs that combine concepts, constructs, techniques and technologies drawn from both SAI and CN invite systematic theoretical as well as experimental exploration in the context of a broad range of problems in perception, knowledge representation and inference, robotics, language, and learning, and ultimately, integrated systems that display what might be considered human–like general intelligence. This chapter examines how today's CN models can be extended to provide a framework for such an exploration.

10.2 A Critical Look at SAI and CN

This section critically examines the fundamental philosophical and theoretical assumptions as well as what appear to be popular conceptions of SAI and CN [Hona90b, Uhr94b, Hona90a, Hona94b]. This examination clearly demonstrates that despite assertions by many to the contrary, the differences between them are less than what they might seem at first glance; and to the extent they differ, such differences are far from being in any reasonable sense of the term, fundamental; and that the purported weaknesses of each can potentially be overcome through a judicious integration of techniques and tools selected from the other [Hona90a, Hona94b, Hona90b, Uhr94b, Uhr90, Bode94].

10.2.1 SAI and CN Approaches to Modelling Intelligence Share the Same Working Hypothesis

The fundamental working hypothesis that has guided most of the research in artificial intelligence as well as the information–processing school of psychology is rather simply stated: *Cognition, or thought processes can, at some level, be modelled by computation.* This has led to the *functional* view of intelligence which is shared explicitly or implicitly by almost all of the work in SAI. Newell's *physical symbol system hypothesis* [Newe80], and Fodor's *language of thought* [Fodo76] are specific examples of this view. In this framework, perception and cognition are tantamount to acquiring and manipulating symbolic *representations*. Representations, in short, are caricatures of an agent's environment that are operationally useful to the agent: operations on representations can be used to predict the consequences of performing the corresponding physical actions on the environment. (See [Newe90, Hona94b, Chan94] for more detailed discussion of the nature of representation and its role in the functional view of intelligence). Exactly the same functional view of intelligence is at the heart of current approaches modelling intelligence within the CN paradigm, as well as the attempts to understand brain function using the techniques of computational neuroscience and neural modelling. This is clearly demonstrated by the earliest work on neural networks by Rashevsky [Rash60], McCulloch and Pitts [Mccu43] and Rosenblatt [Rose62] — from which many of today's CN models [Mccl86, Kung93, Hayk94, Zeid89] as well as the increasing emphasis on *computational models* in contemporary neuroscience are derived [Chur92].

Thus, CN models or theories of intelligence are stated in terms of abstract computational mechanisms just as their SAI counterparts. The abstract descriptions in both cases are usually stated in sufficiently general languages. One thing we know for certain is that such languages are all equivalent (see below). This provides absolute assurance that particular physical implementations of systems exhibiting mind/brain–like behaviour can be described using the language of SAI or CN irrespective of the physical medium (biological or silicon or some other) that is used in such an implementation. And the choice of the language should be dictated by pragmatic considerations.

10.2.2 SAI and CN Rely on Equivalent Models of Computation

One of the most fundamental results of computer science is the *Turing-equivalence* of various formal models of computation – including Turing's specification of the general-purpose serial stored program computer with potentially infinite memory (the *Turing machine*), Church and Rosser's *lambda calculus*, Post's *production systems*, McCulloch and Pitts' *neural networks* (among others). Turing was among the first to formalise the common sense notion of computation in terms of execution of what he called — an *effective procedure* or an algorithm. In the process, he invented a hypothetical computer — the *Turing machine*. The behaviour of the Turing machine is governed by an algorithm which is realised in terms of a *program* or a finite sequence of instructions. Turing also showed that there exists a *universal* Turing machine (essentially a general purpose stored program computer with potentially infinite memory) — one that can *compute* anything that any other Turing machine could possibly compute — given the necessary program as well as the data and a means for interpreting its programs. The various formal models of computation mentioned above (given potentially infinite memory) were proved exactly equivalent to the Turing machine. That is, any computation that can be described by a finite program can be programmed in any general purpose language or on any Turing-equivalent computer [Cohe86]. However, a program for the same computation may be much more compact when written in one language than in some other; or it may execute much faster on one computer than some other. But the provable equivalence of all general purpose computers and languages assures us that *any* computation — be it numeric or symbolic — can be realised, in principle, by both SAI as well as CN systems.

Given the reliance of both SAI and CN on equivalent formal models of computation, the questions of interest have to do with the identification of particular subsets of Turing-computable functions that model various aspects of intelligent behaviour given the various design and performance constraints imposed by the physical implementation media at our disposal.

10.2.3 Problem Solving as State Space Search

The dominant paradigm for problem solving in SAI is *state space search* [Wins92, Gins93]. States represent snapshots of the problem at various stages of its solution. Operators enable transforming one state into another. Typically, the states are represented using structures of symbols (e.g., lists). Operators transform one symbol structure (e.g., list, or a set of logical expressions) into another. The system's task is to find a path between two specified states in the state space (e.g., the initial state and a specified goal, the puzzle and its solution, the axioms and a theorem to be proved, etc.).

In almost any non trivial problem, a blind exhaustive search for a path will be impossibly slow, and there will be no known algorithm or a procedure for directly computing that path without resorting to search. However, search can be guided by the knowledge that is at the disposal of the problem solver. If the system is highly specialised, the necessary knowledge is usually built into the search procedure (in the form of criteria for choosing among alternative paths, heuristic functions to be used, etc.). However, general purpose problem solvers also need to be able to retrieve problem-specific and perhaps even situation-specific knowledge to be used to guide the search during problem-solving. Indeed, such retrieval might itself entail search (albeit in a different space). Efficient, and flexible representations of such knowledge as well as mechanisms for their retrieval as needed during problem solving are, (although typically overlooked because most current AI systems are designed for very specialised, narrowly defined tasks), extremely important.

State space search in SAI systems is typically conducted using serial programs or production systems which are typically executed on a serial von Neumann computer. However, there is no reason not to use parallel search algorithms or parallel production systems.

The CN system (a network of relatively simple processing elements, neurons, or nodes) is typically presented with an input pattern or initialised in a given starting state encoded in the form of a state vector each of whose elements corresponds to the state of a neuron in the network. It is designed or trained to output the correct response to each input pattern it receives (perhaps after undergoing a series of state updates determined by the rules governing its dynamic behaviour). The input-output behaviour of the network is a function of the network architecture, the functions computed by the individual nodes and parameters such as the weights.

For example, the solution of an optimisation problem (traditionally solved using search) can be formulated as a problem of arriving at a state of a suitably designed network that corresponds to one of minimum energy — which is defined to correspond in some natural way to the optimality of the solution being sought [Hopf82]. Ideally, the network dynamics are set up so as to accomplish this without additional explicit control. However, in practice, state updates in CN systems are often controlled in a manner that is not much different from explicit control as in sequential update of neurons in Hopfield networks, [Hopf82] where only one neuron is allowed to change its state on any update cycle to guarantee certain desired emergent behaviours. Indeed, a range of cognitive tasks do require selective processing of information that often necessitates the use of a variety of (albeit flexible and distributed) networks of controls that is presently lacking in most CN models [Hona90c]. Many such control structures and processes are suggested by an examination of computers, brains, immune systems, and evolutionary processes.

In short, in both SAI and CN systems, problem-solving involves state space search and although most current implementations tend to fall at one end of the spectrum or the other, it should be clear that there exists a space of designs that can use a mix of different state representations and processing methods. The choice of a particular design for a particular class of problems should primarily be governed by performance, cost, and reliability considerations for artificial intelligence applications and psychological and neurobiological plausibility for cognitive modelling.

10.2.4 Symbols, Symbol Structures, Symbolic Processes

Knowledge representation in SAI systems involves the use of symbols at some level. The standard notion of a symbol is that it *stands for* something and when a symbol token appears within a symbolic expression carries the interpretation that the symbol stands for something within the context that is specified by its place in the expression. In general, a symbol serves as a surrogate for a body of knowledge that may need to be accessed and used in processing the symbol. And ultimately, this knowledge includes semantics or meaning of the symbol in the context in which it appears, including that provided by the direct or indirect grounding of the symbol structure in the external environment [Harn90].

Symbolic processes are essentially transformations that operate on symbol structures to produce other symbol structures. Memory holds symbol structures that contain symbol tokens that can be modified by such processes. This memory can take several forms based on the time scales at which such modifications are allowed. Some symbol structures might have the property of determining choice and the order of application of transformations to be applied on other symbol structures. These are essentially the programs. Programs when executed — typically through the conventional process of compilation and interpretation and eventually — when they operate on symbols that are linked through grounding to particular effectors — produce behaviour. Working memory holds symbol structures as they are being processed. Long-term memory, generally speaking, is the repository of programs and can be changed by addition, deletion, or modification of symbol structures that it holds. The reader is referred to [Newe90] for a detailed treatment of symbol systems of this sort.

Such a symbol system can compute any Turing-computable function provided it has sufficiently large memory and its primitive set of transformations are adequate for the composition of arbitrarily symbol structures (programs) and the interpreter is capable of interpreting any possible symbol structure. This also means that any particular set of symbolic processes can be carried out by a CN — provided it has potentially infinite memory, or finds a way to use its transducers and effectors to use the external physical environment to augment its memory (just as humans have in their use of stone tablets, papyrus, and books through the ages).

Knowledge in SAI systems is typically embedded in complex symbol structures such as lists [Norv92], logical databases [Gene87], semantic networks [Quil68], frames [Mins75], schemas [Arbi72, Arbi94], and manipulated by procedures or inferences (e.g., list processing, application of production rules [Wate85], or execution of logic programs [Kowa77] carried out by a central processor that accesses and changes data in memory using addresses and indices).

It is often claimed that the CN systems predominantly perform numeric processing in contrast to SAI systems which manipulate symbol structures. But as already pointed out, CN systems represent problem states using state vectors which are manipulated in a network of processors using typically numeric operations (e.g., weighted sums and thresholds). It is not hard to see that the numeric state vectors and transformations employed in such networks play an essential symbolic role although the rules of transformation may now be an emergent property of a large number of nodes acting in concert. In short, the formal equivalence of the various computational models guarantees that CN can support arbitrary symbolic processes. It is not therefore surprising that several alternative mechanisms for variable binding and logical reasoning using CN have been discovered in recent years. Some of these require explicit use of symbols [Shas89]; others resort to quasi-symbols that have some properties of symbols while not being actually symbols in their true sense [Poll90, Macl94]; still others use pattern vectors to encode symbols [Dola89, Smol90, Sun94a, Chen94]. The latter approach to symbol processing is often said to use subsymbolic encoding of a symbol as a pattern vector each of whose components is insufficient in and of itself to identify the symbol in question (see the discussion on distributed representations below). In any case, most, if not all, of these proposals are implemented and simulated on general purpose digital computers, so none of the functions that they compute are outside the Turing framework.

10.2.5 Numeric Processing

Numeric processing, as the name suggests, involves computations with numbers. On the surface it appears that most CN perform essentially numeric processing. After all, the formal neuron of McCulloch and Pitts computes weighted sum of its numeric inputs. And the *neurons* in most CN models perform similar numerical computations. On the other hand, SAI systems predominantly compute functions over structures of symbols. But numbers are in fact symbols for quantities; and any computable function over numbers can be computed by symbolic processes. In fact, general purpose digital computers have been performing both symbolic as well as numeric processing ever since they were invented.

10.2.6 Analog Processing

It is often claimed that CN perform *analog computation*. Analog computation generally implies the use of dynamic systems describable using continuous differential equations. They operate in continuous time, generally with physical entities such as voltages and currents, which serve as physical *analogs* of the quantities of interest. Thus soap bubbles, servomechanisms, and cell membranes can all be regarded as analog computers [Raja81].

Whether physically realisable systems are truly analog or whether an analog system is simply a mathematical idealisation of (an extremely fine grained) discrete system is a question that borders on the philosophical — are time, space, and matter continuous or discrete?. However, some things are fairly clear. Most CN are simulated on digital computers and compute in discrete steps and hence are clearly not analog. The few CN models can be regarded as analog devices — e.g., the analog VLSI circuits designed and built by Carver Mead and colleagues [Mead89] — are incapable of discrete symbolic computations (because of their inability to make all-or-none or discrete choices) [Macl94] although they can approximate such computations. For example, the stable states or attractors of such systems can be interpreted as identifiable discrete states.

Analog systems can be, and often are simulated on digital computers at the desired level of precision. However, this might involve a time-consuming iterative calculation to produce a result that could potentially be obtained almost instantaneously given the right analog device. Thus analog processing appears to be potentially quite useful in many applications (especially those that involve perceptual and motor behaviour). It is possible that evolution has equipped living systems with just the right repertoire of analog devices that help them process information in this fashion. However, it is somewhat misleading to call such processing *computation* (in the sense defined by Turing) because it lacks the discrete combinatorial structure that is characteristic of all Turing-equivalent models of computation [Macl94].

Whether analog processes play a fundamental role in intelligent systems remains very much an open question. It is also worth pointing out that digital computers can, and in fact do, make use of essentially analog devices such as transistors but they use only a few discrete states to support computation (in other words, the actual analog value is irrelevant so long as it lies within a range that is distinguishable from some other range). And when embedded in physical environments, both SAI and CN systems do encounter analog processes through sensors and effectors.

10.2.7 Compositionality and Systematicity of Representation

It has been argued by many e.g. Fodor and Pylyshyn [Fodo88], that *compositionality* and *systematicity* (structure sensitivity) of representation are essential for explaining mind. In their view, CN are inadequate models of mind because CN representations lack these essential properties. Compositionality is the property that demands that representations must possess an internal *syntactic structure* as a consequence of a particular method for composing complex symbol structures from simpler components. Systematicity requires the existence of processes that are sensitive to the syntactic structure. As argued by Sharkey and Jackson [Shar94], lack of compositionality is demonstrably true only for a limited class of CN representations; and compositionality and systematicity in and of themselves are inadequate to account for cognition (primarily for lack of grounding or semantics). Van Gelder and Port [Vang94] have shown that several forms of compositionality can be found in CN representations.

10.2.8 Grounding and Semantics

Many in the artificial intelligence and cognitive science research community agree on the need for *grounding* of symbolic representations through sensory (e.g., visual, auditory, tactile) transducers and motor effectors in the external environment on the one hand and the internal environment of needs, drives, and emotions of the organism (or robot) in order for such representations (which are otherwise devoid of any intrinsic meaning to the organism or robot) to become imbued with meaning or semantics [Harn90]. Some have argued that CN systems provide the necessary apparatus for grounding [Harn94]. It is important to realise that CN as computational models do not provide physical grounding for representations any more than their SAI counterparts. It is only the physical systems with their physical substrate on which the representations reside that are capable of providing such grounding in physical reality when equipped with the necessary transducers and effectors. This is true irrespective of whether the system in question is a prototypical SAI system, or a prototypical CN system, or a hybrid or integrated system.

10.2.9 Serial Versus Parallel Processing

As pointed out earlier, most of today's SAI systems are serial programs that are executed on serial von Neumann computers. However, serial symbol manipulation is more an artefact of most current implementations of SAI systems than a necessary property of SAI. In parallel and distributed computers, memory is often locally available to the processors and can even be almost eliminated in data flow machines which model functional or applicative programs where data is transformed as it flows through processors or functions. Search in SAI systems can be, and often is, parallelised by mapping the search algorithm onto a suitable network of computers [Uhr84, Uhr87b, Hewi77, Hill85] with varying degrees of centralised or distributed control. Many search problems that arise in applications such as temporal reasoning, resource allocation, scheduling, vision, language understanding and logic programming can be formulated as constraint satisfaction problems which often lend themselves to solution using a mix of serial and parallel processing [Tsan93].

Similarly, SAI systems using production rules can be made parallel by enabling many rules to be matched simultaneously in a data flow fashion — as in RETE pattern matching networks [Forg82]. Multiple matched rules may be allowed to fire and change the working memory in parallel as in parallel production systems [Uhr79] and classifier systems [Holl75] — so long as whenever two or more rules demand conflicting actions, arbitration mechanisms are provided to choose among the alternatives or resolve such conflicts at the sensory-motor interface. Such arbitration mechanisms can themselves be realised using serial, parallel (e.g., winner-take-all mechanism), or serial-parallel (e.g., pyramid like hierarchies of decision mechanisms) networks of processes.

CN systems with their potential for massive fine grained parallelism of computation offer a natural and attractive framework for the development of highly parallel architectures and algorithms for problem solving and inference. Such systems are considered necessary by many researchers [Uhr80, Feld82] for tasks such as real-time perception. But SAI systems doing symbolic inference can be, and often are, parallelised, and certain inherently sequential tasks need to be executed serially. On any given class of problems, the choice of decomposition of the computations to be performed into a parallel–serial network of processes and their mapping onto a particular network of processors has to be made taking the cost and performance trade-offs into consideration.

10.2.10 Knowledge Engineering Versus Knowledge Acquisition Through Learning

The emphasis in some SAI systems — especially the so-called knowledge-based expert systems [Wate85] — on knowledge engineering has led some to claim that SAI systems are, unlike their CN counterparts, incapable of learning from experience. This is clearly absurd as even a cursory look at the current research in machine learning [Shav90, Buch93] and much earlier work in pattern recognition [Uhr73, Fu82, Micl86] shows. Research in SAI and closely related systems indeed have provided a wide range of techniques for deductive (analytical) and inductive (synthetic) learning. Learning by acquisition and modification of symbol structures almost certainly plays a major role in knowledge acquisition in humans who learn and communicate in a wide variety of natural languages (e.g., English) as well as artificial ones (e.g., formal logic, programming languages). While CN systems with their micro-modular architecture offer a range of interesting possibilities for learning, for the most part, only the simplest parameter or weight modification algorithms have been explored to date [Mccl86, Kung93, Gall93, Hayk94]. In fact, learning by weight modification alone appears to be inadequate in itself to model rapid and irreversible learning that is observed in many animals. Algorithms that modify networks through structural changes that involve the recruitment of neurons [Hona89a, Hona90a, Hona89b, Hona89c, Hona92a, Hona93, Kung93, Gros82] appear promising in this regard.

Most forms of learning can be understood and implemented in terms of structures and processes for representing and reasoning with knowledge (broadly interpreted) and for memorising the results of such inference in a form that lends itself to retrieval and use at a later time [Mich93]. Thus any CN or SAI or some hybrid architecture that is capable of performing inference and has memory for storing the results of inference for retrieval and use on demand can be equipped with the ability to learn. The interested reader is referred to [Hona94a] for a detailed discussion of systems that learn using multiple strategies and representations. In short, SAI systems offer powerful mechanisms for manipulation of highly expressive structured symbolic representations while CN offer the potential for robustness, and the ability to fine-tune their use as a function of experience (primarily due to the use of tuneable numeric weights and statistics). This suggests a number of interesting and potentially beneficial ways to integrate SAI and CN approaches to learning. The reader is referred to [Uhr73, Holl75, Hona92b, Hona93, Carp94, Shav94, Gall93, Gold94, Book94a] for some examples of such systems.

10.2.11 Associative as Opposed to Address-Based Storage and Recall

An often cited distinction between SAI and CN systems is that the latter employ associative (i.e., content-addressable) as opposed to the address-and-index based storage and recall of patterns in memory typically used by the former. This is a misconception for several reasons: Address-and-index based memory storage and retrieval can be used to simulate content-addressable memory and vice versa and therefore unless one had access to the detailed internal design and operation of such systems, their behaviour can be indistinguishable from each other. Many SAI systems conventional computers use associative memories in some form or another (e.g., hierarchical cache memories). While associative recall may be better for certain tasks, address (or location-based) recall (or a combination of both) may be more appropriate for others. Indeed, many computational problems that arise in symbolic inference (pattern matching and unification in rule-based production systems or logic programming) can take advantage of associative memories for efficient processing [Chen94].

In prototypical CN models, associative recall is based on some relatively simple measure of proximity or closeness (usually measured by Hamming distance in the case of binary patterns) to the stored patterns. While this may be appropriate in domains in which related items have patterns or codes that are close to each other, it would be absurd to blindly employ such a simple content-addressed memory model in domains where symbols are arbitrarily coded for storage (which would make hamming distance or a similar proximity measure useless in recalling the associations that are really of interest). Establishing (possibly context-sensitive) associations between otherwise arbitrary symbol structures based on their meanings and retrieving such associations efficiently requires complex networks of learned associations more reminiscent of associative knowledge networks, semantic networks [Quil68], frames [Mins75], conceptual structures [Sowa84], schemas [Arbi94], agents [Mins86] and object-oriented programs of SAI [Norv92] than today's simple CN associative memory models. This is not to suggest that such structures cannot be implemented using suitable CN building blocks — see [Arbi94, Dyer94, Miik94, Book94b, Barn94] for some examples of such implementations. Indeed, such CN implementations of complex symbol structures and symbolic processes can offer many potential advantages (e.g., robustness, parallelism) for SAI.

10.2.12 Distributed Storage, Processing, and Control

Distributed storage, processing, and control are often claimed to be some of the major advantages of CN systems over their SAI counterparts. It is far from clear as to what is generally meant by the term *distributed* when used in this context [Oden94].

Perhaps it is most natural to think of an item as distributed when it is coded (say as a pattern vector) whose components by themselves are neither sufficient to identify the item nor have any useful semantic content. Thus, the binary code for a letter of the alphabet is distributed. Any item thus distributed eventually has to be reconstructed from the pieces of its code. This form of distribution may be in space, time, or both. Thus the binary code for a letter of the alphabet may be transmitted serially (distributed in time) over a single link that can carry 1 bit of information at a time or in parallel (distributed in space) using a multi-wire bus. If a system employs such a mechanism for transmission or storage of data, it also needs decoding mechanisms for reconstructing the coded item at the time of retrieval. It is easy to see that this is not a defining property of CN systems as it is found in even the serial von Neumann computers. In any event, both CN as well as SAI systems can use such distributed coding of symbols.

And, as pointed out by Hanson and Burr [Hans90], 'distributed coding in and of itself, offers no representational capabilities that are not realisable using a non-distributed coding.

In the context of CN, the term *distributed* is often used to refer to storage of parts of an item in a unit where parts of other items also stored (for example, by superposition). Thus, each unit participates in storage of multiple items and each item is distributed over multiple units. (There is something disconcerting about this particular use of the term distributed in a technical sense: Clearly, one can invent a new name for whatever it is that a unit stores — e.g., a number whose binary representation has a '1' in its second place. Does the system cease to be distributed as a result?) It is not hard to imagine an analogous notion of distribution in time instead of space but it is also fraught with similar semantic difficulty.

The term *distributed* when used in the context of *parallel and distributed processing*, generally refers to the decomposition of a computational task into more or less independent pieces that are executed on different processors with little or no inter-processor communication [Uhr84, Uhr87b, Alma89]. Thus many processors may perform the same computation on pieces of the data (as in single-instruction-multiple-data or SIMD computer architectures) or each processor may perform a different computation on the same data e.g., computation of various intrinsic properties of an image (as in multiple-instruction-single-data or MISD computer architectures), or a combination of both (as in multiple-instruction-multiple-data or MIMD computer architectures). Clearly, both CN and SAI systems can take advantage of such parallel and distributed processing. The reader is referred to [Alma89, Uhr84, Uhr87b] for examples.

Today's homogeneous CN models can be viewed as MIMD computers if the weights or parameters associated with the nodes are viewed as data — which is a reasonable characterisation of what occurs during learning. On the other hand, the weights are stored locally in the processors and can be viewed as part of the instruction or the program executed by the nodes (especially if the weights are not modified) in which case, such CN can be thought of as SIMD computers in which each processor executes the same instruction (typically to compute the threshold or sigmoid function).

The control of execution of programs in CN is normally thought of as being distributed with little or no centralised control. While this appears to be true of *some* CN models (e.g., Hopfield networks operating with asynchronous parallel update), it is an arguable point in the case of *most* CN models. In most cases, the elaborate control structures that are necessary in many cases (e.g., the sequential update of Hopfield networks, synchronisation of neurons in each layer of a multilayer backpropagation network during processing and learning phases) are generally left unspecified. In any event, a wide range of centralised or (to various degrees distributed) controls can be embedded in both SAI as well as CN systems [Hona90c].

10.2.13 Redundancy and Fault Tolerance

Often the term *distributed* is used more or less synonymously with *redundancy* and hence fault tolerance in the CN literature. This is misleading because there are many ways to ensure redundancy of representation, processing and control. One of the simplest involves storing multiple copies of items and/or using multiple processors to replicate the same computation in parallel, and using a simple majority vote or more sophisticated statistical evidence combination processes to pick the result. Redundancy and distributivity are orthogonal properties of representations. And clearly, SAI as well as CN systems can be made redundant and fault tolerant using the same techniques.

10.2.14 Statistical, Fuzzy, or Evidential Inference

Many SAI systems represent and reason with knowledge using (typically first-order) logic. If sound inference procedures are used, such reasoning is guaranteed to be truth-preserving. In contrast, it is often claimed that CN models provide noise tolerant and robust inference because of the probabilistic, fuzzy, or evidential nature of the inference mechanisms used. This is largely due to combination and weighting of evidence from multiple sources through the use of numerical weights or probabilities. It is possible to establish the formal equivalence inference in certain classes of CN models with probabilistic or fuzzy rules of reasoning. But fuzzy logic [Zade75, Yage94] operates (as its very name suggests), with *logical* (hence symbolic) representations. Probabilistic reasoning is an important and active area of research in SAI as well (See [Pear88] for details). Heuristic evaluation functions that are widely used in many SAI systems provide additional examples of approximate, that is, not strictly truth-preserving inference in SAI systems. At the same time, it is relatively straightforward to design CN implementations of logical inference at least in restricted subsets of first-order logic. The reader is referred to [Sun94a, Pink94] for examples of such implementations.

In many SAI systems, the requirements of soundness and completeness of inference procedures are often sacrificed in exchange for efficiency. In such cases, additional mechanisms are used to verify and if necessary, override the results of inference if they are found to conflict with other evidence. Much research on human reasoning indicates that people occasionally draw inferences that are logically unsound [John91]. This suggests that although people may be capable of applying sound inference procedures, they probably take shortcuts when faced with limited computational or memory resources. Approximate reasoning under uncertainty is clearly an important tool that both SAI and CN systems can potentially employ to effectively make rapid, usually reliable and useful, but occasionally fallible inferences in real time.

10.2.15 SAI and CN As Models of Minds/Brains

Some of the SAI research draws its inspiration from (rather superficial) analogies with the mind and mental phenomena and in turn contributes hypotheses and models to the study of minds, Similarly, many CN models draw their inspiration from analogies with the brain and neural phenomena and in turn contribute models that occasionally shed light on some aspects of brain function [Chur92].

Today's CN models are at best, extremely simplified caricatures of biological neural networks [Shep89, Shep90, Mcke94]. They lack the highly structured modular organisation displayed by brains. The brain appears to perform symbolic, numeric, as well as analog processing. The pulses transmitted by neurons are digital; the membrane voltages are analog; the molecular level phenomena that involve closing and opening of channels appears to be digital; The diffuse influence of neurotransmitters and hormones appear to be both analog and digital.

Changes in learning appear to involve both gradual changes of the sort modelled by the parameter changing or weight modification algorithms of today's CN as well as major structural changes involving the recruitment of neurons and changes in network topology [Gree88, Hona89a]. Also missing from most of today's CN models are elaborate control structures and processes of the sort found in brains including networks of oscillators that control timing. Perception, learning and control in brains appear to utilise events at multiple spatial and temporal scales [Gros82]. Additional processes not currently modelled by CN systems include processes that include networks of markers that guide neural development, structures and processes that carry information that might be used to generate other network structures, and so on [Hona90a].

Clearly, living minds/brains are among the few examples of truly versatile intelligent systems that we have today. So even those whose primary interests are in constructing artificial intelligence systems can ill afford to ignore the insights offered by a study of biological intelligence [Mcke94]. (This does not of course mean that such an effort cannot exploit alternative technologies to accomplish the same functions, perhaps even better than their natural counterparts.) But it is a misconception to assume that today's CN model brains any more than today's SAI programs model minds. In short, the processes of the minds appear to be far less rigidly structured and far more flexible than today's SAI systems and the brains appear to have a lot more structure, organisation, and control than today's homogeneous networks of simple processing elements that we call CN. A rich space of designs that combine aspects of both within a well-designed architecture for intelligence remains to be explored.

10.3 Integration of SAI and CN

It must be clear from the discussion in the previous sections that at least on the surface it looks like SAI and CN are each appropriate, and possibly even necessary for certain problems, and grossly inappropriate, for others. But of course each can do anything that the other can. The issues are ones of performance, efficiency, and elegance, and not theoretical capabilities as computational models.

This is a common problem in computing. One computer or programming language may be extremely well suited for some problems but awful for others, while a second computer or language may be the opposite. This suggests several engineering possibilities [Uhr94, Hona94b], including:

1. Try to re-formulate and re-code the problem to better fit the computer or language. This may, for example, entail execution of parallelised CN equivalents or near-equivalents (possibly improvements) serial SAI programs.
2. Use one computer or language for some parts of the process and an other for reminder. Thus, one may embed black-boxes that execute certain desired

functions (e.g., list processing) within a larger network that includes CN modules for certain other functions (e.g., associative storage and recall of patterns).

3. Build a new computer or language that contains constructs from each, and use these as appropriate.
4. Try to find as elegant as possible a set of primitives that underlie both computers or languages, and use these to build a new system.

The term *hybrid* is beginning to be used for systems that in some way try to combine SAI and CN. But usually hybrid refers to systems of type (2) or (3). Types (3) and (4) would appear to be better than (2) although harder to realise, since they would probably be more efficient and more elegant. Thus the capabilities of both SAI and CN should be combined by tearing them apart to the essential components of their underlying processes and integrating these as closely as possible. Then the problem should be re-formulated and re-coded to fit this new system. This restates a general principle most people are coming to agree on with respect to the design of multi-computer networks and parallel and distributed algorithms: the algorithm and the architecture should be designed to fit together as well as possible, giving algorithm-structured architectures and architecture-structured algorithms [Uhr84, Uhr87b, Alma89]. The remainder of this chapter explores one approach to the design of such architectures for intelligent systems — by generalising the rather overly restrictive definition of most of today's CN models while retaining their essential advantages.

10.4 Connectionist Networks (CN) Narrowly Defined

Connectionist networks or *artificial neural networks* are massively parallel, shallowly serial, highly interconnected networks of relatively simple computing elements or *neurons* [Gall93, Kung93, Hayk94, Zeid89, Mccl86]. More precisely, a CN can be thought of as a directed graph whose nodes apply relatively simple numerical transformations to the numerical inputs that they receive via their input links and transmit the resulting output via their output links.

10.4.1 Nodes in a CN Perform Simple Numerical Computations

The input to an n-input node or neuron n_j in a CN is a pattern vector $\mathbf{X}_j \in \Re^n$ or in the case of binary patterns, by a binary vector $\mathbf{X}_j \in [0,1]^n$. Each neuron computes a relatively simple function of its inputs and transmits outputs to other neurons to which it is connected via its output links. A variety of neuron functions are used in practice. The most commonly used are the *linear*, the *threshold*, and the *sigmoid*. Each neuron has associated with it a set of parameters which are modifiable through learning. The most commonly used parameters are the so-called *weights*. The weights associated with an n-input neuron n_j are represented by an n-dimensional weight vector $\mathbf{W}_j \in \Re^n$. In the case of a linear neuron, the output o_j in response to an input pattern $\mathbf{X}_j$ on its input links is given by the vector dot product $\mathbf{W}_j \cdot \mathbf{X}_j$. In the case of a threshold neuron, $o_j =$

1 if $\mathbf{W}_j \cdot \mathbf{X}_j > 0$ and $o_j = 0$ otherwise. For a sigmoid neuron, $o_j = 1 / (1 + e^{-\mathbf{W}_i \cdot \mathbf{X}_j})$. In each of these cases, usually, one of the n inputs is held constant and the corresponding weight is called the threshold.

The functions listed above by no means exhaust the possible choices for neuron functions. A wide range of other possibilities exist — including replacing the simple neurons with digital and analog microcircuits that perform the needed symbolic as well as numeric computations [Shep89, Uhr94a].

10.4.2 Representation and Computation in CN

The representational and computational power of such networks depends on the functions computed by the individual neurons as well as the architecture of the network (e.g., the number of neurons and how the neurons are connected). In general, a layered *feedforward* network with n inputs and m outputs can represent some subset of the possible mappings from $\Re^n$ to $\Re^m$. Such networks are used in data compression, feature extraction, pattern classification and function approximation. A great deal is known about the representational power of different classes of feedforward networks. For example, a 2-layer feedforward network (not counting the input neurons that simply transmit the components of the input vector) with a *sufficiently large* (finite) number of sigmoid neurons in the first layer and linear neurons in the output layer can approximate to arbitrary accuracy, arbitrary continuous functions on bounded closed subsets of $\Re^n$. When feedback loops are included (thereby allowing the network's output at a given time step to be influenced by its output at a previous time step), it can produce complex temporal dynamics. Such networks find use in applications such as robot motion control and approximation of regular grammars from examples. Connectionist networks become Turing-equivalent in their computational power if we allow the networks to grow arbitrarily large.

Only a small subset of possible network topologies have been studied to date. A much broader range of alternatives exist — including those suggested by the brain and contemporary parallel computer architectures.

10.4.3 CN Learn by Modifying Parameters

Much of the research on learning in connectionist networks has tended to focus on algorithms for changing the modifiable parameters (weights) in networks with a certain *a priori* chosen network architecture using samples of the function to be approximated or patterns to be classified. In other words, the task of the learning algorithms is to find a set of weights that yields a satisfactory approximation of the unknown function on the given samples (the expectation is that the network will generalise well on samples not seen during training). This is fundamentally a search or optimisation problem and so a variety of linear and non-linear optimisation methods (gradient-descent, simulated annealing, etc.) can be used in this context. For details of such algorithms, the reader is referred to [Kung93, Hayk94].

Many connectionist learning algorithms use some form of error-guided search e.g., changing each modifiable parameter in the direction of the negative gradient of a suitably defined error measure with respect to the parameter of interest. A commonly used error measure in function approximation applications is the *mean squared error* between the desired and actual network outputs. For pattern classification tasks, a number of different error measures motivated by statistics and information theory (e.g., loss functions, cross-entropy) are possible candidates. For data compression or feature extraction, other error measures can be used to map high-dimensional input patterns to feature vectors in a lower dimensional space with minimal loss of information. Common examples of such methods include *competitive learning networks* and *principal component analysis*.

Almost all such learning algorithms involve searching for a set of weights (a solution to the pattern classification problem) in a high-dimensional *weight space* that meet the desired performance criteria (e.g., classification accuracy on a set of training patterns). In order for this approach to succeed, such a solution must in fact lie within the space being searched and the search procedure (e.g., gradient descent) must in fact be able to find it. This means that unless adequate *a priori* problem-specific knowledge can be brought to bear upon the problem of choosing an adequate network topology, it is reduced to a process of trial and error. This strongly argues for the need for more powerful learning algorithms — including those that incrementally construct the necessary network structures.

10.4.4 CN are Typically Only Partially Specified

A CN is typically specified in terms of the transformations performed by the individual nodes, the topology of the graph linking the nodes, and (if the network is designed to learn) the algorithm used to modify the parameters (typically weights) associated with the nodes. It is important to note that the additional network structures necessary to synchronise the nodes in the network, and to switch between behaving (producing outputs as a function of input) and learning (modifying the parameters as dictated by the learning algorithm) are generally left unspecified. The functions of such network structures are typically performed by programs that simulate CN on a traditional stored program computer. It is not hard to see that if all such functions were to be performed by the CN itself, it would need to be augmented with adequate co-ordination and control structures as well as learning structures to carry out the corresponding tasks [Hona90c].

The next section explores an evolving framework for a broader (and more complete) specification of CN [Hona90a, Hona90b, Uhr90] that not only makes explicit the various structures and functions that need to be built into today's CN but also suggests a broad range of potentially promising alternatives for the design of integrated SAI/CN architectures for versatile, powerful, robust, and adaptive intelligent hybrid systems.

10.5 Connectionist Networks Broadly Defined

The following rather general definition [Uhr90, Hona90b] makes clear that a large variety of CN architectures (other than the ones commonly used today) can be constructed. This definition is not meant to be exhaustive or complete in every respect.

However, it is intended to facilitate the exploration of a vast space of designs for intelligent systems.

A CN is a graph (of linked nodes) with a particular topology Γ. The total graph can be partitioned into three functional subgraphs – $\Gamma_{\mathbf{B}}$ (the *behave / act* subgraph), Γ_{Λ} (the *evolve/learn* subgraph), and $\Gamma_{\mathbf{K}}$ (the *co-ordinate/control* subgraph). The motivation for distinguishing among these three functions will become clear later. It primarily has to do with making explicit everything that is necessary to completely specify the structure and behaviour of a CN.

The nodes in a CN compute one or more different type(s) of functions: **B** (behave/act); Λ (evolve/learn); and **K** (co-ordinate/control). Thus we have a CN $\Omega = (\Gamma, \mathbf{B}, \Lambda, \mathbf{K})$; where $\Gamma = \Gamma_{\mathbf{B}} \cup \Gamma_{\Lambda} \cup \Gamma_{\mathbf{K}}$. Each behave, learn, or control function, has associated with it a suitable subgraph of nodes on which it is executed (to avoid cluttering the notation, this assignment is not explicitly specified in the definition).

It should be immediately clear that today's CN are specified (typically only partially) as follows:

- The topology, $\Gamma_{\mathbf{B}}$ of the subnetwork that behaves. (And much of the total graph, including the entire subgraphs needed to handle learning and control, is usually left unspecified).
- The behave functions computed by each node n_j in $\Gamma_{\mathbf{B}}$. These usually take the form of simple (numerical) transformations $\beta_j(\mathbf{W}_j, \mathbf{X}_j)$ where $\mathbf{W}_j$ is the weight vector associated with node n_j and $\mathbf{X}_j$ is the input vector to node n_j. And common choices are, as already pointed out, threshold, sigmoid, linear, and radial basis functions. Typically, the same node function is used with each of the nodes in the network, or at least, for each node in a given layer of a multilayer network.
- A learning algorithm that translates to a learning function λ_j that modifies the parameters typically the weight vector $\mathbf{W}_j$ at each node n_j in $\Gamma_{\mathbf{B}}$. The actual subnetworks that are needed to actually compute and make these changes or to switch between learning and behaving are left unspecified.

The broad definition of CN given above when superimposed with the critical examination of SAI and CN systems given in section 10.2 above suggests several potentially powerful extensions to today's CN — including more powerful structures and processes for behaving, control, and learning — in other words, a rich space of integrated SAI/CN architectures for intelligent hybrid systems.

10.6 Integrated SAI/CN Designs for Intelligent Hybrid Systems

The broad definition of CN outlined above suggests a general approach to the design of hybrid or integrated architectures for intelligent systems — one that essentially entails extending the range of behave, learn, and control structures and processes so that the capabilities typically associated with SAI systems (e.g., list processing, deductive inference, etc.) can be incorporated in relatively natural and well integrated ways into such systems. This section outlines a range of such potentially useful extensions to today's CN.

10.6.1 More Powerful Behaviour Structures and Functions

As already pointed out, there is no compelling reason to limit ourselves to the simple behave functions (e.g., threshold, linear, sigmoid) computed by the nodes in today's CN. That would be tantamount to arguing that all computer programs must be written using a minimal set of Turing machine instructions. A much broader range of behave structures and functions may be used as appropriate (along with learning and control structures and functions that are designed to work hand-in-hand with such functions). Many such analog, digital, as well as hybrid analog-digital behave structures and functions are suggested by digital and analog computers, microcircuits found in brains, and the design of algorithms and computer programs as well as some of the hybrid SAI/CN architectures that have been explored to date. The reader is referred to [Shep89, Uhr94a] for a detailed discussion of several such microcircuits. The short list that follows is suggestive of the wide range of possibilities that can be beneficially incorporated into CN:

- Microcircuits for comparison of numeric, symbolic, or iconic patterns to determine similarities and differences between two or more patterns (including stored templates).
- A broad range of flexible pattern matching functions — including those for matching highly structured patterns of symbols — e.g., strings, lists, labelled trees and graphs — with stored patterns. A variety of metrics may be used to compute the degree of match — including weighted *edit distances* [Gold94, Hona94a] of the sort used in structural pattern recognition systems.
- Microcircuits for computing minimal generalisations over two or more patterns or minimal specialisation of a stored template to prevent match with a given pattern.
- Automata for parsing symbol structures — preferably fault tolerant CN implementations [Chen94].
- Modiable temporary memories — including CN implementations of such structures using recurrent networks.
- Encoders/decoders for transforming symbol structures for efficient, fault tolerant transmission between spatially separated modules of CN
- Oscillators, clocks, and variable delay circuits.
- Specialised microcircuits for specific perceptual tasks — e.g., spot and oriented edge detectors of the sort found in the mammalian visual system; circuits for computing different intrinsic image properties such as depth, shape, and motion from visual images.
- Shifters for alignment and comparison of multiple sources of input — e.g., from two spatially separated visual sensors.
- Building blocks for components of semantic memories, associative networks, frames, and schema.
- Expandable fault tolerant content-addressable memories.

- Serial-to-parallel (or temporal to spatial) mapping networks and their generalisations.
- Winner-take-all networks to facilitate choice among several competing alternatives.
- Microcircuits for performing dynamic variable binding and unification necessary for matching complex symbolic expressions, performing logical inference, and simulating production systems.
- Dynamically configurable microcircuits for solving optimisation problems through the minimisation of suitably defined energy functions.
- Microcircuits for performing operations on lists such as insertion, deletion, and substitution of elements — analogous to the operations supported by a conventional symbol processing language like LISP.
- Microcircuits that perform simple set operations such as union, intersection, and difference.
- Microcircuits that perform simple spatial and temporal inferences on input spatial patterns or temporal event sequences (e.g., determining whether a particular object is contained within another object in a visual scene; or deciding if a particular event preceded another in time).
- Microcircuits that perform simple statistical computations — e.g., computing histograms, estimating relative frequencies and probability density functions, statistical means, etc.

10.6.2 More Appropriate, and When Indicated, Modular Network Structures

As already pointed out, CN can use virtually any graph topology but only a small subset has been used to date. These include simple one–layer or multilayer feedforward networks with complete connectivity between layers of nodes (typically used for function approximation, associative storage, recall, or classification of pattern vectors, and data compression); recurrent networks with feedback connections from higher layers to lower layers (typically used for temporal pattern recognition and sequential prediction tasks). However, a wide range of other graph topologies may be used as dictated by considerations of efficiency and design constraints imposed by the technology (e.g., VLSI) used for the physical realisation of such networks. A variety of such network topologies are suggested by contemporary developments in the design of parallel and distributed architecures and algorithms [Uhr84, Alma89]. A few such possibilities are enumerated below:

- Near-neighbour connectivity of the sort used in two-dimensional arrays or N-dimensional hypercubes of processors (such topologies have been especially useful in performing a variety of intrinsic image (shape, motion, texture, depth, etc. from gray scale images) computations in image processing and computer vision [Ball82, Uhr87a, Uhr87b, Wech90].
- Near-neighbour tree or pyramid like converging connectivity from one layer to the next of the sort used in several computer vision architectures [Hona89b, Hona89c, Uhr87a, Tani80].

- Network topologies that are specifically designed to expedite certain computations (e.g., Hough transforms that detect lines or parameterised curves in images [Ball82]).
- Dynamically configurable network topologies that facilitate certain computations on demand — e.g., structures used to construct logic proofs [Pink94].
- Specialised networks that carry context sensitive control and co-ordination signals — e.g., to focus or control attention.
- Modular network structures that reflect natural decomposition of tasks (e.g., object identification and object location in a visual image) — including decompositions discovered through learning or evolutionary processes.

10.6.3 More Powerful Learning Structures and Functions

As already pointed out, the only major forms of learning that have been studied in any detail in the context of CN are rote learning (which essentially involves storing patterns for future retrieval as in associative memories) and inductive learning (or learning from examples). A variety of other powerful forms of learning including deductive, abductive, and discovery techniques developed in the context of SAI systems do not have any CN counterparts at present. In today's CN is almost always limited to processes that change modifiable parameters (typically the weight vectors) within an otherwise *a priori* fixed network topology. The list that follows is meant to be merely suggestive of a few of the much wider range of potentially powerful alternatives that exist:

- Learning that modifies the behave functions β_j associated with each of the nodes n_j in the behave subgraph $\Gamma_{\mathbf{B}}$ of a CN or selects among a set of candidate functions based on the characteristics of the problem at hand.
- Learning that modifies the control functions κ_j associated with each of the nodes n_j in the control subgraph $\Gamma_{\mathbf{K}}$ of a CN. For example, such modifications might involve control of the locus, rate, and type of learning.
- Learning that modifies the learning functions λ_j associated with each of the nodes n_j in the learning subgraph Γ_{Λ} of a CN or selects from among a set of candidate learning rules.
- Learning that involves modification of the topology of behave, learn, or control subgraphs by addition and deletion of individual nodes, links, microcircuits, layers, or more generally necessary subgraphs. Given suitable constructive learning mechanisms, CN can adaptively search for and assume whatever connectivity is appropriate for the tasks at hand within the specific design constraints (e.g., local connectivity for VLSI implementation). Only some of the simplest constructive or generative algorithms that dynamically modify the network topology by recruiting nodes as necessary for a particular pattern classification or function approximation task have been examined to date [Hona89b, Hona89c, Hona93, Gall93, Kung93]. Such algorithms extend the search for solution to a suitably constrained space of network topologies. In this context, efficient randomised search techniques such as genetic algorithms are worth exploring [Bala94]. Complementary processes of

deletion of nodes and links are also of interest and a few such mechanisms have been explored to date [Kung93].

- Learning that fine tunes the symbolic representations such as rules and grammars used in SAI systems using the parameter modification processes of CN systems [Gall93, Shav94, Hona92b, Gold94].
- Learning that uses forms of inference such as deduction and abduction which are rarely used in today's CN systems which rely primarily on induction.

10.6.4 More Powerful Control Structures and Functions

As already pointed out, despite assertions to the contrary, some subtle control mechanisms are used (without ever being made explicit — because they are handled by the programs that simulate such CN). On the other hand, perhaps one of the most important limitations of today's CN is their relatively impoverished repertoire of control structures and processes. Many such mechanisms are suggested by an examination of computers and programs, biological organisms, and brains [Hona90c]. The list that follows is suggestive of the wide variety of such potentially powerful control structures and processes that may be beneficially incorporated into CN:

- Control can be introduced by building in particular structures and processes as needed — as in specifying particular microcircuits like winner-take-all nets or decision trees that make choices and selectively transmit the appropriate information.
- Partial control can be built in globally — e.g., through the use of global clocks that synchronise subsets of processors.
- The topology of the network, subnet, and local microcircuits can exercise major control functions — as when a tree of processors successively transforms, combines, and reduces information, or a pipeline of arrays sequentially applies a series of transformations to the input.
- Complete control of the sort found in conventional computers and multi-computers (e.g., execution of instructions in order; controlled sequences of state transitions) can be built in either centrally or locally at each node or small sets of nodes in the CN.
- A variety of co-ordination and control structures (e.g., message passing, blackboard structures for messages, instruction broadcasting, multiplexing, conflict resolution) used in multi-computer networks can be built into the CN.
- A host of control mechanisms of the sort found in conventional programming language constructs (conditional execution, loops, etc.) can be built into local or global microcircuits embedded in the CN.
- CNs may be provided with compact (gene-like) encodings of their structural and functional properties (e.g., the sizes of receptive fields, general topological constraints on connectivity, etc.). Such encodings may be transformed through genetic operators (e.g., crossover, mutation) to yield variant CN specifications. Environmental rewards and punishments may be used as means of guiding whole CN populations to evolve so as to perform better at tasks presented to them by the environment.

- DNA-like encodings can be incorporated, along with the capability to make copies, linking networks over which these copies can be sent, and decoders to transform these encodings into specifications for network structures and processes to be realised.
- Gene-like information can be used to dynamically specify different types of functional units in CN (analogous to the mechanisms of cellular differentiation). Controls can activate or suppress the expression of different functional properties in CN nodes or node ensembles.
- Local interactions of the sort found on the surface of cells that are bound by cell adhesion molecules might be used to determine how to build single units and larger microcircuits into successively larger structures (e.g., through a specification of how the CN microcircuits are to be assembled together).
- The immune system's rapid, evolution like adaptations suggest the possibility of triggering massive proliferation of a range of variant nodes or microcircuits to serve a variety of control functions under specific environmental conditions.
- Mechanisms analogous to chemical markers and pilot cells can — guided by information of the sort found in chemical gradients and lock-in-key-type templates — combine to build complex network structures.
- The complex interactions between chains of enzymes suggest the possibility of mechanisms that can modify, build, and recycle network structures in a self-sustaining manner.
- Multi-messenger pathways analogous to those supported by intra as well as inter-cellular communication (e.g., axonal transport mechanisms, membrane proteins) can be built into and among individual CN units and microcircuits of units.
- Several different types of global controls subnetworks can be used, tailored to and serving different purposes — much like the brain's neurotransmitters, global electrical waves, and chemical messenger systems. Such systems can be embedded in the CN using token-passing networks or distributed encoder–decoder structures.
- Neuromodulator like influences can be achieved by incorporating linking subnetworks that contain links with different amounts of delays, transmit information to changes thresholds, to alter network plasticity.
- Modulatory networks analogous to hormones can now have relatively diffuse, slow acting effects on such global processes as memory storage, memory modification or selective recall of stored memories.
- Regulatory subnetworks can also be used to initiate, modulate, and terminate plasticity of specific CN modules in a controlled fashion during learning.
- Specific control subnetworks can be embedded into the CN to instantiate processes like attention — selectively enhancing or attenuating the relative contributions of different aspects of the environmental stimuli.
- A variety of simple controls that regulate various aspects of learning (including the form and content of the learned representations), alter vigilance, and choose between different learning strategies can be incorporated into CN
- A variety of contextually driven switching mechanisms can be built into the CN to alter, in a dynamic fashion, the functions of different CN modules or the interactions among modules.

- Oscillators can be used to handle a variety of important problems — e.g., to switch between processes; to decide when to initiate, or to terminate, a process; to sample the environmental input at a desired rate.
- Clocks or networks of clocks can execute the equivalent of the *while* loop of a conventional programming languages in CN. For example, information can cycle through several interior layers until a decision network is triggered — which in turn fires into nodes, switching them on so that they in turn fire out in synchrony to other regions of the network.
- Synchronised sampling of environmental stimuli in different sensory modalities can be used as a means of multi-sensory integration.
- Network structures that initiate, transmit and terminate global wave-like signals to large regions of the network can instantiate arousal-like processes or help prepare entire network modules to better process the incoming signals.
- Multi-level shifter-like structures embedded in the CN can be used for a variety of control functions (e.g., to register inputs from two visual sensors, to compensate for the motion of objects in the scene, to dynamically control the degree of smoothing to suppress the noise in the input).
- Feedback pathways can be built into CN to subserve a number of subtle control functions (e.g., selective modulation of the sensory signals as a function of some assessments made at the higher levels, selective attention to specific aspects of the environmental stimuli ordered by their salience).
- Control structures that dynamically link subnetworks that are activated by different features in the environmental stimuli into transient network assemblies can serve a host of useful functions (e.g., dynamic variable binding necessary for complex inferences) in CN.
- Specific subnetworks can be built into the CN that ensure a proper balance in the allocation of different network resources (e.g., nodes, links, long-range communication networks) among different functions as the network learns and evolves.
- Network controls can dynamically alter the incentives for CN units, microcircuits or functional modules to cooperate (as opposed to compete) with each other.

10.7 Summary

SAI and CN are generally regarded as two disparate and perhaps fundamentally different approaches to modelling cognitive processes and engineering intelligent systems. A closer examination of the two approaches clearly demonstrates that there can be no fundamental incompatibility between them. They essentially offer two different (but general-purpose) description languages for modelling systems in general, and intelligent systems in particular.

The choice between the two — much like deciding between two general-purpose programming languages (or equivalently, virtual computer architectures) is primarily a matter of convenience, efficiency and elegance — given a set of design and performance constraints. SAI and CN each demonstrate at least one way of performing certain tasks naturally and thus pose the problem to the other of doing the same thing perhaps more elegantly, efficiently, or robustly than the other.

Living minds/brains offer an proof of existence of at least one architecture for general intelligence. SAI and CN paradigms together offer a wide range of architectural choices. Each architectural choice brings with it some obvious (and some not so obvious) advantages as well as disadvantages in the solution of specific problems using specific algorithms, given certain performance demands and design constraints imposed by the available choices of physical realisations of the architecture. Together, the cross-product of the space of architectures, algorithms, and physical realisations constitutes a large and interesting space of possible designs for intelligent systems. Examples of systems resulting from a judicious integration of concepts, constructs, techniques and technologies drawn from both traditional artificial intelligence systems and artificial neural networks clearly demonstrate the potential benefits of exploring this space. And, perhaps more importantly, the rather severe practical limitations of today's SAI and CN systems strongly argues for the need for a systematic exploration of such design space.

This suggests that it might be fruitful to approach the choice of architectures, implementations, and their physical realisations using the entire armory of tools drawn from the theory and practice of computer science — including the design of programming languages (and hence virtual architectures), computers, algorithms, and programs. Our primary task is to identify subsets of Turing-computable functions necessary for general intelligence, an appropriate mix of architectures for supporting specific subsets of these functions, as well as appropriate realisations of such architectures in physical devices. In the short-term, this entails the design and analysis of *hybrid* systems that use SAI and CN modules to perform different but well co-ordinated sets of functions in specific applications. In the long-term, a coherent theoretical framework for analysis and synthesis of such systems needs to be developed. One way to approach this task is to place both SAI and CN systems within a common framework and identify the various significant dimensions that characterise the resulting space of designs for intelligent systems. The attempt to define CN (and SAI) systems in broader than usual terms in this chapter is an attempt in this direction. This makes explicit some of the dimensions of the design space for intelligent systems that CN (broadly defined) offer — in terms of a variety of behave, control, and learning structures and functions as well as the attendant design/performance trade-offs among: parallel versus serial processing; localised versus distributed representation, processing, and control (in space as well as over time); symbolic, numeric, and analog representation and processing/inference structures and processes; and related issues. It also suggests a number of ways to extend today's CN models — by providing them with more powerful building blocks and functional microcircuits, more appropriate modular topologies, more powerful learning structures and processes, and a rich variety of control structures.

Acknowledgements

The authors would like to thank Drs Suran Goonatilake and Sukhdev Khebbal for their invitation to contribute this chapter to their book. This work was partially supported by the National Science Foundation grant (IRI–9409580) to Vasant Honavar. Vasant Honavar would like to thank his students (especially Rajesh Parekh, Chun–Hsieh Chen and Karthik Balakrishnan) and Professors Bill Robinson and Gregg Oden for many useful discussions on the topics addressed in this chapter.

References

[Alma89] Almasi, G. S. and Gottlieb, A., "*Highly Parallel Computing.*", New York: Benjamin–Cummings, 1989.

[Arbi72] Arbib, M. A. "*The Metaphorical Brain.*" New York: Wiley–Interscience, 1972.

[Arbi94] Arbib, M. A., "Schema Theory: Co-operative Computation for Brain Theory and Distributed AI." In: *Artificial Intelligence and Neural Networks: Steps Toward Principled Integration.* Honavar, V. and Uhr, L. (ed.) San Diego, CA: Academic Press, 1994.

[Bala94] Balakrishnan, K. and Honavar, V., "Evolutionary Approaches to the Exploration of the Design Space of Artificial Neural Networks." In preparation, 1994.

[Ball82] Ballard, D. and Brown, C. "*Computer Vision.*", Englewood Cliffs, NJ: Prentice Hall, 1982.

[Barn94] Barnden, J. A. "How Might Connectionist Networks Represent Propositional Attitudes?" In: *Artificial Intelligence and Neural Networks: Steps Toward Principled Integration.* Honavar, V. and Uhr, L. (ed.) San Diego, CA: Academic Press, 1994.

[Bode94] Boden, M., "Horses of a Different Colour?", In: *Artificial Intelligence and Neural Networks: Steps Toward Principled Integration.* Honavar, V. and Uhr, L. (ed.) San Diego, CA: Academic Press, 1994.

[Book94a]Booker, L. E., Riolo, R. L., and Holland, J. H., "Learning and Representation in Classifier Systems." In: *Artificial Intelligence and Neural Networks: Steps Toward Principled Integration.* Honavar, V. and Uhr, L. (ed.) San Diego, CA: Academic Press, 1994.

[Book94b]Bookman, L., "A Framework for Integrating Relational and Associational Knowledge for Comprehension." In: *Computational Architectures Integrating Symbolic and Neural Processes.* Sun, R. and Bookman, L. (ed.) Boston: Kluwer, 1994.

[Buch93] Buchanan, B. G. and Wilkins, D. C., "*Readings in Knowledge Acquisition and Learning.*", San Mateo, CA: Morgan Kaufmann, 1993.

[Carp94] Carpenter, G. and Grossberg, S. "Integrating Symbolic and Neural Processes in a Self–Organizing Architecture for Pattern Recognition and Prediction." In: *Artificial Intelligence and Neural Networks: Steps Toward Principled Integration*. Honavar, V. and Uhr, L. (ed.) San Diego, CA: Academic Press, 1994.

[Chan94] Chandrasekaran, B. and Josephson, S. G., "Architecture of Intelligence: The Problems and Current Approaches to Solutions.", In: *Artificial Intelligence and Neural Networks: Steps Toward Principled Integration*. Honavar, V. and Uhr, L. (ed.) San Diego, CA: Academic Press, 1994.

[Chen94] Chen, C., and Honavar, V., "Neural Networks for Inference." In preparation, 1994.

[Chur92] Churchland, P. S. and Sejnowski, T. J., "*The Computational Brain*." Boston, MA: MIT Press, 1992.

[Cohe86] Cohen, D., "*Introduction to Computer Theory*.", New York: Wiley, 1986.

[Dola89] Dolan, C. P. and Smolensky, P., "Tensor product production system: a modular architecture and representation.", *Connection Science*, 1:53–58, 1989.

[Dyer94] Dyer, M., "Grounding Language in Perception." In: *Artificial Intelligence and Neural Networks: Steps Toward Principled Integration*. Honavar, V. and Uhr, L. (ed.) San Diego, CA: Academic Press, 1994.

[Feig63] Feigenbaum, E. A., In: "*Computers and Thought*." Feigenbaum, E. A. and Feldman, J. (ed.) New York: McGraw–Hill, 1963.

[Feld82] Feldmann, J. A. and Ballard, D. H., "Connectionist models and their properties.", *Cognitive Science*, 6:205–264, 1982.

[Fodo76] Fodor, J. "*The Language of Thought*." Boston, MA: Harvard University Press, 1976.

[Fodo88] Fodor, J. and Pylyshyn, Z. W., "Connectionism and cognitive architecture: a critical analysis." In: *Connections and Symbols*. Pinker, S. and Mehler, J. (ed.) Cambridge, MA: MIT Press, 1988.

[Forg82] Forgy, C. L., "RETE: A Fast Algorithm for the Many Pattern/Many Object Pattern Match Problem." *Artificial Intelligence*, 19:17–37, 1982.

[Fu82] Fu, K. S. "*Syntactic Pattern Recognition and Applications*" Englewood Cliffs, NJ: Prentice Hall, 1982.

[Gall93] Gallant, S. "*Neural Networks and Expert Systems*" Cambridge, MA: MIT Press, 1993.

[Gene87] Genesereth, M. R., and Nilsson, N. J., "*Logical Foundations of Artificial Intelligence*." Palo Alto, CA: Morgan Kaufmann, 1987.

[Gins93] Ginsberg, M., "*Essentials of Artificial Intelligence*" San Mateo, CA: Morgan Kaufmann, 1993.

[Gold94] Goldfarb, L. and Nigam, S., "The Unified Learning Paradigm: A Foundation for AI", In: *Artificial Intelligence and Neural Networks: Steps Toward Principled Integration*. Honavar, V. and Uhr, L. (ed.) San Diego, CA: Academic Press, 1994.

[Goon92] Goonatilake, Suran and Khebbal, Sukhdev, "Intelligent Hybrid Systems ", *Proceedings of First International Conf. on Intelligent Systems, Singapore*, pp. 207-212, Sept 1992.

[Gree88] Greenough, W. T. and Bailey, C. H., "The Anatomy of Memory: Convergence of Results Across a Diversity of Tests", *Trends in Neuroscience* 11, 142–147, 1988.

[Gros82] Grossberg, S., "*Studies of Mind and Brain*", Boston, MA: Reidel, 1982.

[Hans90] Hanson, S. J. and Burr, D. "What Connectionist Models Learn: Learning and Representation in Connectionist Networks", *Behaviour and Brain Sciences*, 13:1–54, 1990.

[Harn90] Harnad, S. "The Symbol Grounding Problem" *Physica D*, 42:335–346, 1990.

[Harn94] Harnad, S., Hanson, S. J., and Lubin, J., "Learned Categorical Perception in Neural Nets: Implications for Symbol Grounding." In: *Artificial Intelligence and Neural Networks: Steps Toward Principled Integration*. Honavar, V. and Uhr, L. (ed.) San Diego, CA: Academic Press, 1994.

[Hayk94] Haykin, S., "*Neural Networks.*" New York: Macmillan, 1994.

[Hewi77] Hewitt, C., "Viewing Control Structures as Patterns of Passing Messages." *Artificial Intelligence*, 8:232–364, 1977.

[Hill85] Hillis, D., "*The Connection Machine*", Cambridge, MA: MIT Press, 1985.

[Holl75] Holland, J. H. "*Adaptation in Natural and Artificial Systems*", Ann Arbor, MI: University of Michigan Press, 1975.

[Hopf82] Hopfield, J. J. "Neural Networks and Physical Systems With Emergent Collective Computational Abilities", *Proceedings of the National Academy of Sciences*, 79:2554–2558, 1982.

[Hona89a]Honavar, V., "Perceptual Development and Learning: From Behavioral, Neurophysiological and Morphological Evidence to Computational Models." Tech. Rep. 818. Computer Sciences Dept., University of Wisconsin, Madison, Wisconsin, 1989.

[Hona89b]Honavar, V. and Uhr, L. "Brain–Structured Connectionist Networks that Perceive and Learn." *Connection Science*, 1:139–159, 1989.

[Hona89c]Honavar, V. and Uhr, L., "Generation, Local Receptive Fields, and Global Convergence Improve Perceptual Learning in Connectionist Networks" In: *Proceedings of the 1989 International Joint Conference on Artificial Intelligence*, San Mateo, CA: Morgan Kaufmann, 1989.

[Hona90a]Honavar, V., "*Generative Learning Structures for Generalised Connectionist Networks*" Ph.D. Dissertation. University of Wisconsin, Madison, Wisconsin, 1990.

[Hona90b]Honavar, V. and Uhr, L. "Symbol Processing Systems, Connectionist Networks, and Generalised Connectionist Networks" Tech. Rep. 90–23. Department of Computer Science, Iowa State University, Ames, Iowa, 1990.

[Hona90c]Honavar, V. and Uhr, L., "Co-ordination and Control Structures and Processes: Possibilities for Connectionist Networks." *Journal of Experimental and Theoretical Artificial Intelligence*, 2:277–302, 1990.

[Hona92a]Honavar, V., "Some Biases For Efficient Learning of Spatial, Temporal, and Spatio–Temporal Patterns." In: *Proceedings of the International Joint Conference on Neural Networks*. Beijing, China, 1992.

[Hona92b]Honavar, V. "Inductive Learning Using Generalised Distance Measures." In: *Proceedings of the SPIE Conference on Adaptive and Learning Systems*. Orlando, Florida, 1992.

[Hona93] Honavar, V. and Uhr, L.,"Generative Learning Structures and Processes for Generalised Connectionist Networks", *Information Sciences*, 70:75–108, 1993

[Hona94a]Honavar, V., "Toward Learning Systems That Use Multiple Strategies and Representations." In: *Artificial Intelligence and Neural Networks: Steps Toward Principled Integration*. Honavar, V., and Uhr, L. (ed.) San Diego, CA: Academic Press, 1994.

[Hona94b]Honavar, V., "Symbolic Artificial Intelligence and Numerical Artificial Neural Networks: Towards a Resolution of the Dichotomy.", In: *Computational Architectures Integrating Symbolic and Neural Processes*. Sun, R., and Bookman, L. (ed.) New York: Kluwer, 1994.

[Hona94c]Honavar, V. and Uhr, L., (ed.) *Artificial Intelligence and Neural Networks: Steps Toward Principled Integration.* San Diego, CA.: Academic Press, 1994.

[John91] Johnson-Laird, P. and Byrne, J., "*Deduction*." New York: Lawrence Erlbaum, 1991.

[Klir85] Klir, G. J., "*Architecture of Systems Problem Solving*." New York: Plenum, 1985.

[Kowa77]Kowalski, R. A., "*Predicate Logic as a Programming Language*." Amsterdam: North–Holland, 1977.

[Kung93] Kung, S. Y., "*Digital Neural Networks*." New York: Prentice Hall, 1993.

[Levi94] Levine, D. S. Aparicioiv, M., (ed.) "*Neural Networks for Knowledge Representation*." New York: Lawrence Erlbaum, 1994.

[Macl94] Maclennan, B. J., "Image and Symbol – Continuous Computation and the Emergence of the Discrete." In: *Artificial Intelligence and Neural Networks: Steps Toward Principled Integration*. Honavar, V. and Uhr, L. (ed.) San Diego, CA: Academic Press, 1994.

[Mccu43] McCulloch, W. S. and Pitts, W., "A Logical Calculus of the Ideas Imminent in Neural Activity." *Bulletin of Mathematical Biophysics*, 5:115–137, 1943.

[Mccl86] McClelland J., Rumelhart, D. *et al.*, (ed.) *"Parallel Distributed Processing."* Boston, MA: MIT Press, 1986.

[Mcke94] McKenna, T., "The Role of Inter–Disciplinary Research Involving Neuroscience in the Design of Intelligent Systems." In: *Artificial Intelligence and Neural Networks: Steps Toward Principled Integration*. Honavar, V. and Uhr, L. (ed.) San Diego, CA: Academic Press, 1994.

[Mead89] Mead, C. *"Analog VLSI and Neural Systems."* Reading, MA: Addison–Wesley, 1989.

[Meds94] Medsker, L. *"Hybrid Neural Network and Expert Systems."* Boston, MA: Kluwer, 1994.

[Miik94] Miikkulainen, R., "Integrated Connectionist Models: Building AI Systems on Subsymbolic Foundations." In: *Artificial Intelligence and Neural Networks: Steps Toward Principled Integration*. Honavar, V. and Uhr, L. (ed.) San Diego, CA: Academic Press, 1994.

[Mich93] Michalski, R. S. "Toward a Unified Theory of Learning: Multi-Strategy Task–Adaptive Learning." In: *Readings in Knowledge Acquisition and Learning*. Buchanan, B. G., and Wilkins, D. C. (ed.) San Mateo, CA: Morgan Kaufmann, 1993.

[Micl86] Miclet, L., *"Structural Methods in Pattern Recognition."* New York: Springer–Verlag, 1986.

[Mins63] Minsky, M., "Steps Toward Artificial Intelligence." In: *Computers and Thought*. Feigenbaum, E. A. and Feldman, J. (ed.) New York: McGraw–Hill, 1963.

[Mins75] Minsky, M. "A Framework for Representing Knowledge." In: *The Psychology of Computer Vision*. Winston, P. H. (ed). New York: McGraw–Hill, 1975.

[Mins86] Minsky, M. *"Society of Mind."*, New York: Basic Books, 1986.

[Newe63] Newell, A., Shaw, J. C., and Simon, H., In: *Computers and Thought*. Feigenbaum, E. A., and Feldman, J. (ed.) New York: McGraw–Hill, 1963.

[Newe80] Newell, A., "Symbol Systems.", *Cognitive Science*, 4:135–183, 1980.

[Newe90] Newell, A. *"Unified Theories of Cognition."* Cambridge, MA: Harvard University Press, 1990.

[Norm86] Norman, D. A., "Reflections on Cognition and Parallel Distributed Processing.", In: *Parallel Distributed Processing*. McClelland, J., Rumelhart. D. *et al.* (ed.) Cambridge, MA: MIT Press, 1986.

[Norv92] Norvig, P. "*Paradigms in Artificial Intelligence Programming.*" Palo Alto, CA: Morgan Kaufmann, 1992.

[Oden94] Oden, G. C., "Why the Difference Between Connectionism and Anything Else is More Than You Might Think But Less Than You Might Hope." In: *Artificial Intelligence and Neural Networks: Steps Toward Principled Integration*. Honavar, V. and Uhr, L. (ed.) San Diego, CA: Academic Press, 1994.

[Pear88] Pearl, J., "*Probabilistic Reasoning in Intelligent Systems.*" Palo Alto, CA: Morgan Kaufmann, 1988.

[Pink94] Pinkas, G., "A Fault–Tolerant Connectionist Architecture for the Construction of Logic Proofs." In: *Artificial Intelligence and Neural Networks: Steps Toward Principled Integration*. Honavar, V. and Uhr, L. (ed.) San Diego, CA: Academic Press, 1994.

[Poll90] Pollack, J., "Recursive Distributed Representations." *Artificial Intelligence*, 46:77–105, 1990.

[Quil68] Quillian, M. R., "Semantic Memory." In: *Semantic Information Processing.* Minsky, M. (ed.) Cambridge, MA: MIT Press, 1968.

[Raja81] Rajaraman, V., "*Analog Computation and Simulation.*" Englewood Cliffs: Prentice Hall, 1981.

[Rash60] Rashevsky, N., "*Mathematical Biophysics.*" New York: Dover, 1960.

[Rose62] Rosenblatt, F., "*Principles of Neurodynamics.*" Washington, DC: Spartan, 1962.

[Schn87] Schneider, W., "Connectionism: is it a paradigm shift for psychology?" *Behaviour Research Methods, Instruments, and Computers*, 19:73–83, 1987.

[Self63] Selfridge, O. G. and Neisser, U. "Pattern Recognition by Machine." In: *Computers and Thought*. Feigenbaum, E. A. and Feldman, J. (ed.) New York: McGraw–Hill, 1963.

[Shar94] Sharkey, N. and Jackson, S. J., "Three Horns of a Representational Dilemma." In: *Artificial Intelligence and Neural Networks: Steps Toward Principled Integration*. Honavar, V. and Uhr, L. (ed.) San Diego, CA: Academic Press, 1994.

[Shas89] Shastri, L. and Ajjanagadde, V., "Connectionist System for Rule-Based Reasoning with Multi-Place Predicates and Variables." Tech. Rep. MS–CIS–8906. Computer and Information Science Dept. University of Pennsylvania, Philadelphia, PA, 1989.

[Shav90] Shavlik, J. W., and Dietterich, T. G., (ed). "*Readings in Machine Learning.*" San Mateo, California: Morgan Kaufmann, 1990.

[Shav94] Shavlik, J. W., "A Framework for Combining Symbolic and Neural Learning." In: *Artificial Intelligence and Neural Networks: Steps Toward Principled Integration*. Honavar, V. and Uhr, L. (ed.) San Diego, CA: Academic Press, 1994.

[Shep89] Shepherd, G. M., "The significance of real neuron architectures for neural network simulations." In: *Computational Neuroscience*. Schwartz, E. (ed.) Cambridge, MA: MIT Press 1989

[Shep90] Shepherd, G. M. (ed.), "*Synaptic Organisation of the Brain.*" New York, NY: Oxford University Press, 1990.

[Smol90] Smolensky, P., "Tensor Product Variable Binding and the Representation of Symbolic Structures in Connectionist Systems." *Artificial Intelligence*, 46:159–216, 1990.

[Sowa84] Sowa, J. F., *Conceptual Structures: Information Processing in Mind and Machine*. Reading, Massachusetts: Addison–Wesley, 1984.

[Sun94a] Sun, R., "Logic and Variables in Connectionist Models: A Brief Overview." In: *Artificial Intelligence and Neural Networks: Steps Toward Principled Integration*. Honavar, V. and Uhr, L. (ed.) San Diego, CA: Academic Press, 1994.

[Sun94b] Sun, R. and Bookman, L. (ed.) "*Computational Architectures Integrating Symbolic and Neural Processes.*", New York: Kluwer, 1994.

[Tani80] Tanimoto, S. L. and Klinger, A. "*Structured Computer Vision: Machine Perception Through Hierarchical Computation Structures.*" New York: Academic Press, 1980.

[Tsan93] Tsang, E. "*Foundations of Constraint Satisfaction.*" New York: Academic Press, 1993.

[Uhr63] Uhr, L. and Vossler, C., "A Pattern Recognition Program that Generates, Evaluates, and Adjusts its Own Operators." In: *Computers and Thought*. Feigenbaum, E. and Feldman, J. (ed). New York: McGraw–Hill, 1963.

[Uhr73] Uhr, L., "*Pattern Recognition, Learning, and Thought.*" New York: Prentice–Hall, 1973.

[Uhr79] Uhr, L., "Parallel–serial production systems with many working memories." In: *Proceedings of the Fifth International Joint Conference on Artificial Intelligence,* 1979.

[Uhr80] Uhr, L., In: "*Structured Computer Vision*" Tanimoto, S., and Klinger, A. (ed.) New York: Academic Press, 1980.

[Uhr84] Uhr, L. "*Algorithm–Structured Computer Arrays and Networks.*" New York: Academic Press, 1984.

[Uhr87a] Uhr, L., *Parallel Computer Vision*. New York: Academic Press, 1987.

[Uhr87b] Uhr, L., "*Multi-computer Architectures for Artificial Intelligence.*" New York: Wiley, 1987.

[Uhr90] Uhr, L., "Increasing the Power of Connectionist Networks by Improving Structures, Processes, Learning." *Connection Science*, 2:179–193, 1990.

[Uhr94a] Uhr, L., "Digital and Analog Subnet Structures for Connectionist Networks." In: *Artificial Intelligence and Neural Networks: Steps Toward Principled Integration*. Honavar, V. and Uhr, L. (ed.) San Diego, CA: Academic Press, 1994.

[Uhr94b] Uhr, L. and Honavar, V., "Artificial Intelligence and Neural Networks: Steps Toward Principled Integration." In: *Artificial Intelligence and Neural Networks: Steps Toward Principled Integration*. Honavar, V. and Uhr, L. (ed.) San Diego, CA: Academic Press, 1994.

[Vang94] Van Gelder, T. and Port, R., "Beyond Symbolic: Prolegomena to a Kama–Sutra of Compositionality." In: *Artificial Intelligence and Neural Networks: Steps Toward Principled Integration*. Honavar, V. and Uhr, L. (ed.) San Diego, CA: Academic Press, 1994.

[Wate85] Waterman, D. A., "*A Guide to Expert Systems*", Reading, MA: Addison–Wesley, 1985.

[Wech90] Wechsler, H., "*Computational Vision.*" New York: Academic Press, 1990.

[Wins92] Winston, P. H., "*Artificial Intelligence.*" Boston, MA: Addison–Wesley, 1992.

[Yage94] Yager, R. R. and Zadeh, L. A., (ed.) "*Fuzzy Sets, Neural Networks, and Soft Computing.*" New York: Van Nostrand Reinhold, 1994.

[Zade75] Zadeh, L. A. "Fuzzy Logic and Approximate Reasoning." *Synthese*, 30:407–28, 1975.

[Zeid89] Zeidenberg, M. "*Neural Networks in Artificial Intelligence.*" New York: Wiley, 1989.

11

Reasoning with Rules and Variables in Neural Networks

Venkat Ajjanagadde and Lokendra Shastri

11.1 Introduction

The serial von Neumann computer architecture figured predominantly in artificial intelligence and cognitive science research during the last few decades. However, a growing number of people are now convinced that both from the point of view of realising intelligent machines as well as understanding human cognition, a more promising approach is the type that typically goes under the labels of "connectionist models", "parallel distributed processing", and "neural networks". If we abandon the von Neumann architecture and embrace the connectionist architecture, a crucial question regarding the research strategy arises: do we also abandon all that work done over decades within a von Neumann architectural framework, and start afresh? Or, is there a way that we can build upon that work yet adopting a connectionist architecture? Many works on intelligent hybrid systems can typically be viewed as attempts to build upon such earlier work yet incorporating the features of the connectionist paradigm.

There is a spectrum of possibilities with respect to how much one borrows from the earlier work. Towards one end of the spectrum are works which borrow problem description, representations and algorithms entirely from the existing AI work and implement it in a straightforward way in a neural network. For example, an attempt towards a neural network implementation of PROLOG might be of this kind. An effort of that kind is certainly useful in that it may bring in the advantages of speed, graceful degradation etc. It may also provide ideas about how the brain might be carrying out

Intelligent Hybrid Systems, Edited by S. Goonatilake and S. Khebbal.

some tasks of interest. However, overall, such an effort will be more like an implementation of an existing model/theory rather than the development of a new model/theory.

Near the other end of the spectrum are those works that borrow very little from existing AI research. Examples are the use of unstructured neural networks that put the entire burden of formulating the right representations and solving a problem, on a learning algorithm. While the importance and usefulness of learning is undeniable, it seems that it will be prudent to make use of the insights and understandings about "intelligent activity" hard earned by decades of research effort.

The work described in this chapter takes an intermediate path between these two extremes. The connectionist model of reasoning being presented here was developed by bringing together. (1) Fundamental insights borrowed from the artificial intelligence and cognitive science work on reasoning with rules. (2) Psychological observations on the strengths and limitations of human reasoning. (3) The constraints suggested by the requirement that the mechanisms employed are neurologically plausible.

11.2 Reasoning with Rules

A fundamental component of human cognition is the ability to generalise from the experiences and then, make use of that "general knowledge" in dealing with the specific situations encountered in the future. How is such general knowledge represented and how is that knowledge used in specific novel situations?

Rules constitute a form of representing such general knowledge. Applying this general knowledge to specific situations involves *reasoning* with those rules.

For example, consider the rule

$$hit(x,y) \Rightarrow hurt(y)$$

This rule says that "if x hits y, then, y gets hurt". This piece of knowledge might have been learned after witnessing many instances of hitting in the past, and observing that the "hittee" in general gets hurt in an instance of hitting. Now, suppose that we see John hitting Bill. We immediately infer that Bill is getting hurt. That is, starting from the fact $hit(john, bill)$, using the rule $hit(x,y) \Rightarrow hurt(y)$, we infer $hurt(bill)$.

The question we addressed in our research was how to represent a collection of rules in a neural network and realise reasoning as an automatic spreading of activation (i.e., there is no central controller guiding the spreading of activation). We required that the reasoning system should meet the following two rather tight requirements on space and time complexities:

- Space complexity: The number of nodes and links used should be only linear in the size of the knowledge base (as measured by the number of rules and facts present).

- Time complexity: The system should be able to perform reasoning extremely fast. Specifically, the time taken to perform a deductive inference should be linear in the length of the shortest proof.

These two requirements ensure that the system scales well with increases in the problem size.

In this chapter, we will not be discussing the entire reasoning system with all its extensions and details. Instead, we will restrict our focus to a simple deductive forward reasoning task. This discussion will provide a flavour of the overall approach being taken. It will also bring out some key issues such as the representation and propagation of variable bindings. Later, in section 11.5, we will briefly mention some extensions and provide references to the relevant publications.

11.3 Forward Reasoning

For describing the forward reasoning system, let us consider a small knowledge base consisting of the following rules:

$$take-bus(x,y,z) \Rightarrow go(x,y,z)$$
$$drive(x,y,z) \Rightarrow go(x,y,z)$$
$$take-bus(x,y,z) \Rightarrow pay-busfare(x)$$
$$go(x,y,z) \Rightarrow reach(x,z)$$

The intuitive meanings of these rules are obvious. For example, the first rule corresponds to the following knowledge: The fact "x took a bus from place y to place z" implies that "x went from y to z".

First, let us discuss how these rules can be encoded as a connectionist network and then see how reasoning takes place in that network.

Corresponding to every predicate, there exists a *predicate node* (depicted as pentagons in all figures) in the network. Corresponding to each argument of a predicate, there exists an *argument node* (depicted as diamonds in all figures). Thus, in Figure 11.1, corresponding to the tertiary predicate P, there are three argument nodes a1, a2, and a3. Figure 11.1 also shows the encoding of an example rule[1]:

$$P(x,y,z) \Rightarrow Q(z,x,y)$$

To represent this rule, there is a link from the predicate node corresponding to P to the predicate node corresponding to Q. Also, there are links representing the correspondence between the arguments of P and the arguments of Q. For example, as in the above rule, the second argument of Q is to be bound to the same individual that binds the first argument of P. This is indicated by a link from a1 to a5.

[1]Unless mentioned otherwise, variables occurring in a rule are to be assumed as universally quantified.

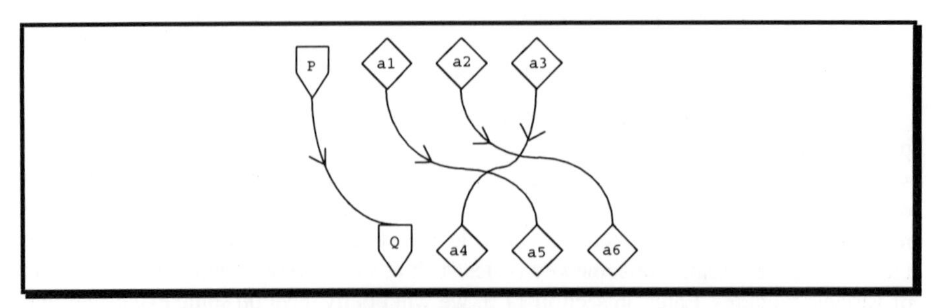

Figure 11.1 — Encoding of the rule $P(x,y,z) \Rightarrow Q(z,x,y)$.

Figure 11.2 shows the encoding corresponding to the example set of rules listed earlier. In addition to the encoding of the rules, corresponding to every "individual" the agent already knows about, there exists a *constant node* in the network. In the network of Figure 11.2, the constant nodes (shown as circles) correspond to the individuals *place1*, *place2*, *john*, and *mary*.

Having discussed how rules are encoded, let us consider the encoding of facts. We distinguish between two kinds of facts: *long-term facts* and *dynamic facts*. Long-term facts are the facts the agent remembers for long time. These facts get represented by appropriate interconnections in the network. Since long-term facts do not take part in the reasoning discussed in this chapter, we will skip the details of how they are encoded in the network. The second kind of facts are dynamic facts. These are the facts that remain in the agent's memory only for a short duration, typically for the duration of one reasoning episode. Intermediate facts generated during reasoning are good examples of dynamic facts. The representations of such facts have to be created dynamically during reasoning and those representations remain only for the duration of the particular reasoning episode.

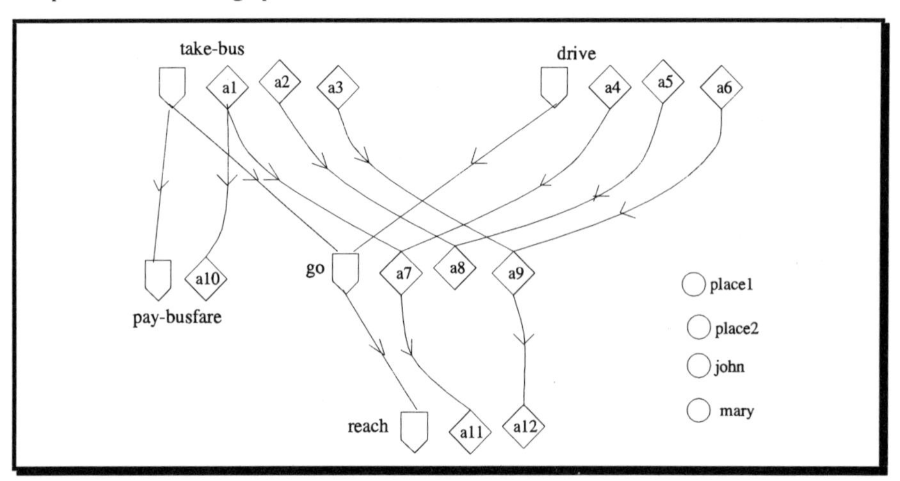

Figure 11.2 — An example network.

11.3.1 Representation of Dynamic Facts

Let us discuss the details of representing dynamic facts by taking the fact *take-bus(john, place1, place2)* as an example. Representing this fact essentially amounts to binding the first, second, and third arguments of the predicate *take-bus* respectively to *john, place1*, and *place2*. In order to achieve such bindings, we have to represent the association between the nodes representing the arguments and the nodes representing the constants that get bound to them. Thus, in the case of *take-bus(john, place1, place2)*, the associations between the argument nodes a1, a2, and a3 respectively with the constant nodes *john, place1*, and *place2* have to be represented. Also, these associations have to be created dynamically: we need to be able to create such associations quickly during the reasoning process and such associations typically last only for short periods of time[2]. The ideas of representing such dynamic associations by having links between the corresponding nodes, using addresses etc., all pose major difficulties [Shas93]. The problem of finding a feasible way to represent dynamic associations between nodes is what has come to be known as the *connectionist variable binding problem* [Feld82, Mals86].

The approach we have taken is to represent dynamic bindings by the synchrony of activation of the corresponding nodes. In its simplest form (used in this chapter and in all our previous publications [Ajja94a]), the idea is to represent bindings by having the argument nodes bound to the same object to be active at the same time and the argument nodes bound to different objects to be active at different times. The approach is fundamentally tied to the observation that at any time, human reasoning involves only a small number objects. Let us label the objects involved in reasoning at a particular time as $obj_1, obj_2, \ldots, obj_n$.. The idea is to multiplex the time among those objects, in a round-robin fashion. Imagine the time as divided into chunks of size T each, where T is the time taken to cycle through all the objects currently being reasoned about, by attending to them one by one. Each such chunk can thus be viewed as further divided into n parts, wherein during each part one object is attended to. Let us refer to the chunks of duration T as *cycles* and the n parts of each cycle as *phases*[3]. The n phases within a cycle will be referred to as "first phase", "second phase", "nth phase".

We select (arbitrarily) different phases to represent different individuals participating in a reasoning episode[4]. Once different phases have been chosen

---

[2]Permanently retaining the bindings amounts to retaining all the facts inferred. This would put infeasibly high requirement on memory – solving this problem is one of the uses of reasoning to begin with.

[3]Note that the terms *cycles* and *phases* are introduced here just to simplify discussion. Specifically, it is not meant that there is a global clock. Also, the duration of the "cycle" can change from one episode of reasoning to another depending on the number of objects involved in the particular reasoning episode.

[4]During different reasoning episodes, a particular individual can be assigned different phases. Also, during different reasoning episodes, a particular phase can be associated with different individuals. But, within a reasoning episode each phase corresponds to only one individual.

corresponding to the different individuals participating in a reasoning episode, representations of the dynamic bindings is done as follows.

Let p_i denote the individual to whom ith phase corresponds during the reasoning episode. The argument nodes corresponding to the arguments bound to p_i will be active in the ith phase of the cycles. Also, if there exists a constant node in the network corresponding to the individual p_i, then, that constant node will also be active in the ith phase of the cycles.

To make things concrete, let us consider the representation of the dynamic fact *take-bus(john, place1, place2)*. In this fact, there are three constants — *john*, *place1*, and *place2*. We choose three different phases to represent them — say, we choose the first, second, and third phases to represent *john*, *place1* and *place2* respectively.

In *take-bus(john, place1, place2)*, the first argument is bound to *john*. This will be represented by the argument node a1 and the constant node *john*. Figure 11.2 being active in the first phase (i.e., the phase corresponding to *john*) of each cycle in Figure 11.3. In *take-bus(john, place1, place2)*, the second argument is bound to *place1*. So, the argument node a2 and the constant node *place1* will be active in the second phase. Similarly, the argument node a3 and the constant node *place2* will be active in the third phase.

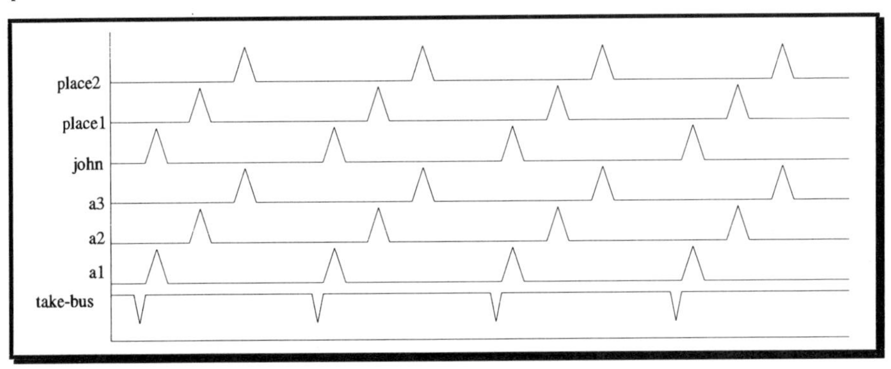

Figure 11.3 — Representation of the dynamic fact *take-bus(john, place1, place2)*.

The activities mentioned in the previous paragraph represent the bindings in the fact *take-bus(john, place1, place2)*. Together with the representation of these bindings, if we also have the predicate node corresponding to *take-bus* active, then that completes the representation of the dynamic fact *take-bus(john, place1, place2)*. The complete activity pattern representing this fact is shown in Figure 11.3.

More than one argument node can be active in the same phase of a cycle. That would represent that the individual to which that phase corresponds binds all of those arguments. Thus, in the network shown in Figure 11.2, if the argument nodes a1, a7, and a10 all happen to be active in the first phase of the cycles during a reasoning episode, that represents that all arguments are bound to the individual to whom the first phase has been assigned in that reasoning episode (i.e., *john*, in our example). This suggests how multiple dynamic facts can be simultaneously represented in the network.

Figure 11.4 depicts the simultaneous representation of the facts *take-bus(john, place1, place2)*, *go(john, place1, place2)*, and *pay-busfare(john)*. Note that *john* is bound to the arguments a1, a7 and a10; similarly, *place1* is bound to a2 and a8; *place2* is bound to a3 and a9.

11.3.2 Forward Reasoning in the Network

Having seen how dynamic facts are represented, let us now discuss how forward reasoning is realised. In forward reasoning, the reasoning system is given some fact; the system then derives other facts that follow from it. In connectionist terms, this amounts to the following : The activity pattern representing the given fact will be clamped onto the network; then, the network has to automatically generate the patterns of activity representing the facts that can be inferred from the given fact. We now discuss how this takes place in the network by taking *take-bus(john, place1, place2)* as the starting fact. The pattern of activity representing this fact is shown in Figure 11.3.

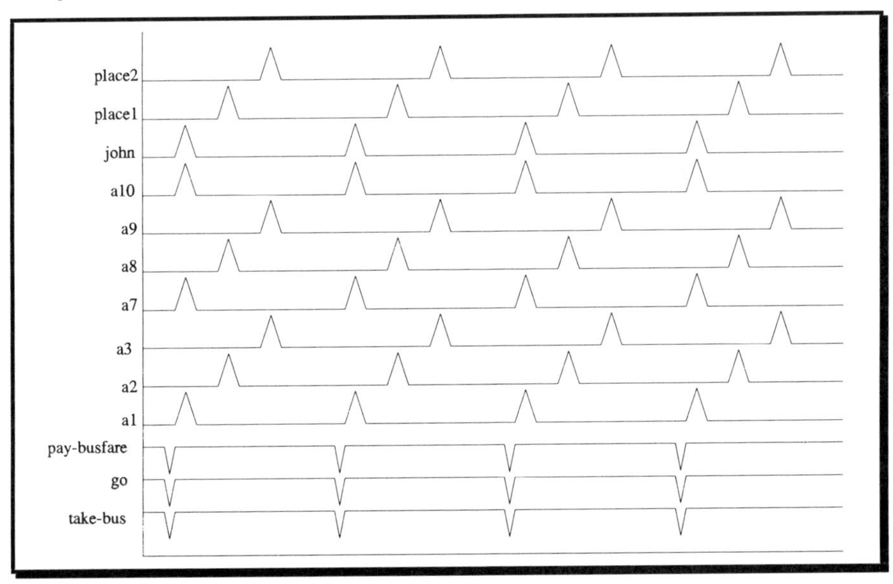

Figure 11.4 — Representation of the dynamic facts *take-bus(john, place1, place2)*, go *(john, place1, place, pay-busfare(john)*.

Let us see how reasoning takes place once this pattern of activity is clamped onto the network[5]. What we desire is that the activity patterns corresponding to *go(john, place1, place2)*, *pay-busfare(john)*, and *reach(john, place2)* should automatically get created. These three are the facts that follow from the given starting

[5]Clamping the pattern of Figure 11.3 would amount to causing the seven nodes involved in this pattern to become active as shown in Figure 11.3.

fact *take-bus(john, place1, place2)* using the rules encoded in the network of Figure 11.2.

Consider the activations of the nodes when the pattern of Figure 11.3 corresponding to *take-bus(john, place1, place2)* is clamped onto the network. The nodes a1 and *john* will be active in the first phase of every cycle; nodes a2 and *place1* will be active in the second phase; nodes a3 and *place2* will be active in the third phase. Also, the predicate node of *take-bus* will be active (predicate nodes remain active throughout the cycle.).

Now, activation starts spreading from the nodes of *take-bus*. The link from a1 to the first argument node of *go*, i.e., a7, causes a7 to become active in synchrony with a1. So, a7 also starts firing in the first phase of every cycle, thereby representing that the first argument of *go* is bound to *john*. The link from a2 to a8 causes a8 to become active in the second phase. That is, the binding of a8 to *place1* gets created. Similarly, the flow of activation along the link from a3 to a9 causes the latter node to become active in the third phase of every cycle, thereby representing that a9 is bound to *place2*. In addition, flow of activation via the link from the predicate node of *take-bus* to that of *go* causes the latter node to become active as well. Now, the entire representation of the inferred fact *go(john, place1, place2)* has been created. Simultaneous with the flow of activation from the nodes of *take-bus* to the nodes of *go*, activations would have also been flowing towards the nodes of *pay-busfare*. Activation flow along the link from a1 to a10 causes the latter node to become active in synchrony with the former. Thus, a10 starts becoming active in the first phase of every cycle, thereby representing that it is bound to *john*. Also, the predicate node of *pay-busfare* becomes active by receiving activation from the predicate node of *take-bus*. Thus, simultaneous with the creation of the pattern representing *go(john, place1, place2)*, the pattern corresponding to *pay-busfare(john)* also gets created.

Now, the activation from the nodes of *go* flow towards the nodes of *reach*. In a manner similar to that described above, the activity pattern corresponding to *reach(john, place2)* gets created.

This concludes a brief description of how forward reasoning involving single antecedent rules (i.e., rules of the form $P(\ldots) \Rightarrow Q(\ldots)$) is achieved.

11.3.3 Multiple Antecedent Rules

The encoding of single antecedent rules described above can be extended to handle multiple-antecedent rules. We illustrate the encoding of such rules with the help of an example. Figure 11.5 depicts the encoding of the rule:

$$P_1(x1, x2) \wedge P_2(x3, x2, x4) \wedge P_3(x2, x4) \Rightarrow Q(x1, x4)$$

Notice that the above rule should fire only if **all** the predicates in the antecedent are 'satisfied'. This also requires that the fillers bound to the roles of the antecedent predicates satisfy certain constraints. For example, the second argument of P_1, the second argument of P_2, and the first argument of P_3 should be bound to the same filler. Similarly, the third argument of P_2 and the second argument of P_3 should also be bound to the same object. The proposed encoding of dynamic role bindings makes it extremely easy to enforce such constraints: checking that two or more roles are bound to the same filler simply involves checking that the two role nodes are firing in synchrony.

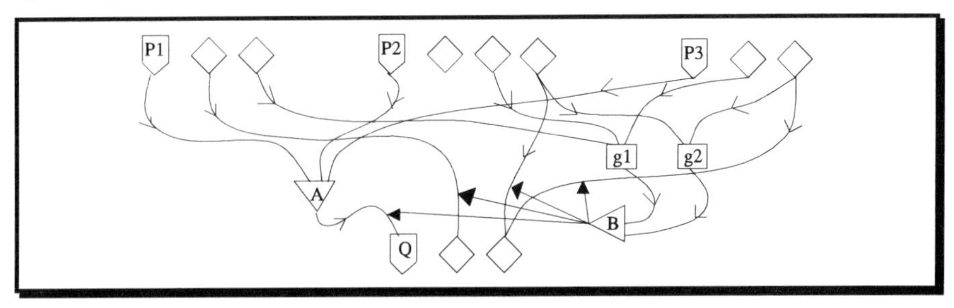

Figure 11.5 — Encoding of a multiple antecedent rule.

With reference to Figure 11.5, the triangular nodes A and B are simple *conjunctive* nodes: a conjunctive node becomes active if it receives activation along *all* its inputs. The inputs need not be in-phase, and it suffices that they arrive within a reasonable window of time. The square shaped nodes g1, g2 are *coincidence* detectors. A coincidence detector becomes active if it receives *synchronous* (or in-phase) activation along all its inputs. Links terminating with dark arrow heads act as 'enabling' modifiers. Any link that is impinged upon by an enabling modifier only propagates activation if the enabling modifier is also active. Observe that the output of the conjunctive node B will be high if and only if the role bindings of the antecedent predicates satisfy the necessary equality constraints. The activation of B enables the propagation of bindings from the roles of the antecedent predicates to the roles of the consequent predicate.

11.4 Related Work

Several connectionist reasoning systems have been developed over the last few years. For a relative comparison of those systems with ours, an interested reader may refer to [Shas93].

There is a growing body of evidence supporting the presence of synchronous activity in the brain. The fact that EEG recordings display rhythmic activity – even though they measure gross electrical activity summed over *millions* of neurons – strongly suggests that there exists significant temporal synchronisation in neuronal activity [Sejn81]. The existence of oscillatory activity in the olfactory bulb, hippocampus, and cerebellum has also been documented by several researchers [Gers70, Free81, Macv80]. The most compelling evidence however, comes from recent findings of stimulus-related synchronous oscillations in the cat visual cortex [Gray89a, Gray89b, Eckh90, Eckh88]. These findings support the hypothesis that the dynamic binding of all features related to a single object may be realised by the synchronous firing of the cells encoding these features. Our work has demonstrated how this widely observed phenomenon of synchronous activation can play a fundamental role in the performance of an important task, namely, reasoning.

11.5 Extensions and Work in Progress

In this chapter, we discussed a connectionist system that performs forward reasoning. Detailed discussion of the realisation of a backward reasoning system can be found in [Shas93]. It should be noted that our system can handle even more expressive rules than the ones discussed here. The cases where (1) the arguments of a predicate appearing in a rule are bound to constants, (2) there are existentially quantified variables in the rule have been discussed in [Shas93]. A way of extending the system to deal with rules containing function symbols has been sketched in [Ajja90].

The reasoning system, as presented in this chapter, has the limitation that during any reasoning episode, the system can represent only one dynamic fact per predicate. Note, however that, *any* number of dynamic facts can be represented simultaneously, as long as they involve different predicates. One possible scheme for extending the system to deal with several, but a bounded number of dynamic facts involving the same predicate has been discussed in [Mani92, Shas93]. That scheme involved rather complicated circuitry that is difficult to learn. An alternative, significantly simpler way of dealing with multiple dynamic instances of a predicate has recently been developed [Ajja94b].

The work described in this chapter and most of the extensions mentioned in the previous two paragraphs, had *deductive reasoning* as their main focus. In deductive reasoning, one performs only those inferences which are guaranteed to be valid with respect to the knowledge base. This is in contrast to the situation in *abductive reasoning* [Char85] — a form of reasoning that pervades a whole variety of intelligent tasks including language understanding, vision and diagnosis. In abductive reasoning, the inferences are not guaranteed to be valid; instead, they are only *plausible*. A close examination of the problem of abductive reasoning involving background knowledge and structured explananda [Ajja93] brings out some interesting issues that need to be addressed. Guided by the analysis of those issues [Ajja93], a connectionist system for abductive reasoning has also been developed [Ajja91].

Currently, our focus is on a critical re-examination of the above mentioned work. This re-examination has involved a careful scrutiny of some of the conceptual distinctions (such as rule/fact distinction, role/filler distinction, type/predicate distinction), the representations employed, and the reasoning strategy used. The result has been a extended system with significant further increases in conceptual elegance, reasoning power, representational efficiency, ease of learning and neurological plausibility.

References

[Ajja90] Ajjanagadde, V., "Reasoning with function symbols in a connectionist system", *Proceedings of Cognitive Science Conference*, 1990, 285-292.

[Ajja91] Ajjanagadde, V., "Abductive reasoning in connectionist networks: Incorporating variables, background knowledge, and structured explananda", Tech. Rept. No. WSI 91-7, Wilhelm-Schickard-Institut, Universitaet Tuebingen, Tuebingen, Germany, 1991.

[Ajja93] Ajjanagadde, V., "Incorporating background knowledge and structured explananda in abductive reasoning: A framework", *IEEE Transactions on Systems, Man, and Cybernetics*, **23**, 1993, 650-664.

[Ajja94a] Ajjanagadde, V., "Revising the representations in a connectionist reasoning system", (in preparation).

[Ajja94b] Ajjanagadde, V., "Goals as soft constraints on reasoning", (in preparation).

[Char85] Charniak, E., and McDermott, D., "*Introduction to Artificial Intelligence.*" Addison Wesley: Reading, MA, 1985.

[Eckh88] Eckhorn, R., Bauer, R., Jordan, W., Brosch, M., Kruse, W., Munk, M., and Reitboeck, H.J., "Coherent oscillations: A mechanism of feature linking in the visual cortex? Multiple electrode and correlation analysis in the cat", *Biological Cybernetics*, **60**, 1988, 121-130.

[Eckh90] Eckhorn, R., Reitboeck, H.J., Arndt, M., and Dicke, P., "Feature linking via Synchronisation among Distributed Assemblies: Simulations of Results from Cat Visual Cortex", *Neural Computation*, **2**, 1990, 293-307.

[Feld82] Feldman, J.A. "Dynamic connections in neural networks", *Biological Cybernetics*, **46**, 1982, 27-39.

[Free81] Freeman, W.J., "A physiological basis of perception", in *Perspectives in Biology and medicine*, **24**, 1981, 561-592.

[Gers70] Gerstein, G.L., "Functional association of neurons: Detection and interpretation", In: F.O. Schmitt (ed.), *The Neurosciences: Second Study Program*, The Rockfeller Univ. Press: New York, 1970.

[Gray89a] Gray, C.M., and Singer, W., "Stimulus specific neuronal oscillations in orientation specific columns of cat visual cortex", *Proceedings of National Academy of Sciences*, **86**, 1989, 1698-1702.

[Gray89b] Gray, C.M., Konig, P., Engel, A.K., and Singer, W., "Oscillatory responses in cat visual cortex exhibit inter-columnar synchronisation which reflects global stimulus properties", *Nature*, **338**, 1989, 334-337.

[Macv80] MacVicar B., and Dudek, F.E., "Dye-coupling between CA3 pyramidal cells in slices of rat hippocampus", *Brain Research*, **196**, 1980, 494-497.

[Mals86] von der Malsburg, C., "Am I thinking assemblies?", In: G. Palm and A. Aertsen (eds), it Brain Theory, 1986, Springer: Berlin.

[Mani92] Mani, D.R., and Shastri, L., "A connectionist solution to the multiple instantiation problem using temporal synchrony", *Proc. of the Cognitive Science Conf.*, 1992, 974-979.

[Sejn81] Sejnowski, T.J., "Skeleton filters in the brain", In: G.E. Hinton and J.A. Anderson (eds), *Parallel Models of Associative Memory*, Erlbaum, 1981.

[Shas93] Shastri, L., and Ajjanagadde, V., "From simple associations to systematic reasoning: A connectionist encoding of rules, variables, and dynamic bindings using temporal synchrony", *Behavioural and Brain Sciences*, **16**, 1993, 417-494.

12

The NeurOagent:A Neural Multi-agent Approach for Modelling, Distributed Processing and Learning

Khai Minh Pham

12.1 Introduction

In AI, three fundamental levels can be distinguished: the *knowledge level* (knowledge modelling methodology) as defined by Newell [Newe82], the *symbol level* (techniques for knowledge representation such as rules, semantic networks, and frames) and the *association level* (connectionist technologies). Usually, the neuro-symbolic hybrid systems focus on the integration of the last two levels. The benefits of neuro-symbolic integration means that knowledge can be acquired via the learning capabilities of the neural component and symbolic problem solving mechanisms can be applied on top of a distributed processing architecture. The *neurOagent*™ is a neural multi-agent approach, based on the *macro-connectionist*™, which proposes a double integration. The first concerns the *association* and *symbol* levels, where a neuro-symbolic *fully-integrated* processing unit provides learning capabilities and distributed inference

™ neurOagent, macro-connectionism are registered trademarks

Intelligent Hybrid Systems, Edited by S. Goonatilake and S. Khebbal. ©1995 John Wiley & Sons Ltd.

mechanisms. Secondly at the *knowledge level,* the *concept operational modelling* (COM) methodology is concerned with knowledge base modelling.

Knowledge acquisition is often considered to be a bottleneck for the development of expert systems, but two different cases must be distinguished: (1) the expertise is complex and a modelling phase is necessary to define the conceptual model of the expertise (knowledge level), and (2) the expertise cannot be formalised, learning capabilities are therefore necessary. The graphic knowledge-based systems shell called intelliSphere©, based on the neurOagent approach, has allowed the development of solutions to different industrial problems (for example, image processing, design by optimisation and constraints satisfaction, and medical diagnosis) where classical approaches have presented significant limits for modelling or learning.

12.2 The neurOagent: A macro-connectionist Model

The neurOagent approach is a fully-integrated model where the basic unit is more complex than the usual neurons in neural networks and where the modelling approach is an intrinsic part of the system.

Usually, the basic functional unit in a artificial neural network corresponds to a single formal neuron. Figure 12.1, shows the structure of the neurOagent which is inspired by the neurobiological macro-connectionist level [Dela87,Chan89], i.e. the basic study unit at this level is the assembly of neurons and not a single neuron.

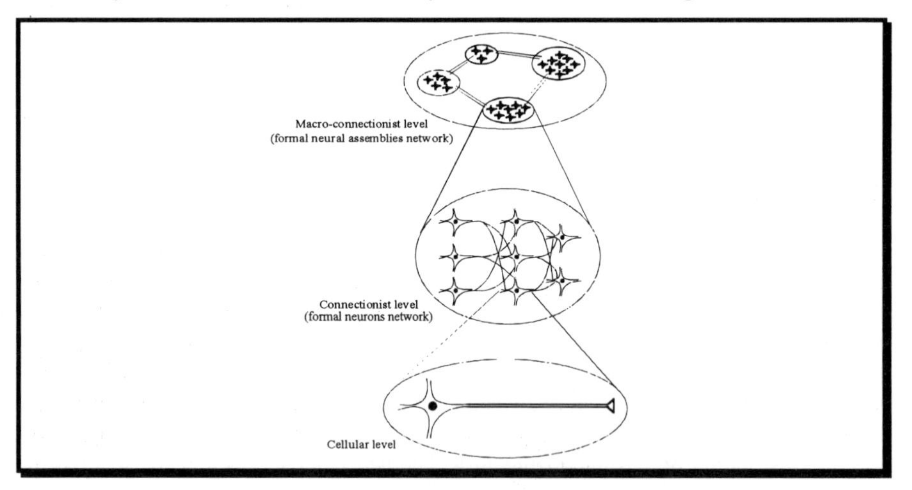

Figure 12.1 — The different neural organisation levels.

From the computer science point of view, the common characteristics between the connectionist and the macro-connectionist levels are:

- an automata network,
- each automaton has a transition function and a memory,
- a communication medium based on the numeric propagation,
- numeric learning capabilities,
- associative memory properties,
- the triggering of the automaton's behaviour is based on the numeric stimulation that the automaton has received.

In comparison to the connectionist level, the macro-connectionist level differs by the following :

- The automaton can be complex.
- The automaton is functionally autonomous.
- The propagation mechanism is more complex.
- The topology of the network presents a semantic organisation.

These characteristics are implemented in the neurOagent approach. The neurOagent is the continuation of research work carried out on Neuronics [Pham88, Pham89, Pham91]. The neurOagent is both an analysis and modelling entity for the knowledge level and an implemented neuro-symbolic processing unit for the development of knowledge-based systems. In this chapter, we are going to focus only on the neuro-symbolic integration aspects.

12.3 The neurOagent's Structure

Figure 12.2, shows the neurOagent's structure which consists of three main elements :

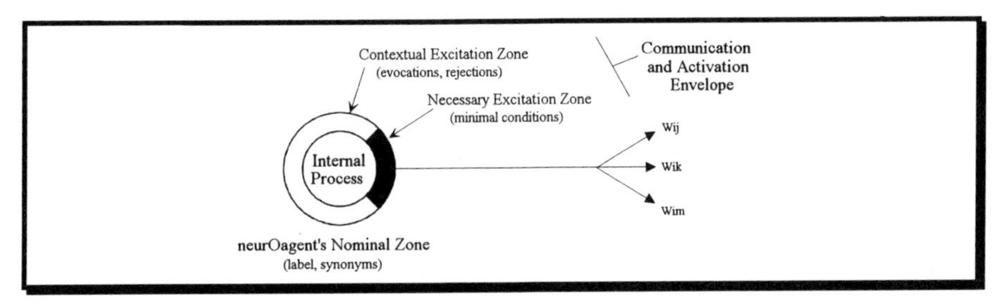

Figure 12.2 — Graphic representation of a neurOagent.

1. the *Communication and Activation Envelope* which ensures a standard communication between the neurOagents, and which controls the neurOagent's state;
2. the *Nominal Zone* which describes the neurOagent (name, synonyms) and

3. the *Internal Process* which determines the neurOagent's functional specifications.

12.3.1 Communication and Activation Envelope

The Communication and Activation Envelope contains :

1. the Necessary Excitation Zone and
2. the Contextual Excitation Zone.

The excitation zones are areas which will receive the connections. With the neurOagent approach, the connection is established between neurOexpressions E_i and a given neurOagent *j*. A neurOexpression is a logic expression of neurOagents. If the neurOexpression is atomic, it corresponds to one neurOagent. In this case, the connection will be established between two neurOagents.

The ***Necessary Excitation Zone*** of neurOagent *j* is the zone where all the conditions, which must be validated, are connected. Indeed, having a necessary condition does not mean it is sufficient, it is a minimal condition. By default, the *Necessary Excitation Zone* is a Minimal Excitation Zone. For example, the concept WHEEL is absolutely necessary for the validation of the concept CAR, but it is not sufficient. However, it is possible to express a minimal and sufficient condition in the *Necessary Excitation Zone*.

The ***Contextual Excitation Zone*** of neurOagent n_i is the zone where one defines what evokes and what rejects the concept of neurOagent n_i. Using connection weights, it is possible to grade evocations and rejections. The evocation is a positive real number that can express a relatively weak (i.e. 20, 40), or strong evocation (i.e. 80, 100). The rejection is a negative real number that can express a relatively weak (i.e. –20, –40), or strong rejection (i.e. –80, –100). The connection weights are expressed in percentages.

There are two ways to establish the connection weights :

1. Explicitly : Five levels can generally be distinguished : (very weakly positive 20, weakly positive 40, positive 60, strongly positive 80, very strongly positive 100) or (very weakly negative –20, weakly negative –40, negative –60, strongly negative –80, very strongly negative –100);
2. By learning from examples.

Contrary to usual neural networks, the connection weights can be dynamically modulated according to the current context, during the propagation process. A connection weight explicitly defined may be modulated :

1. By the Modulation Coefficient: If the *Modulation Coefficient* is modified by a numerical function (fuzzy function, statistical function, or any numerical function), the effective stimulation will be different from the defined connection weight: the connection weight and the stimulation weight will be different. This notion of modulation coefficient is important because it allows external numerical functions to be directly

integrated inside the system. Fuzzy functions are already included inside intelliSphere.

2. By the stimulation function: By default, the stimulation function is linear, but it can be sigmoidal.

12.3.2 Nominal Zone

The *Nominal Zone* contains the neurOagent's name and its synonyms :

- Label: The neurOagent's main identification. This is the neurOagent's main name which will be used to designate it.
- Synonyms: They are used to assign other names to a neurOagent.

12.3.3 Internal Process

There are two types of actions (as shown in Figure 12.3): *Cognitive actions* and *Productive actions*.

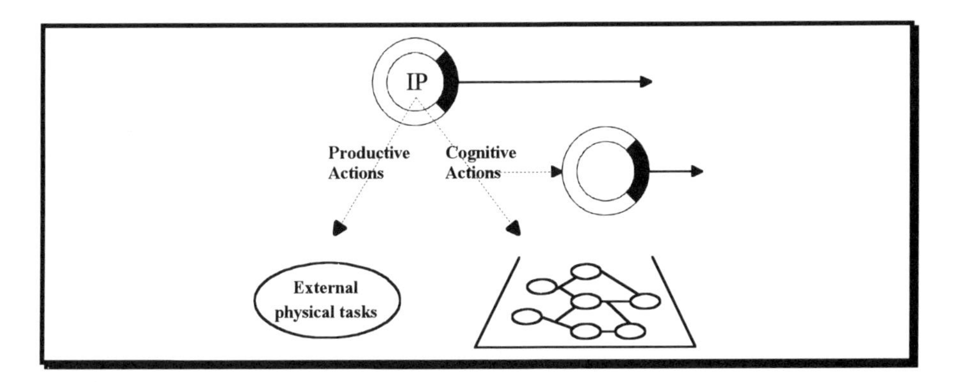

Figure 12.3 — Internal process actions.

Cognitive actions have a direct effect on the inference strategies. For example, functions within intelliSphere which are used to validate, inhibit or activate other neurOagents as well as to dynamically modulate the connection weights or functions used to express fuzzy predicates. Finally, an *internal process* can include an assembly of neurOagents.

Productive actions are functions programmed in C which correspond to external processing (i.e. operate a video camera, execute an SQL request, call an existing library, etc.) or the encapsulation of another neurOagent-based application.

12.4 The neurOagent's Basic Behaviour

The neurOagents's basic behaviour is determined by the *behaviour parameters*, by the neurOagent's state and the inference propagation mechanisms.

12.4.1 The Behaviour Parameters

The neurOagent's inferential behaviour is conditioned through the following *behaviour parameters :*

1. Excitation Threshold
2. Excitation Level
3. Modulation Coefficient
4. Propagation Order
5. (Non) Monotonicity
6. Refractory Period and Propagation Phase
7. Hypothesis
8. Question
9. External Stimulation

The ***Excitation Threshold*** sets the validation threshold for the Contextual Excitation Zone. The default setting for the *Excitation Threshold* parameter is 100%.

The ***Excitation Level*** allows the determination of the state of the *Contextual Excitation Zone*. The *Excitation Level* el_j of neurOagent j is established as follows:

$$el_j = \sum_{i=1}^{n} (\mathrm{f}(S_{Ei,j}) * \mathrm{eval\_exp}(Ei)) \tag{12.1}$$

where f is the stimulation function. This function is linear by default, but it is possible to choose a double sigmoidal function in intelliSphere, $S_{Ei,j}$ is the stimulation between neurOexpression E_i and neurOagent j, eval_exp(E_i) is the logic evaluation function for the neurOexpression E_i.

The default setting for the *Excitation Level* parameter is 0%.

The ***Modulation Coefficient*** is used to dynamically modulate the neurOagent's connection weights. Indeed, the connection weight is a fixed value determined by the expert or by learning from examples. But, neurOagent j will receive a stimulation $S_{Ei,j}$ from neurOexpression E_i if the latter is validated and not the connection weight $W_{Ei,j}$. as shown by Figure 12.4.

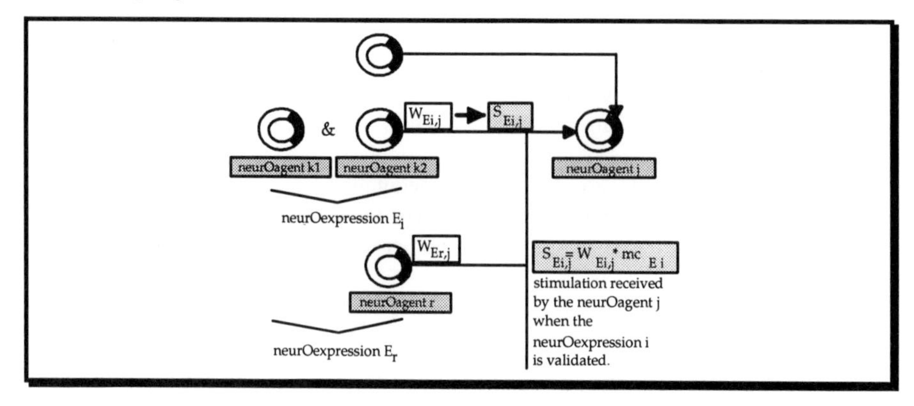

Figure 12.4 — A neurOagent receives stimulation and not connection weights.

The stimulation $S_{Ei,j}$ is calculated as follows:

$$S_{Ei,j} = W_{Ei,j} * mc_{Ei} \tag{12.2}$$

With $S_{Ei,j}$ the stimulation which is effectively received by the neurOagent j from the neurOexpression E_i, $W_{Ei,j}$ the connection weight defined between the neurOexpression E_i and the neurOagent j, mc_{Ei} the *Modulation Coefficient* of the neurOexpression E_i.

$$mc_{Ei} = \min_{h=1}^{n} (mc_h) \tag{12.3}$$

where mc_h is the *Modulation Coefficient* of the neurOagent h which belongs to the neurOexpression E_i. Numeric functions are used to establish the neurOagent's *Modulation Coefficient*: functions for fuzzy predicates, and statistical functions.

The default setting for the *Modulation Coefficient* parameter is 100%.

As Figure 12.5, shows, the default ***Propagation Order*** parameter is set so that when the neurOagent is validated, the system will first execute its *Internal Process* and then propagate the information. However, it is possible to change this order so as to synchronise certain parallel tasks. Once the neurOagent is validated, it is possible to propagate the information first and then execute the neurOagent's *Internal Process.*

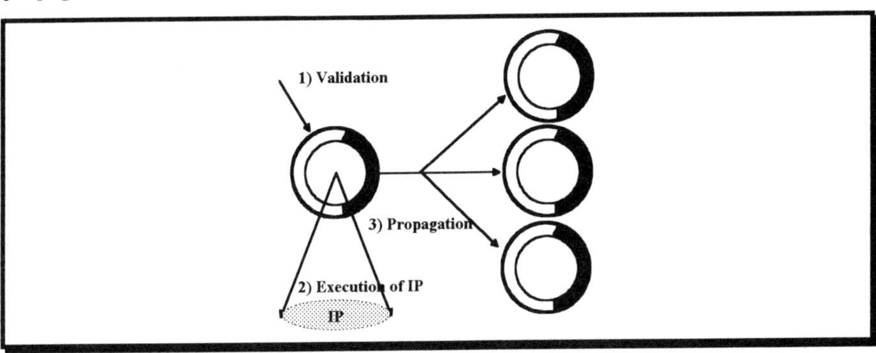

Figure 12.5 — Default propagation order.

A ***monotonic*** reasoning is a type of reasoning based on the assumption that once a fact has been determined, it cannot be altered even if new and more pertinent information comes up. This seriously limits a large number of applications for which the evolution of the various states must be taken into account.

A non-monotonic reasoning is a type of reasoning in which the assumptions are systematically changed in light of new information.

Traditionally, knowledge-based systems consist of two main parts: an inference engine and a static knowledge base. Only the inference engine can manage the reasoning mechanism and therefore the notion of monotonicity or non-monotonicity. This is why a system is usually either monotonic or non-monotonic; it cannot manage both mechanisms at the same time.

In many applications, monotonicity and non-monotonicity are both necessary, particularly in areas where certain decisions cannot be changed. For example, in the defence sector, launching a missile is an action which cannot be changed once it is started, even if the system receives new information opposing the launching. Having a strictly non-monotonic system does not solve the problem, since the missile cannot be called back once it is launched. However, we must keep in mind the fact that the missile has been launched, adapt to the new context and eventually destroy the missile in mid-air. In this example, the concept of missile launching is monotonic but all the reasoning which takes place before this concept is validated must be non-monotonic.

The neurOagent approach is one of the few technologies which can manage both monotonic and non-monotonic reasoning. This is possible because the neurOagent is a knowledge automata which allows a distributed reasoning process. The (non) monotonicity is directly managed by each neurOagent locally. It is possible to dynamically change the value of the *Monotonicity* parameter according to the current context too.

The default setting for the *Monotonicity* parameter is NON.

Although a ***Refractory Period*** exists in neurobiological systems, within artificial connectionist systems this factor is rarely simulated, but could allow an efficient control of loops. When *Refractory* is set to YES, a neurOagent which is already validated or inhibited cannot be validated or inhibited again during the refractory period which starts at the time of the neurOagent's validation or inhibition.

If *Refractory* is set to NO, the neurOagent is sensitive to any new stimulation and does not take into account the refractory period. This is interesting, for example, for optimisation problems to avoid local loops. The notion of "phase" indicates if the neurOagent already participates in a propagation phase. In this case, the neurOagent presents a temporary "refractory period". The "refractory period" and the "phase" participate in the loops' control.

The default setting for the *Refractory* parameter is YES.

When a neurOagent's *Excitation Level* is greater or equal to 80 and less than 100 and its *Necessary Excitation Zone* is validated, the neurOagent spontaneously goes into ***hypothesis***. This is the first of two mechanisms which creates spontaneous *backward propagation*. The other occurs when the *Contextual Excitation Zone* is validated (i.e. the neurOagent's *Excitation Level* is greater than its *Excitation Threshold*) but the *Necessary Excitation Zone* is not yet completely evaluated. Once again the neurOagent will go into hypothesis and therefore try to validate itself. Once a neurOagent goes into hypothesis, it seeks additional information which will enable it to validate itself. This means that the neurOagent will go back (backward propagation) to the neurOagents which are connected to it and are not yet validated. These neurOagents will then go into hypothesis themselves.

If the *Hypothesis* parameter is set to NO, then the neurOagent will not be able to go into hypothesis.

The *Hypothesis* parameter can have a third value: REACTIVE. When the neurOagent goes into hypothesis, it will go through a procedure where it will seek the information it is missing to validate itself instead of asking the user, or it will display the question in a different format.

The default setting for the *Hypothesis* parameter is YES.

During the reasoning process, it may be wise to determine the state of a neurOagent. If ***Question*** is set to YES, the neurOagent's state will be determined ;

- directly using a dialogue box in which the user will have to give a "YES/NO" answer.
- indirectly using the "reactive hypothesis".

The default setting for the *Question* parameter is YES.

External Stimulation means that the user can directly validate or inhibit the neurOagent. If the *external stimulation* parameter is set to NO, the neurOagent can only be validated or inhibited by other neurOagents.

The default setting for the *External Stimulation* parameter is YES.

12.4.2 The neurOagent's State Determination

At any given time, each neurOagent is characterised by its state which is established as the result of all the stimulation it receives on its *Activation and Communication Envelope.*

The neurOagent's *Activation and Communication Envelope,* as shown in Figure 12.6, can have different configurations according to the connections it receives on its excitation zones:

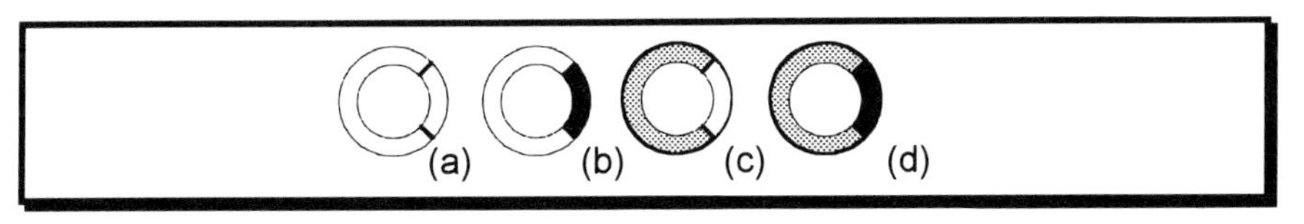

Figure 12.6 — NeurOagent's activation & communication envelope

Figure 12.6(a) No connection on the excitation zones.
Figure 12.6(b) Connections only on the *Necessary Excitation Zone*.
Figure 12.6(c) Connections only on the *Contextual Excitation Zone*.
Figure 12.6(d) Connections on the *Necessary* and the *Contextual Excitation Zones*.

12.4.2.1 Defining the neurOagent's State

Three states are possible for a neurOagent:

1. undetermined,
2. validated,
3. inhibited.

Table 12.1, summarises all of the neurOagent's validation states :

neurOagent's Configuration	Necessary Excitation Zone	Contextual Excitation Zone	**neurOagent's general state**
	Validated	Abs.	**Validated**
	Abs.	Validated	**Validated**
	Validated	Validated	**Validated**

Table 12.1 — Validation states for the neurOagent, Abs. means that there is no connection.

Table 12.2 summarises the neurOagent's inhibition states :

neurOagent's Configuration	Necessary Excitation Zone	Contextual Excitation Zone	**neurOagent's general state**
	Inhibited	Abs.	**Inhibited**
	Abs.	Inhibited	**Inhibited**
	Inhibited	Any	**Inhibited**
	Any	Inhibited	**Inhibited**

Table 12.2 — Inhibitation states for the neurOagent.

In all other configurations, the neurOagent is in an undetermined state.

12.4.2.2 Defining the State of the Necessary Excitation Zone

The *Necessary Excitation Zone* is validated when the neurOexpressions connected to this zone, i.e. the logic conditions described with neurOagents, are validated. This zone is inhibited in the opposite situation.

If there is no connection on the *Necessary Excitation Zone* then it is considered to be validated.

The evaluation of the *Necessary Excitation Zone* of the neurOagent *j*, nez_j, is formulated as follows:

$$nez_j = \begin{cases} 1 & if\,(nb\_nez_j > 0) \text{ and } (\text{eval\_nez}(j) = True) \\ 1 & if\,(nb\_nez_j = 0) \\ 0 & otherwise \end{cases} \quad (12.4)$$

with nb_nez_j the number of connections on the *Necessary Excitation Zone* of the neurOagent j, eval_nez(j) the logic evaluation function of the neurOexpressions connected on the *Necessary Excitation Zone* of the neurOagent j.

12.4.2.3 Defining the State of the Contextual Excitation Zone

The *Contextual Excitation Zone* is validated when the sum of positive and negative stimulation is greater or equal to the neurOagent's *Excitation Threshold*.

The *Contextual Excitation Zone* is inhibited when the sum of the positive and negative stimulation is inferior or equal to the negative value of the *Excitation Threshold*. Usually, the value of the *Excitation Threshold* is equal to 100%. The *Contextual Excitation Zone* is then inhibited when the sum of the positive and negative stimulation is inferior or equal to -100%.

If there is no connection on the *Contextual Excitation Zone* then it is considered as validated. The evaluation of the *Contextual Excitation Zone* of the neurOagent $_j$, cez_j is formulated as

$$cez_j = \begin{cases} 1 & if\,(nb\_cez_j > 0) \text{ and } (el_j \geq et_j) \\ 1 & if\,(nb\_cez_j = 0) \\ 0 & otherwise \end{cases} \qquad (12.5)$$

with nb_cez_j the number of connections on the *Contextual Excitation Zone*, el_j the *Excitation Level* of the neurOagent$_j$, and et_j the *Excitation Threshold* of the neurOagent$_j$.

12.4.3 The Inference Propagation Mechanisms

The propagation's mechanisms are :

- forward propagation,
- backward propagation,
- spontaneous backward propagation,
- retropropagation of necessities.

The propagation mechanisms are asynchronous, i.e. the update of neurOagents depends on events. The spontaneous backward propagation mechanism is managed by the *Hypothesis* parameters.

Figure 12.7 shows that different inference propagation mechanisms can be obtained within the same system.

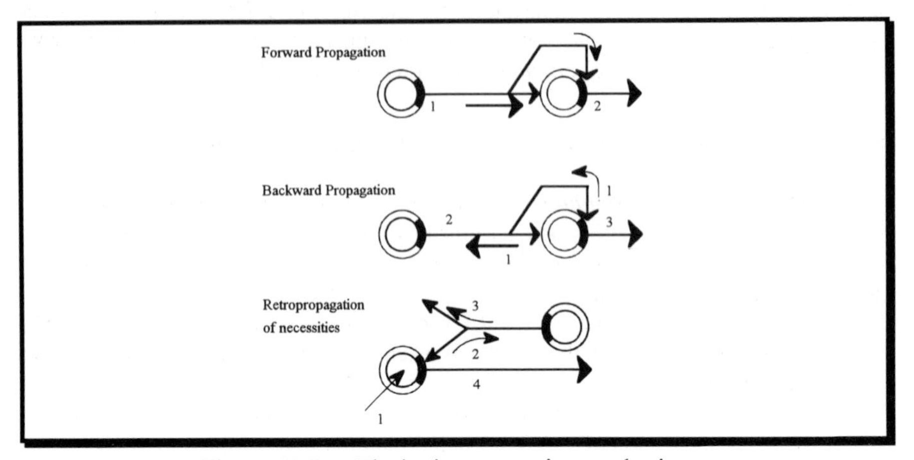

Figure 12.7 — The basic propagation mechanisms.

12.4.4 Reasoning Mechanisms

Thanks to the different features which were described above, the neurOagent approach provides very rich properties for the reasoning process (see Table 12.3):

Reasoning Mechanisms	Method
Deduction and/or reasoning by hypothesis	Forward, backward and mixed propagation
Triggering of spontaneous hypothesis	Spontaneous backward propagation
Inheritance, implicit deduction	Retropropagation of necessities
Constraint satisfaction	Constraint propagation
Selective challenging of deductions	Distributed monotonicity and non-monotonicity
Processing of non-perfect information	Inference through numeric propagation
Fuzzy predicates	Fuzzy primitives
Event behaviour	Distributed inference
Parallel tasks processing	Execution/propagation order

Table 12.3 — Reasoning mechanisms using the neurOagent

12.4.5 Learning

With the neurOagent approach, it is possible to design a knowledge base through explicit modelling, learning or both. Of course this enhances the quality of the knowledge bases, since in many cases, explicit modelling is not sufficient nor is learning from examples.

The learning provided has two objectives :

1. to automatically establish the connection weights as in usual connectionist models, and
2. to automatically establish the topology of the network.

Due to the neurOagent's connectionist architecture, the system will not provide a "black box" at the end of the learning, it will be able to reach semantic conclusions: necessary minimal conditions for the validation of outputs, simultaneous presence of certain features, specificity of certain features.

Let's say we had to study a medical database in order to design a knowledge base on pathologies diagnosed in various patients. At the end of the learning process, we would obtain the connection weights established by the system, but we would also find out the following :

- which symptoms are minimal in order to diagnose a given pathology,
- which symptoms are always found together, and
- which symptoms are specific to a given pathology.

The neurOagent architecture, has the following characteristics :

- The topology of the neurOagents network is built during the learning period.
- The input parameters can be qualitative and / or quantitative.
- Input parameters can be missing.
- The neurOexpressions are built during the learning period.
- It can be mixed directly with explicit knowledge.
- The order in which the examples are presented has no importance.
- The order of the input data included in the examples has no importance.

The learning algorithm used with neurOagents is similar to Probabilistic Neural Networks (PNN) [Mack92, Werb91, Lee91]. However, it presents a number of differences (the neurOexpressions, the excitation zones, and the topology building). Figure 12.8, shows how the topology of a neurOagent is built up during learning.

During the learning period, the parameters which are necessary for the calculation of the connection weights are established:

- $N(E_i|O_j)$: The number of examples where the neurOexpression E_i was present when the output is O_i.
- $N(n_i)$: The number of examples where the neurOagent n_i was present during the learning period.
- $N(O_j)$: The number of examples where the output is O_j.

- The total number of examples which were presented during the learning period.
- *Nb_Class*: The number of output classes which were presented during the learning period.

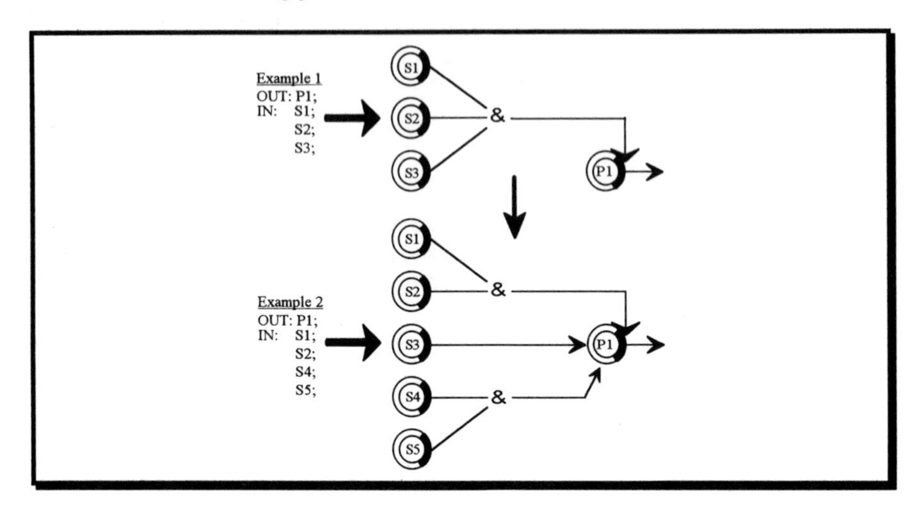

Figure 12.8 — The building of the topology of neurOagents during the learning period.

The evaluation of the impact of E_i on the different outputs is based on a comparative process. The connection weight $w(E_i,O_j)$ between the neurOexpression E_i and an output-neurOagent O_i is calculated as follows :

$$w(E_i,O_j) = \frac{P(E_i|O_j)}{\pi(E_i)} \tag{12.6}$$

where $P(E_i|Oj)$ is the probability of E_i when O_i is present and $\pi(E_i)$ is the estimated probability of E_i.

$$P(E_i|Oj) = \frac{N(E_i|O_j)}{N(O_j)} \tag{12.7}$$

$$\pi(E_i) = P(E_i|O_j) + ((E_i|\overline{O_j}) * (Nb\_Class - 1)) \tag{12.8}$$

where $\pi(E_i|Oj)$ is the estimated probability of E_i when the class of the output is not Oj.

$$\frac{(\min[N(n_i)] - N(E_i|O_j))}{(N - N(O_j))} \tag{12.9}$$

Note, $\pi(E_i)$ is an estimated probability, the probability of E_i would be:

$$P(E_i) = \frac{\sum_{j=1}^{Nb\_Class} N(E_i|O_j)}{N} \tag{12.10}$$

In the present case, this calculation has no meaning because E_i is a neurOexpression, i.e. it can be a combination of signs, and not a single sign.

To avoid taking irrelevant information into account, a significant threshold ST is defined as follows:

$$ST = \max \left\langle \begin{array}{l} \frac{100}{Nb\_Class} + SD \\ PT \end{array} \right. \tag{12.11}$$

where standard deviation SD and parasite threshold PT are arbitrary constants.

The final value of the connection weight is calculated as follows:

$$w(E_i, O_j) = \left\langle \begin{array}{ll} 0 & if\, w(E_i, O_j) \leq ST \\ w(E_i, O_j) & otherwise \end{array} \right. \tag{12.12}$$

12.5 The neurOagent: a Neural Multi-agent Approach

There are two main ways to integrate the multi-agent and the neural networks approaches: the first concerns the establishment of distributed architecture's and communication protocols between modules where neural networks are embedded, and the second focuses on the definition of an entity which presents both multi-agent and neural properties. The neurOagent approach corresponds to the second method.

12.5.1 The Granularity of the neurOagent

The neural networks are integrated using the concepts of distributed artificial intelligence techniques. The two extreme poles are represented by: *reactive simple agents* and *cognitive complex agents*. Figure 12.9, shows the granularity scale, where connectionist neural networks are extreme reactive systems, i.e. the lowest granularity.

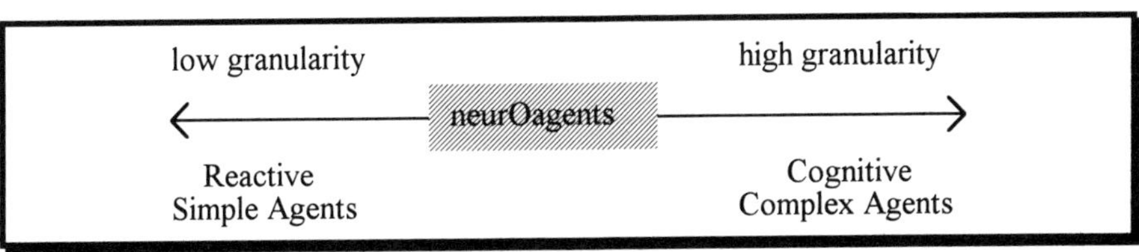

Figure 12.9 — Granularity scale.

According to the type of actions included in the Internal Process of the neurOagent, the latter is a reactive or a cognitive agent. The philosophy of the neurOagent approach is more reactive than cognitive, but sufficiently cognitive to be different to the usual connectionist neural networks which are usually more suitable for signal processing type applications.

12.5.2 The Communication Types

In multi-agent systems, two main communication types can be distinguished as blackboard systems or message passing systems.

By default, there is no blackboard in neurOagent-based systems. The message passing corresponds essentially to the numeric propagation between the neurOagents. The problem solving with the neurOagent approach is based on local interactions.

The problem solving with the neurOagent approach is therefore close to the Eco Problem Solving which is based on the paradigm of "computational ecosystems" [Hube88]. The problem solving is the result of local interactions between simple agents which correspond to an undeterministic evolution of the system. A multi-agent system for the Eco Problem Solving presents a number of laws (law of autonomy, of locality, of memory ordering, of pleasure and of economy) and the notion of tropic behaviour [Gene87] (local satisfaction condition, local rejection condition, survival reaction, cost function, information transmitted as messages). A number of these features are the same as the neurOagent (for example the law of locality, the local satisfaction and rejection conditions, information transmitted as messages). However, the basic neurOagent system has fewer heuristic strategies.

12.6 The neurOagent and Other Approaches

The neural networks and the multi-agent models are the basic paradigms for the definition of the neurOagent approach. However, there are other approaches which can be represented totally or partially by the neurOagents. This allows other technologies to have additional modelling facilities, inference properties and learning capabilities.

12.6.1 The neurOagent and Decision Trees

It is easy to represent decision trees with neurOagents, as seen in Figure 12.10.

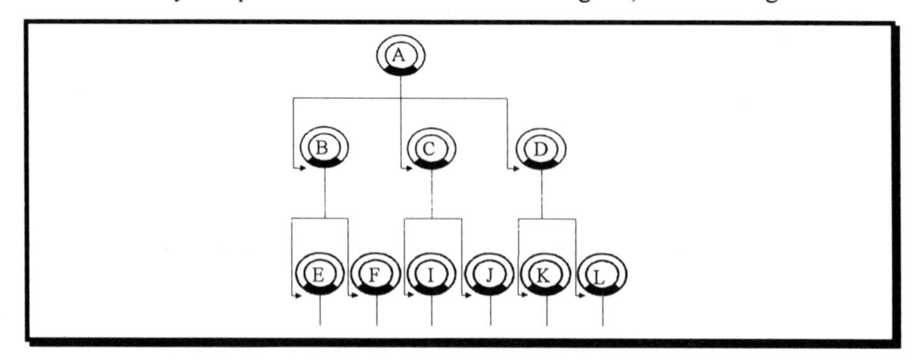

Figure 12.10 — A decision tree with neurOagents.

The connections are on the Necessary Excitation Zones. They represent only the minimal conditions to eventually validate B or C or D. Indeed, to validate the nodes B or C or D, you must validate at least A. When the node A is validated, according to the test which is included in the Internal Process of A, the node B or C or D will be validated.

12.6.2 The neurOagents and Fuzzy Predicates

Thanks to the *Modulation Coefficient*, the neurOagents can be used to implement some features of fuzzy associative memories [Kosk92]. For example, in Figure 12.11, the fuzzy membership set for a traffic density problem is shown.

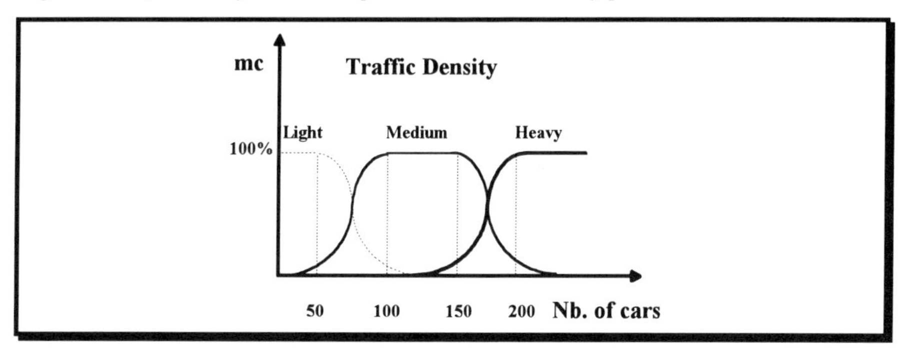

Figure 12.11 — Traffic density.

Figure 12.12, shows a number of fuzzy primitives such as *iS_until, iS_from, iS_between,* that are provided with intelliSphere.

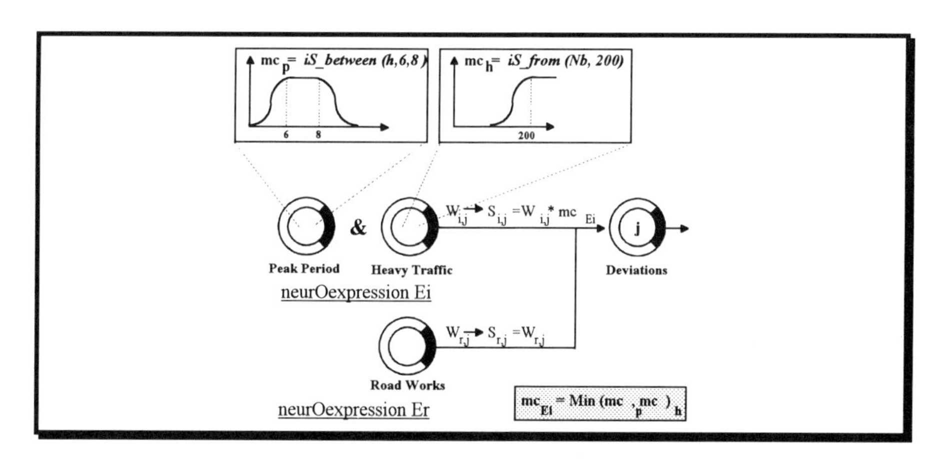

Figure 12.12 — Fuzzy predicates.

12.6.3 Transcription Rules into neurOagents

Figure 12.13, shows that it is possible to directly transcribe a rule-based system into a neurOagent-based system. However, it is better to use the *Concept Operational Modelling* (see section 12.7.) approach by integrating the analysis which would have already been done when the rule-based system was being developed.

The most simple transcription way is as follows:

1. Regroup the rules which have the same conclusion.
2. The common condition(s) of these rules is (are) changed into neurOagent(s) connected on the Necessary Excitation Zone of the conclusion-neurOagent.
3. For each rule that was selected during step 1, the non common condition(s) is (are) changed into neurOagent(s) connected on the Contextual Excitation Zone with a connection weight equal to 100.

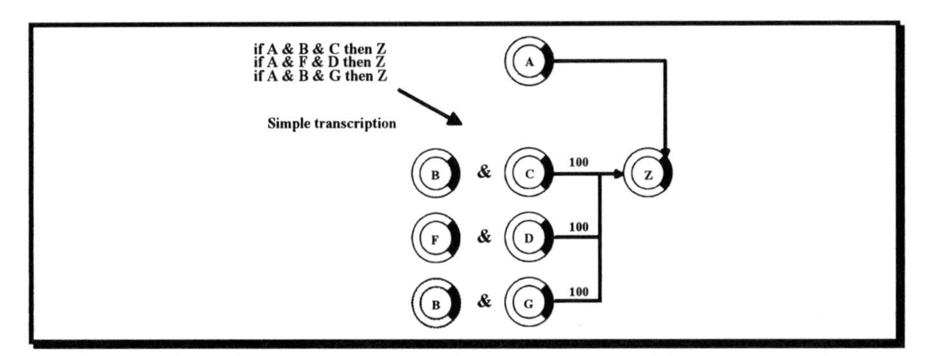

Figure 12.13 — Transcription rules into neurOagents.

12.6.4 The neurOagents and Sets of Rules

The neurOagents can be used to organise an existing rule-based system. They can be considered as meta-rules. Figure 12.14, shows that in the same way as the objects, the neurOagents can encapsulate sets of rules, via their *Internal Process*.

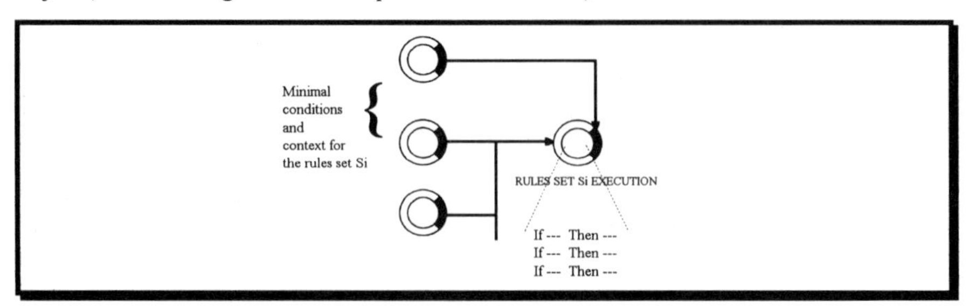

Figure 12.14 — Encapsulation of rules within the neurOagent.

By comparison with objects, the neurOagents present:

- a more flexible organisation because the hierarchy is not mandatory,
- better inference properties, and
- learning capabilities.

12.6.5 The neurOagents and Objects

In knowledge acquisition, two kinds of knowledge can be distinguished: dynamic and static knowledge. Object-oriented systems are adapted to represent static knowledge where knowledge classification is an important aspect. This level corresponds to the domain level, in the sense of KADS [Schr93]. The other levels (inference, task and strategy) are represented by dynamic knowledge. They can be managed by the neurOagents which are more adapted to dynamic knowledge. Figure 12.15, shows how neurOagents can manage objects. This would provide flexible organisation, richer inference properties and better learning capabilities.

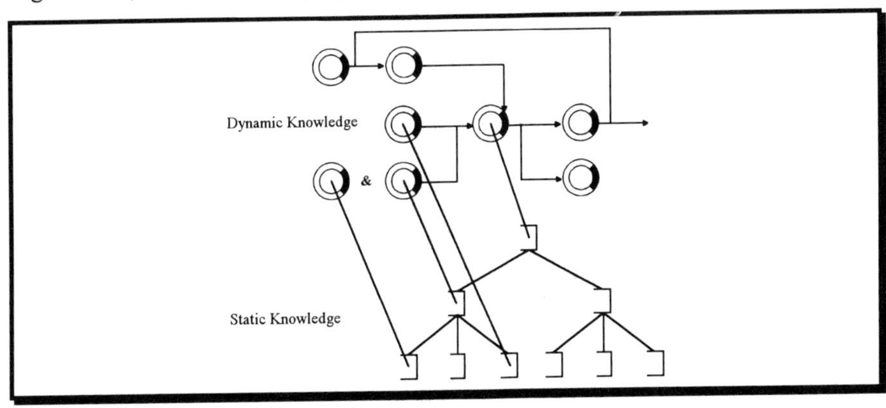

Figure 12.15 — Encapsulation of objects by the neurOagents.

12.7 The Concept Operational Modelling (COM)

Often, the knowledge acquisition is considered to be a bottleneck for the development of expert systems. Indeed, for several years, the AI research has focused on the symbol level, i.e. the techniques of knowledge representation (rules, semantic networks, and frames). The development of complex knowledge-based systems has shown the necessity to have a knowledge acquisition methodology for knowledge modelling. This corresponds to the debate on the knowledge level which was introduced by Newell [Newe82]. Today, there are several investigations into the development of knowledge acquisition methodologies and/or tools (AQUINAS,

GRAPES, KADS, KOD, MOISE, MOLE, and SBF). When the conceptual model of the expertise is determined, the implementation phase can start.

Figure 12.16, shows the *Concept Operational Modelling* methodology, proposed with the neurOagent approach, which allows a directly operational knowledge modelling. There is no interpretation, therefore no introduction of biases. This allows a control of the analysis at any stage and provides a better maintenance of the knowledge base.

In this section, we present the main principles of the *Concept Operational Modelling*. The objective of the *Concept Operational Modelling* is to build the *Generic Knowledge Base* which defines the fundamental principles of the expertise. The instantiate of this *Generic Knowledge Base* provides the *Operational Knowledge Base*.

12.7.1 The Generic Knowledge Base

A *Generic Knowledge Base* is an influence network of *Generic Concepts*. It is the "skeleton" of the knowledge base, it is a flexible "grammar" of expertise. It ensures the coherence of interactions between the different *Generic Concepts*. The knowledge engineer can directly control the coherence of the expertise.

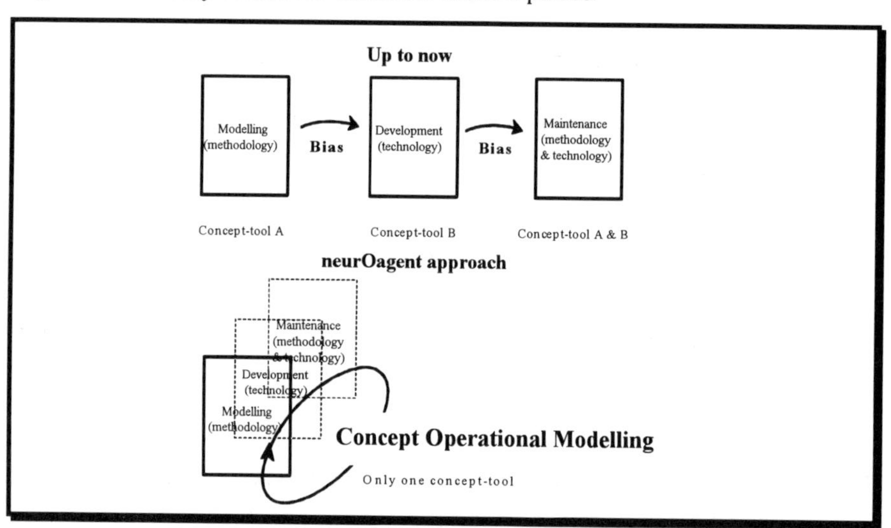

Figure 12.16 — NeurOagent approach.

A ***Generic Concept*** is a generic neurOagent which is a model for the instantiate of neurOagents which belong to the same type. For example Figure 12.17, shows a decision system which proposes a *Generic Network* for DRUG PRESCRIPTION.

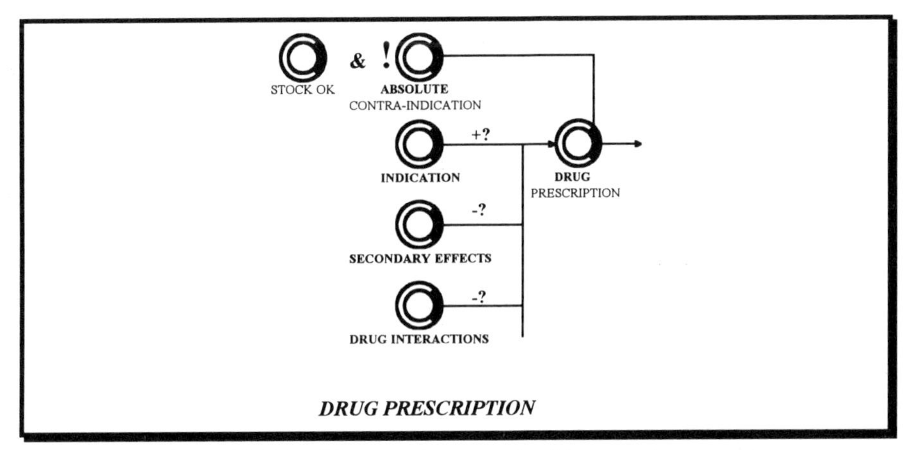

Figure 12.17 — The generic network of the generic concept.

A generic neurOagent is defined by its local network, termed *Generic Network*, and its *Behaviour Parameters*. Figure 12.18, show that in a *Generic Concept*, you can have a recursive definition.

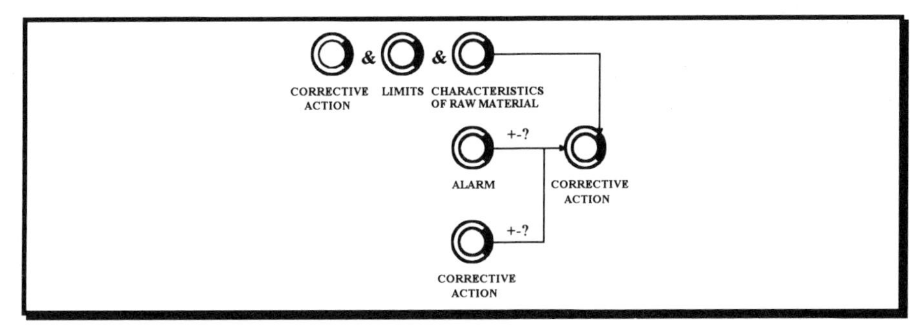

Figure 12.18 — A recursive definition for a generic concept.

12.7.2 The Operational Knowledge Base

Figure 12.19, shows that an instantiate of the *Generic Knowledge Base* allows the building of all or a part of the *Operational Knowledge Base*. The latter is an influence network of *Operational Concepts* which are either *Instantiated Concepts* or *Particular Concepts*. A *Particular Concept* is a concept which is operational but which was not instantiated. The instantiate can be pure, i.e. it is strictly based on the *Generic Concepts*, or degraded, i.e. some *Particular Concepts* can be introduced. The *Particular Concepts* are interesting because, in a number of cases, a complete normalisation knowledge in *Generic Concepts* is not possible.

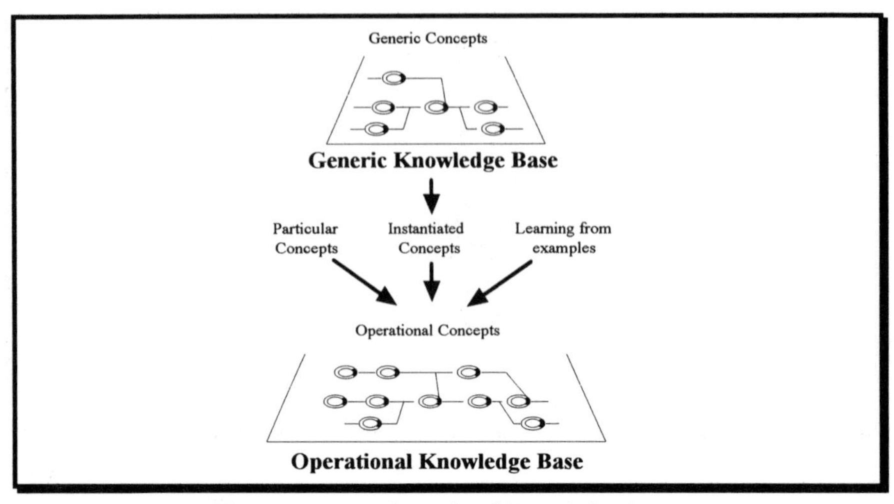

Figure 12.19 — Operational Knowledge Base.

The *Generic Concepts* will allow the instantiate of the *Operational Concepts*. For example, Figure 12.20, shows that ASPIRIN PRESCRIPTION is an instantiation of the *Generic Concept* DRUG PRESCRIPTION described above, i.e. its local network must be coherent with the *Generic Network* of the concept DRUG PRESCRIPTION and it will inherit the values of the *Behaviour Parameters*.

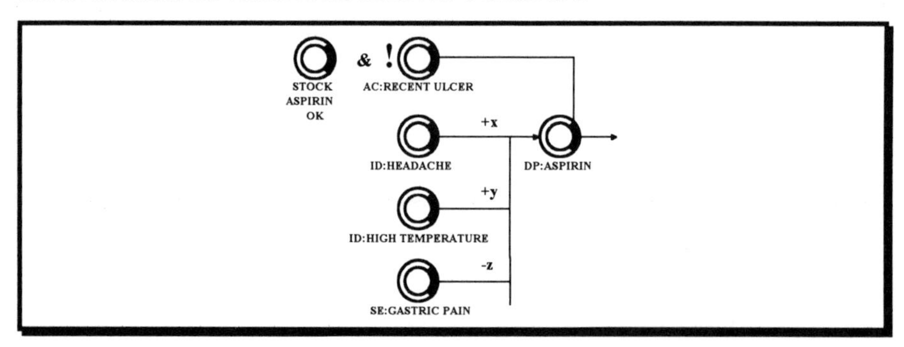

Figure 12.20 — Instantiation of the Generic Network of the Generic Concept DRUG PRESCRIPTION.

The concept of ASPIRIN PRESCRIPTION is directly operational because it is represented by a network of neurOagents. There is no additional interpretation of the analysis representation in another technique which corresponds to the implementation.

Naturally to use and to instantiate *Generic Concepts* is to strongly evoke the Conceptual Graphs approaches [Sowa84, Sowa92] and KL-ONE [Brac85].

The *Generic Concepts* can be used in different applications. You can change a part of the influence network of *Generic Concepts* to adapt it to a new application.

The main job for the knowledge engineer is to define, with the experts, the *Generic Concepts* and to build a library of *Generic Concepts* which is the entities knowledge skeleton.

12.8 Conclusion

The neurOagent approach suggests a possible way for a large integration which includes the knowledge, the symbol and the association levels. The *Concept Operational Modelling* offers a solution to complex knowledge modelling problems which are lacking in the "pure" connectionist approaches. The neurOagent is not a black box, it provides learning capabilities, which produce a valuated network, and provides a semantic organisation.

The research on the neurOagent approach is today an operational technology: intelliSphere is a commercial development shell which is used in several industrial applications (diagnosis, constraints satisfaction and optimisation, prediction). The neurOagent combines the strengths of neural and symbolic techniques. It can be used as a basic building block where existing approaches can be integrated either by the construction of new *Generic Concepts* or by the encapsulation of external technologies.

Acknowledgements

I wish to thank V. Piffero and M. Kieu for the discussions concerning the *Concept Operational Modelling*, T.D. Nguyen for his reflections and tenacity in the development of intelliSphere, and P. Hopkins for my English.

References

[Brac85] Brachman, R. and Schmolze, J., "An overview of the KL-ONE knowledge representation system." *Cognitive Science*, 9, 2, pp 171-216.

[Chan89] Changeux, J.P. and Dehaene, S. "Neural models of cognitive functions." In *Cognition*. 33 (1-2), pp 63-109. 1989.

[Dela87] Delacour, J., "*Apprentissage et Mémoire*." Masson, 1987.

[Gene87] Genesereth, M. and Nilsson, N. "*Logical Foundations of Artificial Intelligence*", Morgan Kaufman, 1987.

[Hube88] Huberman, B.A. and Hogg, T., "The Behaviour of Computational Ecologies.", In *The Ecology of Computation*, B.A. Huberman (eds) North Holland, 1998.

[Kosk92] Kosko, B. "*Neural Networks and Fuzzy Systems*", Prentice-Hall, 1992.

[Lee91] Lee, D.S., Srihari, S.N. and Gaborski,R. "Bayesian and neural network pattern recognition : a theoretical connection and empirical results with hand-written characters", In *Artificial Neural Networks and Statistical Pattern Recognition,* I.K. Sethi and A.K. Jain, editors, Elsevier Science Publishers B.V., pp 89-108, 1991.

[Mack92] MacKay, D.J.C. "Bayesian Model Comparison and Backprop Nets.", In *Neural Information Processing Systems 4.* Edited by J.E. Moody, S.J. Hanson and R.P. Lippmann. pp 839-846, Morgan Kaufmann 1992.

[Newe82] Newell, A. "The Knowledge Level." In *Artificial Intelligence, 18 (1) pp 87-127, 1982.*

[Pham88] Pham, K.M. and Degoulet, P., "MOSAIC: a Macro-connectionist Organisation System for Artificial Intelligence Computation.", In *Proc. IEEE ICNN 88. International Conference on Neural Networks.* Vol. II. pp 533-540. 1988.

[Pham89] Pham, K.M. and Degoulet, P., "MOSAIC: Medical Knowledge Processing Based on a Macro-connectionist Approach to Neural Networks", In *Proc. of 6th Congress on Medical Informatics.* MEDINFO 89. Vol. I. pp 82-86. 1989.

[Pham91] Pham, K.M. and Degoulet, P., "MOSAIC: a Macro-connectionist Expert Systems Generator.", In *Expert System With Applications 1991.* Vol II. pp 29-45. 1991.

[Schr93] Schreiber, G., Wielinga, B. and Breuker, J. "KADS A Principled Approach to Knowledge-Based System Development." *Knowledge-Based Systems* 11. Academic Press.

[Sowa84] Sowa, J., "Conceptual Structures.", Information in Mind and Machine, Addison Wesley, Reading Mass, 1984.

[Sowa92] Sowa, J. "Current research issues in Conceptual Graphs", *Actes 7th Annual Workshop on Conceptual Graphs*, New Mexico, 1992.

[Werb91] Werbos, P.J. "Links between Artificial Neural Networks (ANN) and Statistical Pattern Recognition.", In : *Artificial Neural Networks and Statistical Pattern Recognition*, edited by I.K. Dethi and A.K. Jain. pp 11-31.North-Holland 1991.

13

Genetic Programming of Neural Networks: Theory and Practice

Frederic Gruau

13.1 Introduction

Cellular encoding (CE) is a scheme to encode neural networks on rooted point labelled trees. We prove that this scheme verifies seven properties: completeness, compactness, closure, modularity, scalability, power of expression, abstraction. These properties prove that CE is a machine language for a neural computer that is suited to the genetic algorithm. The genetic algorithm is used to evolve cellular code, and synthesise neural networks for the symmetry function of arbitrarily large size. This problem had not been solved yet with neural networks and illustrates the properties of CE. Genetic programming (GP) uses the genetic algorithm to evolve LISP computer programs. The seven properties show that the genetic search performed with CE is an hybridisation of GP and neural networks which is referred to as the genetic programming of neural networks.

Most classical neural learning algorithms aim at finding weights for a neural network whose architecture is frozen. They usually operate by making small adjustments in the direction given by the computation of a gradient vector. Recently, some algorithms have been proposed that optimise the architecture by pruning or adding nodes one by one during the process of learning [Fahl90]. These algorithms also use "small adjustments" starting from an initial solution. A small adjustment searching method presents two drawbacks. It depends heavily on the initial solution, usually randomly generated, and it has difficulties escaping from local minima. The genetic

Intelligent Hybrid Systems, Edited by S. Goonatilake and S. Khebbal.

algorithm (GA) is a different kind of search method without these drawbacks. A GA handles a population of solutions and makes a global walk through the search space. Therefore it can explore many local minima. A GA works by finding a set of features which describes a good solution to the problem [Holl75]. Because of these fundamental differences with the "small adjustments" method, applying GA to neural network optimisation should enable a more thorough exploration of the neural networks landscape.

In [Muhl89], Muhlenbein proposes a general framework for "genetic neural networks". In fact, the GA does not directly handle neural networks. Instead, it handles structures called chromosomes which encode a neural network. Therefore, in order to apply a GA to neural network optimisation, there is only one problem to solve: how to encode a neural network in a chromosome. The GA does not "see" any neural networks directly, it sees the landscape of codes. If this landscape does not have good properties, the GA will not be able to make a successful search in it.

Walsh functions have recently been applied to the analysis of the GAs for estimating the suitability of a given encoding [Gold89]. Since the finding of the Walsh coefficients needs more computation than that is required to yield the solution, the method is not of practical interest to evaluate encoding in real world problems.

In the particular case of neural network optimisation, we will show that it is possible to define verifiable properties of good encoding. In [Whit90, Mill89] the connectivity matrix of the network is directly encoded. This direct encoding allows one to encode any architecture having a fixed number of neurons. This gives very long chromosomes of size n^2 for a network including n neurons. Only small networks can be found, otherwise the search space becomes too large. In order to be able to search a wider space, the encoding has to be more compact. Moreover, if the weights are also encoded, GA cross-over operations on the connections can produce non-functional offsprings. This is known as the structural/functional problem [Whit90]. A solution can be to introduce some abstraction in the representation and encode in the chromosome a collection of abstract features more likely to provide building blocks.

In [Harp89] the chromosome is a concatenation of parameters that describe the number of layers, the size of the layers, and how layers are interconnected. This method provides abstraction and compactness. Big nets can be coded with small chromosomes, but only within a restrictive range of architectures. It cannot, for example, reach modular architectures with well defined and repeated groups of neurons.

Kitano's work in [Kita90] begins to implement modularity and scalability. In his method, the chromosome encodes a matrix grammar. This grammar allows one to get in $\log_2(n)$ steps a $n \times n$ matrix of characters which is then transformed into a $n \times n$ connectivity matrix. One starts with the 1 x 1 matrix which single character is the axiom of the grammar. At each time step, each character of the current matrix is rewritten into a 2 x 2 matrix of characters, so that the size of the current matrix doubles up. With a grammar encoding, the same piece of code that represents a subset of rules can be used repeatedly during the decoding phase to produce the same pattern of connections at different places in the neural net. Still, although there can be repeated patterns in the connectivity matrix, repeated subnetworks do not appear in Kitano's experimental results. A more direct property of modularity is desired (see section 13.5).

Secondly, [Kita90] characterises scalability as the ability of the GA to find big neural networks for a parameterised problem of big size. However, its goodness is tested on an atypical case: a degenerated version of the encoder/decoder problem where the number of neurons in the hidden layer is bigger than the number of input or output neurons. In the original problem, the number of hidden neurons must be the logarithm of the number of input neurons and this is the point that makes the problem challenging. In section 13.5, a theoretical rather than experimental property of scalability is proposed.

Thirdly, it is not clear that with a matrix-grammar, every architecture can be expressible, in a short code. In fact, Kitano's encoding scheme requires artificial conventions that very likely decrease the compactness of the code. A $m \times m$ matrix of bits must be developed for a network of n neurons where m is the smallest power of two bigger than n. One must extract an $n \times n$ matrix, and use only 25% of the bits in some cases. In order to get an acyclic graph for a feedforward network, one must consider only the upper right triangle. Moreover a grammar produced by the GA is not clean. It may have many rules that rewrite the same character, and rules that are not used. Only a small percentage of the coded rules are used and the efficiency is rather poor.

In this paper we present an encoding scheme called cellular encoding (CE) based on a graph grammar instead of a matrix grammar and explain why CE is more efficient than the existing encoding. Seven theoretical properties to qualify an encoding are defined, and an extended version of CE is shown to verify all of them. The CE is like a machine language for neural networks. We show that increasing the size of the instruction set allows for a growing set of properties. The seven properties show that CE is like a programming language. We present a result of simulation which clearly shows that the combination GA+CE is able to generate modular neural networks where the same neuronal group is repeated many times, as a procedure which is used many times. Because of the seven discussed properties, using the GA+CE method can be viewed as the genetic programming of neural networks. The conclusion will compare pure genetic programming (GP) and the genetic programming of neural networks.

13.2 The Cellular Encoding

A GA user has to provide a decoding algorithm from a chromosome to a solution of the optimisation problem that he or she is solving. However, for optimising neural networks, this decoding can be very long and complex [Kita90]. By analogy with nature, the corresponding algorithm is called a developmental process, as proposed in [Bele90]. The developmental process used in CE is inspired from Kitano's one, because a chromosome also encodes a rewriting grammar. Nevertheless, the interpretation and the encoding of the grammar are different.

Instead of rewriting a matrix, the grammar of CE rewrites neurons. In fact, rewriting neurons is akin to a "neural network machine language". This language can describe network architectures as well as weight patterns in a natural and clear way. During the developmental process, instead of a global rewriting of the connectivity matrix, each node of the graph is processed separately, one after the other. The final product of the rewriting is exactly the encoded neural network and there is no need for an additional process. On the contrary, Kitano's developmental process produces a raw matrix of characters, which needs further processing with some artificial conventions before yielding a neural network.

In CE, an attribute grammar is used. The grammar is not encoded as a set of rules but as a rooted, point labelled tree (or grammar tree) with ordered branches, where the labels refer to program symbols that compute the attributes of the symbols. Figure 13.1 shows that this encoding scheme can be more compact. This encoding as a tree ensures that the GA always produces deterministic grammars where all the rewriting rules are used and which yield finite networks.

Special attributes allows one to encode recursive grammars that produce infinite networks. Instead of stopping the development of the network when the size of the connectivity matrix reaches a given bound, a simple mechanism allows one to apply the grammar rules for a fixed number of loops *l*. Thus, a grammar can develop an entire family of neural networks parameterised by *l*.

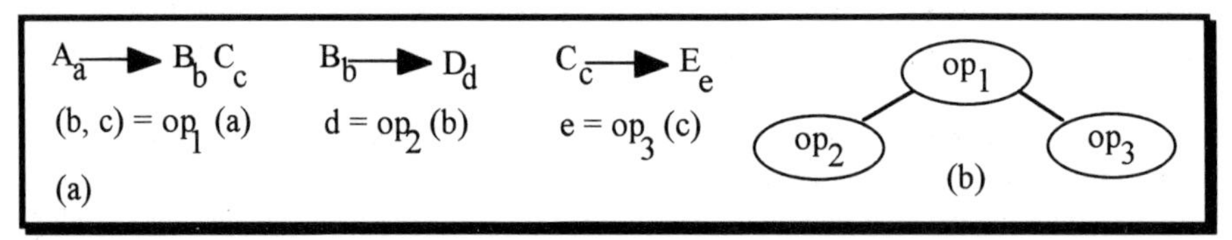

Figure 13.1 — The encoding of grammar (a) a set of rules,(b) a rooted, point labelled tree with ordered branches, upper case are symbols of the grammar. Capital "A" is the axiom, Capitals "B" and "C" are non terminal, "D", "E" are terminals. Lower case letters are attributes, op_i are program symbols that compute the attributes.

We now present the CE framework. A neuron is a node of an oriented graph with ordered connections, which is called the network graph. A neuron manages a set of input weights and some internal registers. A rewriting grammar encompasses terminal and non terminal symbols. With the formalism of CE, non terminal neurons called reading cells are neurons that have a pointer (reading head) to a node of the grammar tree. The attributes of a neuron encode the connectivity and the value of some internal registers. The program symbols that modify the attributes are called rewriting program symbols. During a rewriting step, a reading cell of the current network graph executes the rewriting program symbol that labels the node pointed to by its reading head. A rewriting program symbol has the arity of the node of the grammar tree it labels. A reading cell that executes a program symbol will behave differently depending on the arity. If the arity is 0 the cell loses its reading head and become a terminal neuron. If the arity is 1, the cell places its reading head on the unique subtree. The program symbols with arity 1 can modify the value of an input weight, and also of an internal register of the cell. If the arity is 2 the cell divides into two child cells, so increasing the size of the network graph. Each child places its reading head on one of the two subtrees. When a cell divides, the values of the internal registers and of the weights are recopied in the child cells. With the formalism of CE, the axiom of the grammar corresponds to the initial network graph. The basic initial network graph is a single non-terminal cell with no links, and a reading head pointing to the root of the grammar tree. The internal registers of this reading cell are initialised with predefined default values. At an intermediate step, some neurons of the network graph are terminal neurons, others are reading cells and are still rewriting themselves. When the rewriting process ends, all the neurons are terminal neurons, the network graph is the derived neural network.

The order in which the cells rewrite is determined as follows. Once a cell has been rewritten, it enters a First In First Out (FIFO) queue. If the cell divides, the child which reads the left subtree enters the FIFO queue first. The next cell to be rewritten is the head of the FIFO. This order of execution tries to model what would happen if the cells were rewriting themselves in parallel.

The initial graph can be completed so as to implement various features of neural networks. In order to specify a particular list of neurons, special cells called pointer cells are used. Figure 13.2(a) shows the initial network graph completed with two pointer cells.

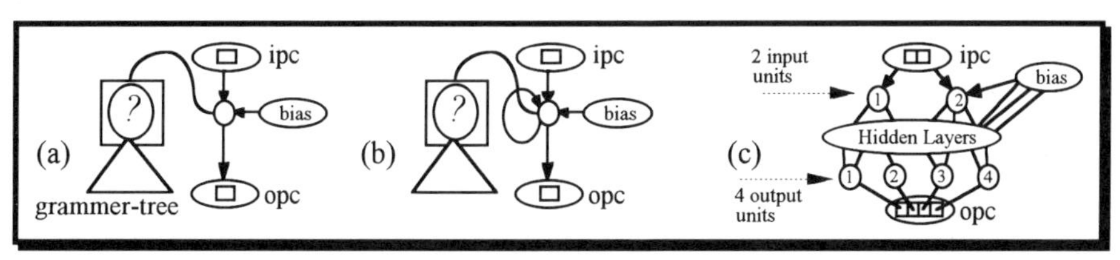

Figure 13.2 — The initial network graph (a) for a feedforward neural network, (b) for a recurrent neural network, (c) the final graph. "ipc" stands for Input Pointer Cell, and "opc" stand for Output Pointer Cell.

The input and output pointer cells are pointers that define the list of input and output units. At the end of the rewriting steps, the neighbours of the input pointer cell are the input units. The ordered output links of the input pointer cell define an order in the input units (the same for the output pointer cell). Moreover, a special cell called bias unit, with no input links and a constant activation of +1 is added in the initial graph. The threshold of a neuron is encoded as the weight of the connection from this bias unit. Last the unique reading cell of the initial network graph can have a recurrent link from its output to its input. This recurrent link is used to develop recurrent networks. The initial network graph without the recurrent link is *ACYC* and the one with a recurrent link, *CYC*. A particular CE with r rewriting program symbols is determined by a r+1 - tuple of symbols. The first symbol refers to the initial network graph and the last r symbols refer to the rewriting program symbols. In the notation, the first symbol will be slanted. In the next sections, it will be shown that increasing the number of rewriting program symbols allows one to obtain a variety of useful properties.

13.3 Architecture Encoding

In this section, rewriting program symbols that develop the architecture are studied. An architecture is a directed graph. Each vertex stands for a neuron. The architecture also includes two ordered sets of vertices. These lists specify the input and output neurons. First a minimal set of three program symbols encoding any architecture is presented.

The first program symbol is the parallel division noted **PAR**. It is a binary program symbol. In the parallel division both child cells inherit the input and output links from their parent cell. If the parent cell has a recurrent link, the two children connect to each other and to themselves. Figure 13.3 shows how the links have to be reordered once the new cell has been added.

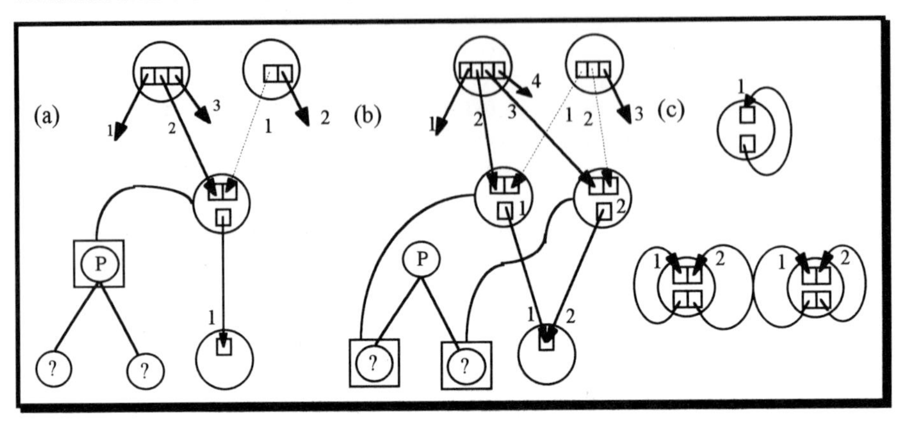

Figure 13.3 — The parallel division (a) before the execution, (b) after the execution, (c) the case of a recurrent link. Positive weights are represented by — and negative weights by - - - .

The second program symbol noted **CUT** cuts an input link. It has an integer argument which specifies the number of the input link to cut. The third program symbol noted END makes the cell lose its reading head and become a terminal neuron.

Figure 13.4 shows how to encode an arbitrary graph.

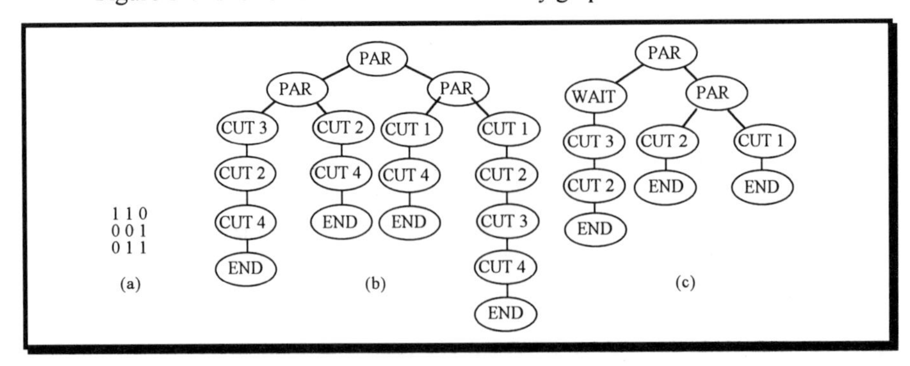

Figure 13.4 — Encoding of an arbitrary recurrent network (a) the target connectivity matrix, (b) code that uses two program symbols **PAR CUT**, (c) code that uses three program symbols **PAR CUT WAIT**.

Since all the leaves and only the leaves are labelled with **END**, the program symbol **END** will not be represented in the next figures. The development of a given graph having n nodes is done in two stages. During the first stage, a fully connected graph of m nodes is developed, where $m = 2^{[\log_2(n)]}$ and $[x]$ is the ceiling function. This is done with a binary tree of depth $[\log_2(n)]$ labelled with t **PAR**. During the second stage, each cell cuts its supernumerary input links. All the links to the supernumerary n - m cells are cut. The final graph has n nodes, connected as desired. The following property is convenient for the normalisation:

Property 1 (completeness) *An encoding scheme is complete with respect to a set of architectures (neural networks) if any element of the set can be encoded.*

The preceding construction yields:

Proposition 1 {*CYC*, **END, PAR, CUT**} *is complete with respect to the set of all architectures.*

Because of the program symbol **CUT**, the topology of the final network depends on which cell is rewritten first. In the preceding construction, the cells start to cut their input links only when the growth of the net is finished. This is why a net whose size is an exact power of two must be developed. This development can be avoided by means of an unary waiting program symbol **WAIT**, which makes the cell wait for the next rewriting step. The program symbol **WAIT** is necessary for a rewriting order to be genetically programmed by generating appropriate delays. Figure 13.4 shows how **WAIT** can be used to encode a neural network with a shorter chromosome. There is no need to create more cells than the final number of cells. Every cell starts to cut at the same time. The cells that have finished their division earlier are just delayed with the program symbol **WAIT**. Clearly, this construction makes the code shorter. Let us define the property of compactness as follows:

Property 2 (compactness) *An architecture encoding scheme $\mathcal{E}$ is said to be topologically more compact than another encoding F if every architecture encoded by F with a code of size s, can also be encoded by $\mathcal{E}$ with a code of size* O(s). *If the schemes also encode weights, $\mathcal{E}$ is said to be functionally more compact than F if for every neural network N encoded by F with a code of size s, there exists a neural network encoded by $\mathcal{E}$ with a code of size* O(s) *that implements the function associated to N.*

The size of a code is the number of bits necessary to store it. Since less than 31 rewriting program symbols will be used, 5 bits suffice to encode each program symbol plus a **NULL** character. Hence, the size of the code needed to store a tree of α nodes is $5\alpha + 1$. We want to show compactness with respect to the classical representations of graphs. First consider the direct encoding $C1$ of the connectivity matrix. Recall that it produces a code of size n^2 for a network of n neurons. Provide the reading cell with an internal register called link register that contains a link number. Define the unary program symbol **INCLR** incrementing the link register and the unary program symbol **CLIP** cutting the link pointed by the link register. As shown if Figure 13.5, each cell of the fully connected graph can select its subset of connections with the execution of less than n **CLIP** alternated with n **INCLR**.

Since the fully connected graph has n cells, the size of the total code is $O(n^2)$. The following holds:

Proposition 2 {*CYC*, **END, CLIP, INCLR, PAR, WAIT**} *is topologically more compact than C*1.

Since the program symbol **CUT** can be implemented using **CLIP**, it will not be used any more, except for the sake of clarity, in the figures. **CLIP** will be denoted as a **CUT** with no arguments.

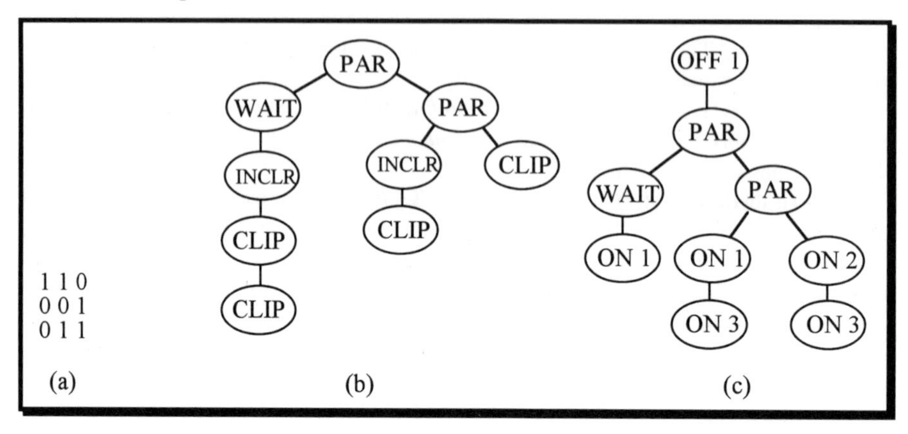

Figure 13.5 — Compact encoding of an arbitrary recurrent network (a) the target connectivity matrix, (b) compactness with respect to *C*1, (c) compactness with respect to *C*2.

Second, consider the representation *C*2 where the list of connections is encoded. The neuron numbering can be encoded in log (n) bits. For each connection, the numbers of the two connected neurons are specified. Therefore *C*2 produces an encoding of size O(l log (n)) for a network with l connections. This encoding is interesting when there are few connections. Assume a link has two states: **on/off**. When a link is duplicated during a cell division, the state of the link is also duplicated. At the end of the rewriting, the links with states **off** are removed. Define the program symbol **ON** that turns on the state of the link specified by an integer argument, and its counterpart **OFF**. Figure 13.5(c) shows how to encode an arbitrary network with the program symbols **ON** and **OFF**. Let the default value for the recurrent link in the initial graph be ***on***. The program symbol **OFF** 1 labelling the root of the grammar tree sets the state of the recurrent link to **off**. The program symbols **PAR** develop a fully connected graph where all the links are **off**. Last, each cell c builds its set of l_c connections by applying l_c program symbols **ON** on the numbers of the desired neighbours. Since each of these numbers are coded with log(n) bits, the size of the total code is O(l log (n)).The following holds:

Proposition 3 {*CYC*, **END, ON, OFF, PAR, WAIT**} *is topologically more compact than C*2.

Last, consider the scheme *C*3 that encodes layered neural networks, with complete interconnections between layers. *C*3 specifies the number of neurons in each layer and a list of pairs of integers pointing at the input layer and the output layer of each bridge. Define two other division program symbols. The sequential division noted

SEQ makes the first child cell inherit the input links of the parent cell, the second child inherit the output links, and the first child connect to the second with a link whose weight is +1 and state is on. An example is shown in Figure 13.6(a) (b).

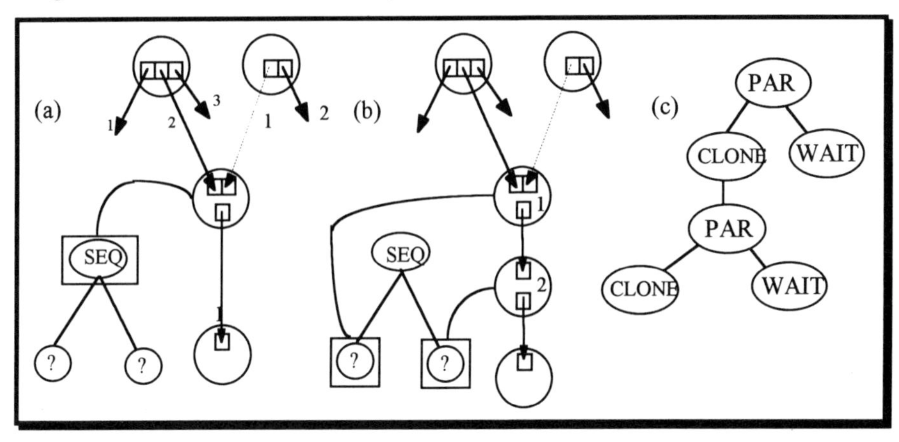

Figure 13.6 — How to efficiently encode a layered neural network (a) before the sequential division, (b) after the sequential division, (c) the code of a layer of 7 cells using the clone program symbol. Weight +1 is represented by — and weight -1 by - - - .

The clone division denoted **CLONE** is like **PAR** except that the two child cells place their reading head on the same node of the grammar tree. Since there is only one subtree to the nodes labelled with **CLONE**, this program symbol is unary. It is called a clone because the two child cells will execute exactly the same code, and therefore give birth to the same subnetwork of neurons. A given neural network with c layers is developed as follows: apply c-1 times **SEQ** to produce a string of c cells. At this stage, each cell stands for a layer. The rank of the cell in the string corresponds to the layer number. Now each cell standing for layer i must execute n_i times **PAR**, where n_i is the number of cells of the ith layer. As shown in Figure 13.6 (c), a layer of n_i neurons can be developed with a combination of less than $3\log_2(n_i)$ **PAR** and **CLONE** program symbols. The total size of the code is: $c+3\sum_{i=1}^{c}\log_2(n_i)$. It is less than three times the size of the code produced by $C3$. Thus:

Proposition 4 {*CYC*, **END, ON, OFF, PAR, SEQ, CLONE, WAIT**} *is topologically more compact than C3.*

We have extended this property to bridged layered neural networks. Let us now define another property.

Property 3 (closure) *An architecture encoding scheme is closed with respect to a set of architectures, if every possible code develops an architecture in this set.*

Whatever the rewriting program symbols are, a given CE is always closed with respect to the set of all architectures. This means that any code generated by the GA can actually develop an architecture. It ensures that whenever the GA generates a chromosome, the GA will be able to develop and evaluate a network, and receive a feed-back.

Suppose that the architecture one is looking for is known to belong to a particular set. It is possible to make the GA use this *a priori* knowledge. One must design an encoding scheme that is closed with respect to the set and the GA will automatically place its individuals in the set. For example one can require acyclic architectures in order to be able to do backpropagation. If one starts to rewrite from an acyclic network graph, it will remains acyclic, thus {*ACYC*, **END, PAR, SEQ, CUT, INCLR, WAIT**} is closed with respect to the set of acyclic architectures, but it is not complete with respect to this same set. In order to obtain completeness, the program symbol merge, denoted **MRG**, is introduced. The program symbol MRG has an integer argument i and rewrites a cell in the following manner: let l be the input link which number is l and c the neighbour cell which is connected through l. The program symbol replaces l by the list of the input links of c, as shown in Figure 13.7 (a)(b). If after this operation, c has no more output links, c is suppressed. Figure 13.7(c) shows how to build the code of an arbitrary acyclic graph.

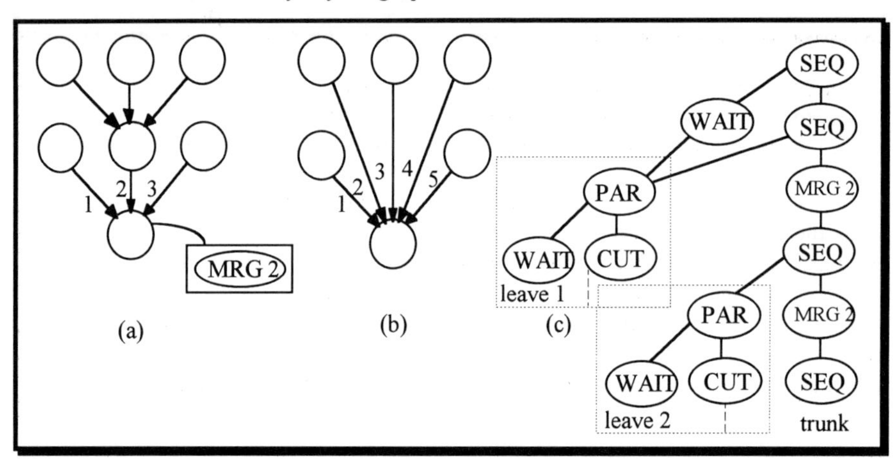

Figure 13.7 — The merge program symbol (a) before the application of the program symbol, (b) after the application, (c) the code of an arbitrary acyclic graph.

In an acyclic graph, it is possible to number the nodes so that each node is connected only to other nodes having a greater number than itself. The proposed code is made of a trunk and n leaves, where n is the number of nodes. At an intermediate step, all the cells of the current network graph are connected to the cell that reads the trunk. Periodically, the cell that reads the trunk divides, and creates a child c that reads leaves i and represents the node of number i. At this stage, all the cells that represent nodes with a number less than i are connected to c. The cell c will read on the ith leave how to prune the supernumerary input links. The corresponding cutting program symbols are located inside the dashed line in Figure 13.7(c).Thus:

Proposition 5 {*ACYC*, **END, PAR, SEQ, CUT, INCLR, MRG, WAIT**} *is closed and complete with respect to the set of acyclic architectures. Adding* **ON, OFF** and **CLONE** *yields the property of compactness with respect to* $C1$, $C2$ *and* $C3$.

Consider the bias unit as a particular input unit which remains with a constant activity of +1. This is modelled by connecting the input pointer cell to the bias unit, in the initial network graph. For each neuron, we want a path from one input unit to this neuron (the output of a neuron not having such a path would be constantly zero). One can also require that there exists a path from each neuron to a given output unit (otherwise the neuron would not be used in the computation of the neural net). An architecture that meets these two requirements is a meaningful architecture. Let us first show the closure property with respect to the set of meaningful architectures that are also acyclic. The semantic of the program symbols that prune links must be slightly modified in the following manner: a reading cell can apply one of the program symbols **CUT, OFF, MRG** on a link whose state is on only if the number of input links which state is on is strictly greater than one. Take *ACYC* as the initial network graph. It is a connected graph that has more than one node. A case analysis shows that the application of any of the rewriting program symbols already defined keeps the network graph connected. Hence the final network graph is connected. Only the input pointer cell has no input links, and the output pointer cell has no output links. From these three properties one deduces easily that the architecture is meaningful and:

Proposition 6 {*ACYC*, **END, PAR, SEQ, CUT, INCLR, MRG, WAIT**} *is closed with respect to the set of meaningful acyclic architectures.*

Now, consider the case of meaningful recurrent architectures. The construction uses the fact that a meaningful recurrent architecture contains a meaningful acyclic architecture. Take the initial network graph *CYC*. Define an order in the cells of the network graph with two rules:

- During a cell division, the child that reads the left subtree is less than the one that reads the right subtree.
- If a cell $c1$ is less than a cell $c2$, any descendant of $c1$ including $c1$ is less than any descendant of $c2$ including $c2$.

Provide each cell c with two input sites, one for the link connected to cells less than c, and one for the link connected to cells greater than or equal to c. When a cell division takes place, the two input sites have to be managed so as to respect their definition. In order to develop a meaningful recurrent architecture, one must respect the preceding restriction in the application of pruning program symbols with the input site connected to inferior cells.

With **ON, OFF** and **CLONE**, one can again derive the property of compactness with respect to $C1$, $C2$ and $C3$. The construction with {**ON, OFF**} is difficult because the new semantics does not allow to set all the links to **off**. At each step, for each cell, at least one input link must be on. Hopefully, this can be ensured without increasing the size of the code more than the desired complexity.

Lastly, if one wishes to impose the number of input units and output units, one can take the kind of initial network graph shown in Figure 13.8.

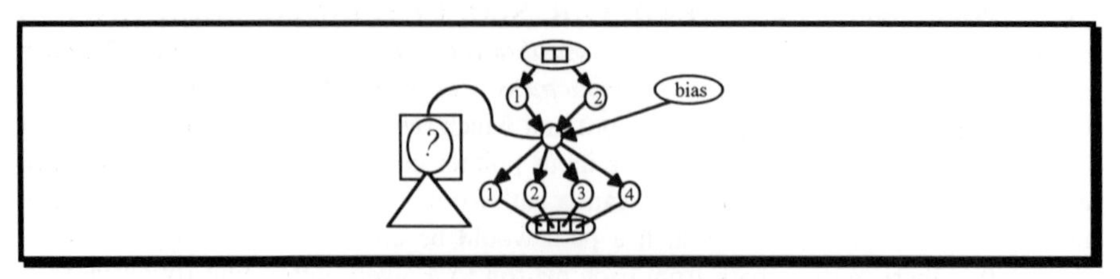

Figure 13.8 — Imposing two input units and four output units.

13.4 Neural Network Encoding

In this section we describe how to encode a neural network. The architecture, the weights, and the kind of sigmoids used by each neuron must be encoded. A simple way of encoding weights is to provide a unary program symbol **CONST** which weights the link pointed by the link register of the cell to the value specified by an argument. This argument can be ± 1, integer or real, depending on the desired neural network. Consider the encoding scheme $C1$ and $C2$ defined in the preceding section. The encoding scheme $D1$ is like $C1$ except that a list of weight values is provided together with the connectivity matrix. The same holds with $D2$ and $C2$. We have the following result:

Proposition 7 {*CYC*, **END, PAR, SEQ, CUT, INCLR, ON, OFF, MRG, WAIT, CONST**} *is complete with respect to the set of all neural networks, where the units have the same sigmoid. It is functionally more compact than D1 and D2.*

To show these properties, a two-stage development is used. First, the architecture is developed, then the weights are set. From a GA point of view, an encoding using the **CONST** program symbol presents some drawbacks. The GA forms its initial population with random individuals. As described by Koza in [Koza92], when a program symbol like **CONST** needs an argument, a random value is generated in a specified range. If there is no mutation, which is the case in Koza's scheme, the possible value of the argument in the future generations are limited to the set of values represented in the initial population. In the case of the program symbol **CONST**, it means that the possible values of the weights are restricted to this finite set. If a particular weight is needed, and is not available in the current population, the GA will not be able to provide it. Moreover, with the program symbol **CONST**, there is no possibility to find values with a progressive refinement, through the formation of building blocks of increasing size. A solution to the problem is to use two other program symbols: **MULT** which multiplies the weight by its argument, and **ADD** which adds its argument to the weight. It is possible to go further in this direction by using a set of four program symbols: **INC, DEC, MULT2** and **DIV2**, which respectively increments, decrements, multiplies by two and divides by two the weight of the input link pointed by the link register. These program symbols are interesting because they have no arguments. If we use only this set without **CONST**, then the formation of weight values will be done by a subtle combination of these four program symbols. The precision to which weights are coded is not fixed *a priori*. If the encoding of a number takes b bits with a fixed point format, an encoding of this number can be built with $O(b)$ program symbols chosen in {**INC, DEC, MULT2, DIV2**}. This allows to have the same property of compactness with respect to $D1$ and $D2$ if the weights are coded in a fixed point format. A similar technique can be used to remove the argument of other program symbols. For example, if this argument is a link number, the program symbol can use the value of the link register. In order to keep the property of compactness, one needs the program symbol **MULT2LR** which multiplies by 2 the value of the link register, and set the value to 0 if it is greater than the number of input links. For a purpose of clarity, in the next figures, program symbols will still use an integer argument.

A neural network is not entirely determined by a weighted architecture. Each neuron may have features on its own, such as a different sigmoid, or a different learning procedure to adjust its weights. To encode this, an internal register of the cell is attributed to each of these features, and program symbols that modify the value of these registers are introduced. For example, assume that there is a choice between the two sigmoids plotted in Figure 13.9.

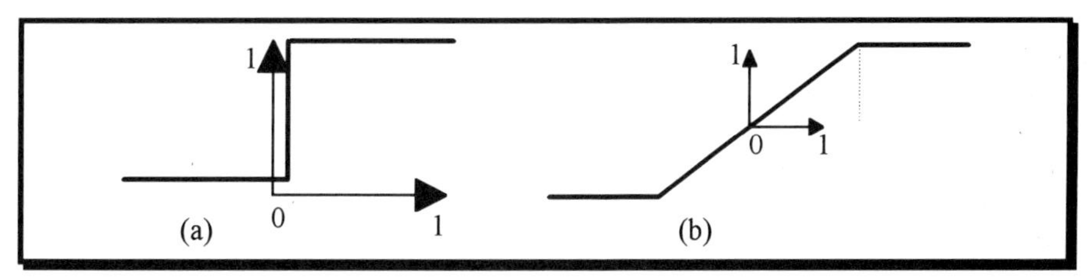

Figure 13.9 — Two sigmoids (a) the sign function, (b) the bounded linear function. The constant *max* is a parameter of the model.

Let a sigmo register store the type of sigmoid. A program symbol **SIGN** sets the value of the sigmo register to **sign**, a program symbol **LIN** sets the value to **lin**. When the cell becomes a terminal neuron, its sigmo register contains the name of its sigmoid.

13.5 Properties of a Programming Language

In this section, we will show the most interesting properties of cellular encoding, that makes it appear like a programming language. We will need to build the encoding of a neural network that simulates a Turing machine. As a first step, we want to achieve the following property.

Property 4 (modularity) *Consider a network N_1 that includes at many different places, a copy of the same subnetwork N_2. The code of N_1 is* ***modular*** *if it includes the code of N_2, a single time.*

This property tries to formalise the intuitive idea of modularity. Modularity should help to reduce the apparent complexity of an initial network by decomposing it into a set of less complex subnetworks connected by a simple structure. The process is repetitive: subnetworks may themselves be decomposed. Modularity is connected with the reusability problem: the aim is to transform the neural network design into a construction box activity, whereby networks would be built by a combination of existing standard elements. Since each module can be separately understood by a human reader, the architectures produced are easily understandable.

In order to reference a subnetwork N_2, one must provide a program symbol that makes the reading head of the cell jump on the subtree defining N_2. We use a method advocated by Ray in [Ray91] called "Addressing by Template", inspired by molecular biology. Ray writes:

> In most machine codes, when the IP jumps to another piece of code, the exact numeric address of the target code is specified in the machine code. Consider that in the biological system, by contrast, in order for protein molecule *A* to interact with protein molecule *B*, it does not specify the exact co-ordinate where *B* is located. Instead molecule *A* presents a template on its surface which is complementary to some surface on *B*. Diffusion brings the two together, and the complementary conformations allow them to interact.

Addressing by template is illustrated by the unary program symbol **JUMP**. Each **JUMP** program symbol is followed by a sequence of **NOP** (no operation) program symbols, of which there are two kinds: **NOP_0** and **NOP_1**. Suppose we have a piece of code with four instructions in this order: **JUMP(NOP_0(NOP_0(NOP_1(..))))**. The cell reading the **JUMP** will search in the tree the nearest occurrence of the complementary pattern: **NOP_1(NOP_1(NOP_0(..)**. The nearest is defined from the natural topology on a graph where the distance between two vertices is the length of the shortest path that connects them. If the pattern is not found, the **JUMP** program symbol is ignored. If the pattern is found, the cell positions it's reading head to the end of the complementary pattern. The branching can loop. In order to avoid the infinite growth of the network the cell is provided with a life register. The life register of the reading cell in the initial graph is initialised with a value l. If the complementary pattern is found, the **JUMP** program symbol decrements the life register and removes the reading head only if the life register is positive. With this algorithm, the number of loops is bounded by l. A code develops a family of neural networks parameterised by l. The size of the network increases with l.

In order to fulfil the property of modularity, one must provide the possibility to encode separately the N_2 subnetwork from the main network N_1. For this purpose a binary program symbol **DEF** is defined. When a cell executes **DEF** it places its reading head on the right subtree. In order to be used, the left subtree must contains some labels of **NOP** program symbols. The left subtree encodes a subnetwork whereas the right subtree corresponds to the main network.

In order to be able to connect an included subnetwork to the rest of the network in a functional way, another program symbol of division called splitting division, and denoted **SPLIT** is required. A cell executes the **SPLIT** division as follows:

First, assume that the number of input links is equal to the number of output links. The cell splits into n children, where n is the number of links. The inheritance of the links is shown in Figure 13.10. The child cell with index i inherits the input link and the output link numbered *i*. Each child cell places its reading head on the same node of the grammar tree.

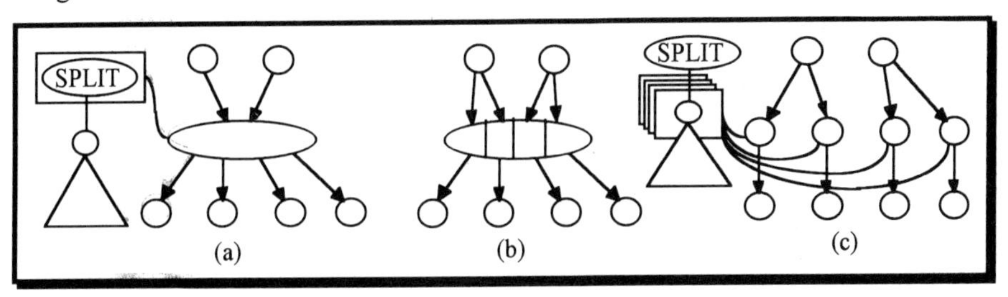

Figure 13.10 — The splitting division (a) before the division, (b) duplication of links in the input sites, in order to make the total number of links equal to four, (c) creation of four neurons.

If now there are less input links than output links, some of the input links are duplicated in an homogeneous manner to make equal the number of input and output links, and apply the splitting. If the cell has less output links than input links, the output links are likewise duplicated.

The CE can be used to design a neural network that does a particular mapping. For example, Figure 13.11 shows a neural network that simulates a Turing machine. The design method is to build networks of higher complexity by specifying how to connect together subnetworks of smaller complexity. The representation scheme uses this decomposition. Each subnetwork of neurons is named and represented separately. To represent a subnetwork *h*, instead of using the subtree *t* that encodes *h*, one represents a graph of cells which is obtained in the following manner. Take as an initial network graph, a reading cell whose reading head is placed on the root of *t*, and whose neighbours are a set of labelled cells. The labelled cells represents the neighbours of the cells that will include *h*, during the development of the complete neural network. Make this initial network graph develop, with two conventions.

- Whenever a cell reads a **JUMP** program symbol, freeze the development of the cell, introduce a new subnetwork *g* whose code is the subtree of the node labelled by **JUMP**, and write the name of the included subnetwork, followed by the name of *g* in the circle representing the frozen cell.
- Whenever a cell reads a **SPLIT** program symbol, freeze the development, introduce a new subnetwork *g* whose code is the subtree of the node labelled by **SPLIT**, write **SPLIT** followed by the name of *g* in the circle representing the frozen cell.

If the subtree of the node labelled by **JUMP** or **SPLIT** is only one node **END**, it is not necessary to introduce a new "virtual" subnetwork.

In the final network graph, the labelled cells indicate how the subnetwork *h* is to be connected to the rest of the network. Recall that the final network depends on the order of execution. In order to finish the representation, the order of execution of the frozen cells must be specified if needed. In the case of the Turing machine, it must be specified that the cells frozen on a **SPLIT** program symbol have to execute this program symbol only when all the neighbours are neurons. Thus, in the final encoding they must be preceded by a certain number of **WAIT** program symbols so as to be correctly delayed. Note that this number is always a linear function of l, the initial value of the life register. These **WAIT** program symbols can be encoded by means of a subnetwork which includes itself and does nothing but wait.

Figure 13.11(a) shows that the neural net is recurrent and has two layers: a processing layer and a mixing layer.

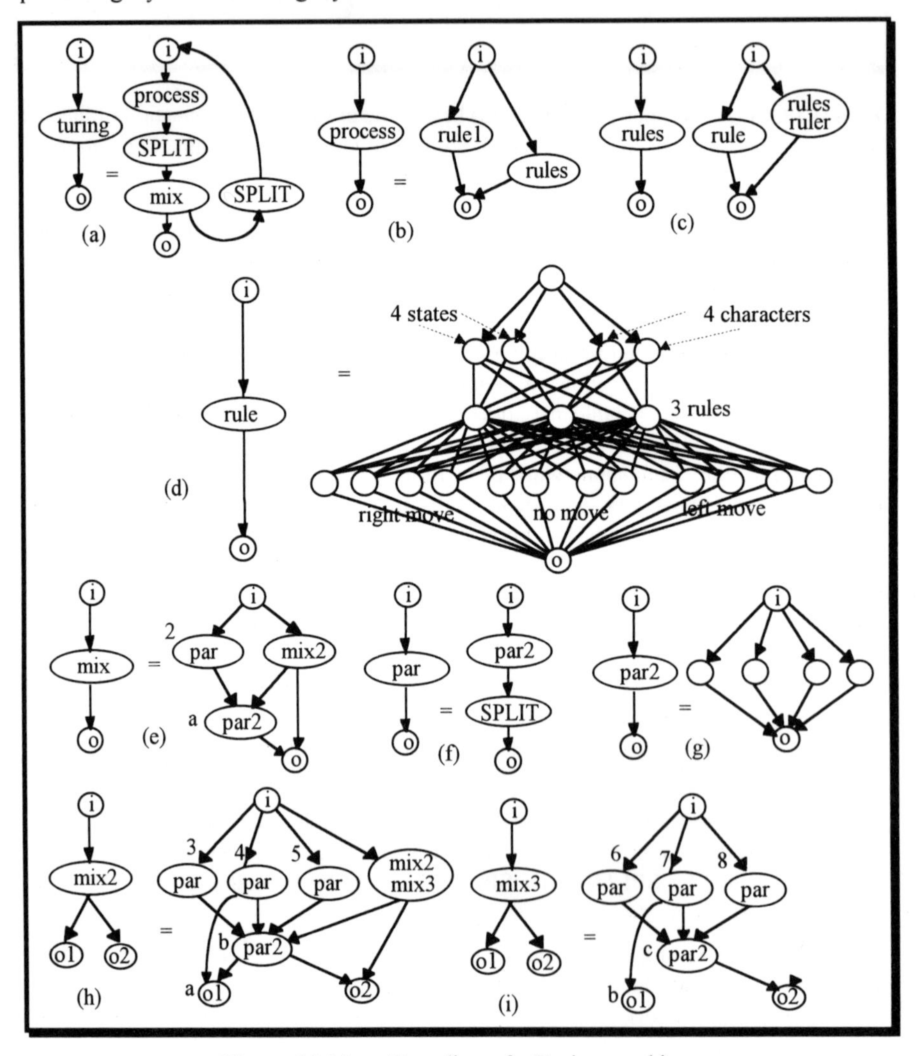

Figure 13.11 — Encoding of a Turing machine.

The layers connect to each other in a one-to-one manner, thanks to two **SPLIT** inserted between the layers. Each time the data flows through the neural net, a transition of the Turing machine is simulated. Figure 13.11(b) and (c) shows that the processing layer applies in parallel l subnetworks of neurons called **rule** to a tape of l registers. Because of a slight difference, the leftmost **rule** is called **rulel**, and the rightmost, **ruler**. The subnetwork **rule** encodes all the transition rules of the Turing machine. Each group **rule** corresponds to a register of the Turing machine's tape. For each register, a character and a state are encoded. If the reading head is positioned on the register, the state encoded is the state of the Turing machine, otherwise the state is 0. On the input of the ith **rule** is applied the binary code of the state and the character of the corresponding register. In the example, two bits are used for the state, and also two bits for the character. Hence if there are q states and an alphabet of c characters, the number of input units of **rule** is $i = [\log_2(q)]+[\log_2(c)]$ The number of output units of **rule** is 3 * i, three rows of i cells. One row represents each of the possible three moves of the reading head of the Turing machine. The subnetworks **ruler** and **rulel** have only 2 * i units. Since they handle a register on the extremity of the tape, only two moves are possible. If the reading head of the Turing machine is not on a register, the state is 0, its code is the concatenation of $[\log_2(q)]$ 0, and the output of **rule** is 0 everywhere except for the $[\log_2(c)]$ units of the middle row whose activities are the same as the $[\log_2(c)]$ input units. This accounts for the fact that the character on the tape remains unchanged. If the reading head is on the register, it moves in one direction: left, standby or right. The corresponding row of output neurons will indicate the next state, and the character to write. The two other rows will output zero values. Each hidden unit of **rule** implements one **rule**.

The mixing layer must compute the next state and character of the tape registers, from the total output of the process layer. Therefore, $3il$ activities must be recombined into il activities. Figure 13.12(a) shows how this must be done. Figure 13.11(e),(f),(g),(h) and (i) show how this is done. Each of the il activities can be computed with the logical **OR** applied to three of the $3il$ activities.

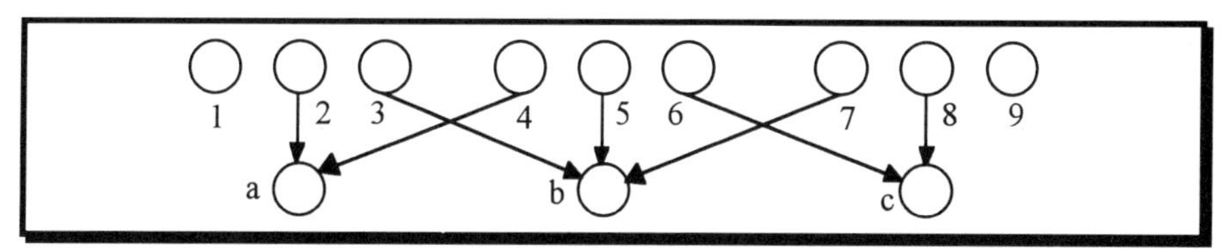

Figure 13.12 — The action of the mixing layer.

Now, let us define precisely what is meant by simulating a program on a machine that can have an arbitrary big memory with a parameterised neural net of finite size. First, define the possible functions implemented by a neural net as follows:

Definition 1 *A neural net N implements a function* $f\!:\{0,1\}^n \rightarrow \{0,1]^n$ *if it has more than n input and p output units, and, if a given vector v of p bits is clamped to the first input units of N, all the other activities are initialised with 0, and N is relaxed* *with a synchronous dynamic, after stabilisation, the activities of the first output units of N will be f(v).*

Now, consider a machine $M(l)$ with a memory of l bits and a program P that makes computations on bit strings. Run on some particular input bit strings, the program P will have enough memory, and will stop after some time, yielding a binary vector. By definition, these particular input bit strings form the input set of the function computed by P on $M(l)$, and the output of this function is given by P.

Definition 2 *A neural network $N(l)$ parameterised by the integer l, simulates a program P on a given type of machine $M(l)$ with a memory of l bits, if $N(l)$ implements the function computed by P on $M(l)$.*

Property 5 (power of expression) *A parameterised neural network encoding has a power of expression greater than a programming language if for every program P written with this language, there exists a code of a parameterised neural net that simulates the program P. Moreover, the size of the code is $O(s(P))$, where $s(P)$ is the number of bits needed to store the source code of P.*

In the case of the Turing machine, a program is just a list of transition rules. The preceding construction can produce a code of a parameterised neural network that simulates an arbitrary Turing machine program if the following adjustment is made: the input units of the processing layer that encode the state must not be connected to the input pointer cell. Their initial activity must be zero, except for the register of the tape which will initially contain the reading head of the Turing machine. Initially the state encoded in this register must be the initial state of the Turing machine q_0. Let us now evaluate the size of the codes. The number of bits needed to encode r transition rules is $s = (\log_2(q)) + (\log_2(c))r$. The size of the code of **par2** is $O(\log_2(q)) + (\log_2(c))$. The size of the code of respectively **rule, ruler, rulel** is $O(s)$. The size of the codes of the other subnetworks are constant. Hence the total size is $O(s)$.

Proposition 8 {*CYC*, **END, PAR, SEQ, SPLIT, CUT, INCLR, MRG, WAIT, CONST, NOP_0, NOP_1, JUMP, DEF**} *has a power of expression greater than the "Turing machine language".*

From the fact that CE has a power of expression greater than the Turing machine results in a sort of maximal property of compactness:

Proposition 9 {*CYC*, **END, PAR, SEQ, SPLIT, CUT, INCLR, MRG, WAIT, CONST, NOP_0, NOP_1, JUMP, DEF**} *is functionally more compact than any other neural network encoding.*

In order to prove this, consider an arbitrary neural network encoding scheme E. Let e be the code of a particular neural network, encoded with this scheme. Call $s(e)$ the size of e. There exists a program with input e and i that builds the neural network encoded by e, runs the network on the input i, and returns the output of the network. This program can be executed by a Turing machine T. If the initial tape of T is the concatenation of e and i, the output of T will be the output of the neural network. Let N be the neural network that simulates T. Let $s(e)$ be the size of e. A string of $s(e)$ neurons must be added to N so that a string of $s(e)$ registers of the tape is automatically initialised with e. Figure 13.13(a) shows what the final neural network looks like. It depicts a neural network that simulates a Turing machine that develops a neural network encoded by e=$\overline{10}$, and simulates it. The simulated neural net has two input units. For simplification only the neurons corresponding to the characters of the tape are represented. The encoding of the final neural network must specify the string of $s(e)$ neurons, therefore, it is a bit more complex than the encoding in Figure 13.11.

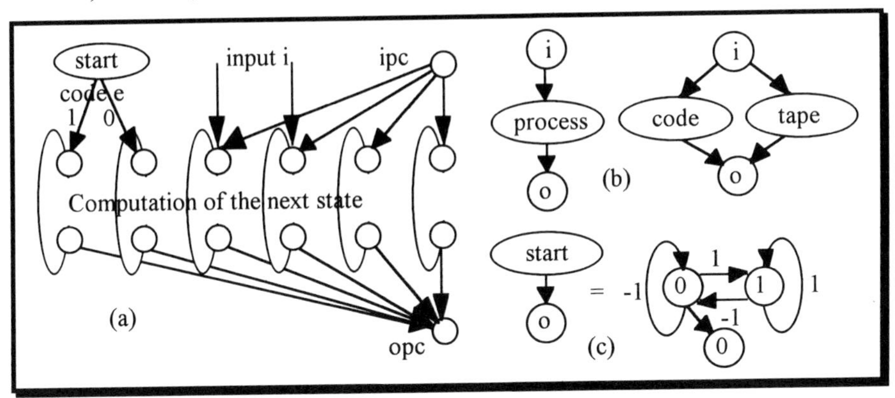

Figure 13.13 — A neural network that simulates a Turing machine that encodes a neural network.(a) outline of the neural network, (b) modification of the process layer, (c) the subnetwork **start.**

Figure 13.13 (b) shows the first part of the code of the process layer. The subnetwork code will develop $s(e)$ subnetwork of the type **rule.** The ith subnetwork is not connected to the input pointer cell. It is connected to the subnetwork **start** through an intermediate neuron with a weight 0 or 1, depending on the value of the ith bit of e. The role of the subnetwork **start** is to send a one and then only zero values. The subnetwork **tape** develops a tape of l registers. The mixing layer must be slightly modified in order to mix $l+s(e)$ registers instead of only l. The size of the total encoding is a linear function of $s(e)$. The total size is therefore $O(s(e))$. A close analysis shows that the slope of the linear function does not depend on T, and therefore, not on E either.

Another interesting property is scalability. Consider a problem of variable size, and a parameterised neural network encoding scheme. Suppose that the same code can be used to solve a problem of size one and problems of big sizes, using a different parameter. If the GA can find a code for a network that solves the problem of small size, it will not be harder for the GA to find a network for the problem of size 1,000,000, as the corresponding code is the same.

Property 6 (scalability) *Let (f_l) be a family of functions on a finite domain, where l is an integer index. A parameterised encoding is scalable with respect to (f_l) if there exists a code such that for any l the neural net encoded with a parameter l computes the function f_l.*

Consider the following definition:

Definition 3 A family of functions f_n on a finite domain is said to be recursive if there exists a Turing machine T that computes $f_n(x)$ if n and x are initially written on its input tape. If f_n is recursive, $TM_{space}(f_n)$ is the minimal number of registers needed by T to compute every value of f_n.

The following holds:

Proposition 10 {*CYC*, **END, PAR, SEQ, SPLIT, CUT, INCLR, MRG, WAIT, CONST, NOP_0, NOP_1, JUMP, DEF**} *is scalable with respect to the set of recursive family of functions with* $TM_{space}(f_n)$ *smaller than* k^n *for a certain k.*

In order to prove this proposition, we have to provide an encoding of an arbitrary Turing machine, such that the network developed with a size parameter l simulates the Turing machine with a finite tape of k^l registers, instead of only l registers as it has already been done. The encoding is similar to the one already proposed. As it is only a technical matter, it is not presented here. The family of languages that can be recognised by a Turing machine using less than k^n registers for a certain k is called PSPACE. PSPACE is so big that it includes almost all known languages [Ullm79]. Therefore, CE is scalable to all "everyday" families of functions. A cellular automaton can also be simulated with a construction similar to the case of the Turing machine. This time, the mixing layer must precede the processing layer, and its task is to reproduce at a location i, the concatenation of the states of the automata neighbours to the ith automaton.

Proposition 11 {*CYC*, **END, PAR, SEQ, SPLIT, CUT, INCLR, MRG, WAIT, CONST, NOP_0, NOP_1, JUMP, DEF**} *has a power of expression greater than cellular automata.*

Turing machine and cellular automata are universal computing devices. Cellular automata are particularly powerful. In general, a given algorithm can be efficiently programmed on a cellular automaton. It is possible to transform a set of transition rules of a cellular automaton into the CE code of a neural net in an efficient way. We can consider even more efficient computing devices than cellular automata. The following last property makes a step forward, and completes the list of seven properties.

Property 7 (abstraction) *An encoding is abstract if it has a power of expression greater than a high-level programming language like Pascal.*

Using an abstract encoding scheme allows us to reduce the problem of genetically finding a neural network to the problem of genetically finding a program that computes the target mapping. When there exists a small program that computes the target mapping, the space of program to search is smaller than the space of neural networks of bounded size, that contains some solution network. So an encoding that is abstract is likely to work better with the GA, if the mapping to learn can be computed with a short program. Since property of abstraction is hard to demonstrate, the proof is not reported here. In fact, the proof is a program, it is the compiler JaNNeT (Just an Automatic Neural Network Translator) that transforms a Pascal program into a CE which length is proportional to the length of the Pascal program. JaNNeT is described in [Grua93a].

13.6 Genetic Programming of Neural Networks

Koza has shown that the tree data structure is an efficient encoding which can be effectively recombined by genetic program symbols. *Genetic programming* has been defined by Koza as a way of evolving computer programs with a GA [Koza92]. In Koza's Genetic Programming paradigm the individuals in the population are LISP *s*-expressions which can be depicted graphically as rooted, point-labelled trees with ordered branches. Since a cellular code has exactly the same structure, the same approach is used for generating and recombining cellular code. The set of alleles is the set of cell program-symbols that label the tree, the search space is the hyperspace of all possible labelled trees. During crossover, a subtree is cut from one parent tree and replaces a subtree from the other parent tree; the result is an offspring tree. The subtrees which are exchanged during recombination are randomly selected.

Objective:	Find the cellular encoding of a neural network family that computes and the symmetry of an arbitrary large number of bits
Terminal set:	**END**
Function set:	**JUMP, BLIFE, PAR, SEQ, CUT, WAIT, INCLR, DECLR, & INCBIAIS, DECBIAIS, CONST +1,CONST -1**
Fitness cases:	For each developed neural net: If the neural net have less than six inputs and the whole input space is tested otherwise 32 symetric and 32 non symetric random patterns of bits are tested.
Hits:	Fitness cases for which the neural net produces the right output
Raw fitness of one neural net:	The number of hits of the neural net.
Standardised fitness of one neural net:	Raw fitness divided by the total number of fitness cases. It ranges between 0 and 1
Standardised fitness	The sum of standardised fitness of the developed neural net. It ranges between 0 and 10
Parameters:	M = 1,024, T = 300 s, t_m = 0.005, l =30
Success predicate:	The standardised fitness equals 10. That is the first 10 networks are developed and score the maximum number of hits

Table 13.1 — Tableau for genetic search of a neural network for the even parity problem up to 11 inputs, using cellular encoding.

There exists two models of parallel GAs (PGA). In the massive parallel model, individuals are located on a 2-D grid, and recombination is done between individuals which are near each other, with respect to the 2-D topology. In the island model, individuals are distributed in sub-population. Each sub-population is on an island, and islands exchange some of their individuals from time to time. Our GA is the Mixed Parallel GA described in [Grua93c]. The massive parallel model is simulated on each processor, whereas the processors between themselves implement an island model of PGA. Hence advantages of both the massive parallel model and the island model are combined. The mixed PGA runs on four processors of an IPSC860 parallel machine.

Müehlenbein [Mueh91] promotes the use of individual hill-climbing. Each individual generated by the GA should try to enhance itself with hill-climbing. Hill-climbing is done by exploring the neighbourhood of the individual, with respect to a natural topology given by the problem. In the case of neural network optimisation, a learning procedure can be used as a hill-climber. The learning procedure defined

[Grua93b] was implemented. This learning is a crossover between backpropagation and the bucket brigade used in classifier systems. It learns boolean weights. The learning is applied for each cellular code produced by the GA, on the first neural network of the family. The learned information is then coded back on the cellular code and can be used to develop the other networks of the family. In [Grua93d] experimental results suggest that it is the best combination between GA and learning on the phenotype. Table 13.1 summarises the key features of the problem for evolving a cellular encoding that generates a family of neural networks for computing the symmetry function of arbitrary large size. New program-symbols are used, their effect is as follows. We use a **JUMP** program-symbol without labels, the cell always places its reading head on the root of the grammar tree. **BLIFE** tests the value of the life, if it is one the cell goes to read the left subtree, else it goes to read the right subtree. **DECLR** decrements the link register. We did not encode the bias as a weight to a bias unit, but as a separate register of the cell, and uses special program-symbols **INCBIAS** and **DECBIAS** to set its value. In fact we develop the 10 first neural networks. So the solutions are tested for the symmetry problem with up to 2*10=20 inputs. Nevertheless, the analysis of the grammar shows that it generates solutions to symmetry of arbitrary large size, because the recurrence has been learned.

Not all the 10 neural networks are developed in the early generation, this saves time. The following rules decide when to continue the development of other networks: If $N_1, N_2, \ldots, N_L$ have already been developed, then, continue to develop N_{L+1} if N_L correctly processes more than a fraction r of the training set, where r is a given threshold. If L > L_{max} stop the development. In the present experiment, $r = 0.7$ and $L_{max} = 10$.

We used the genetic programming algorithm but with some differences from Koza. We do not stop the GA after a given number of generations. Instead we stop when elapsed time is greater than a variable T (five minutes here). We use a mutation rate $t_m = 0.005$, each allele has a probability t_m to be mutated into another allele of the same arity. We found that mutation consistently improves the genetic search. It may be because all the leaves have the same label **END**, and Koza's argument that crossover on leaves mimics mutation is not valid. Koza considers the depth of the individuals in the initial population. We take into account only the number of nodes and generate an initial population of trees having exactly the same number of nodes which is 30 here. After, when exchange of subtree is done during crossover, the size of the tree can increase (not above 50 nodes) or decrease. Here, we did things differently from normal GP only because at the time of implementation we where not aware of the details of Koza's GP.

The average time over 100 experiment for finding a solution is 62 seconds. An example of cellular code found by the GP algorithm is SEQ (SEQ (PAR (END) (BLIFE (PAR (JUMP (END)) (END)) (END))) (PAR (END) (INCBIAS (INCBIAS (BLIFE (CONST -1 (END)) (JUMP (END))))))) (WAIT (WAIT (WAIT (CONST -1 (END)))))), The corresponding 10th neural network that solves the symmetry of 20 inputs is shown in Figure 13.14. It is clearly a combination of 20 times the same neuronal group. To our knowledge, existing paradigms for neural networks do not provide a way to reuse the set of weights discovered in one part of the network in other part of the network were a similar sub-task must be performed on a different set of inputs. The results of this simulation clearly shows that GA+CE provide a way to reuse not only a set of weights, but a complete neuronal group.

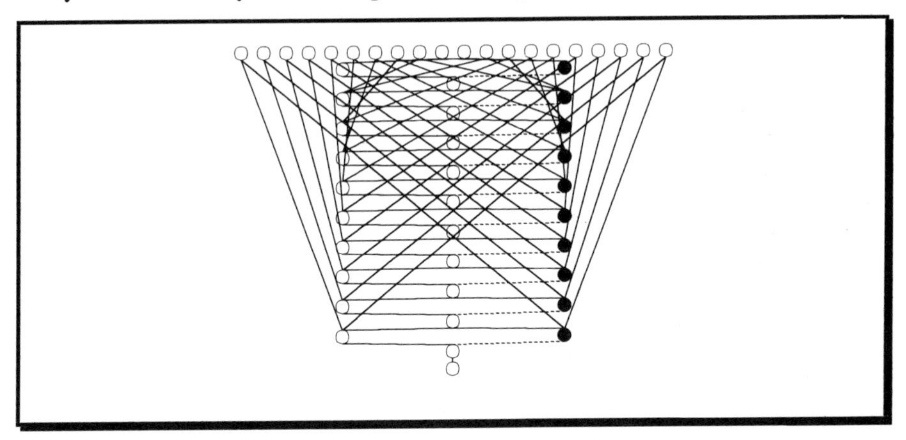

Figure 13.14 — A neural network for the symmetry problem of 20 bits. When the threshold is 1, the neuron are black disks, otherwise the empty circle indicates a threshold 0. Plain lines represent a weight 1, dashed lines represent a threshold -1.

The method can be applied to generate neural networks other than boolean neural networks. It is not a problem to encode integer weights and thresholds, or even real numbers. Cellular encoding is efficient if the function to be computed by the neural network has regularities. Cellular encoding is able to encode regularities in a compact way, but regularity is everywhere in nature.

13.7 Conclusion

When applying genetic algorithms to neural network synthesis, a key consideration is the choice of encoding scheme. We have proposed seven theoretical properties that rate the efficiency of a neural network encoding. A neural network encoding which has all the desired properties has been studied. This encoding, called cellular encoding(CE) uses an alphabet of 20 alleles. Each allele corresponds to a program symbol that slightly modifies the topology and labels of a graph describing a network. A given combination of such program symbols develops an initial graph of a few cells, into a neural network. The program symbols are: **END, PAR, CLONE, SEQ, SPLIT, MRG, CUT, ON, OFF, INC, DEC, MULT2, DIV2, INCLR, MULT2LR, WAIT, NOP_0, NOP_1, JUMP, DEF**.

The properties of CE can be summarised as follows:

Completeness Any neural network can be encoded, thus the GA can reach the solution neural network if it exists at all.

Compactness The CE is topologically more compact than classical representations of neural networks, and functionally more compact than any other representation. This ensures that the codes manipulated by the GA are minimal in size.

The shorter the codes are, the shorter the search space is, and the less effort the GA has to do.

Closure The CE always develops meaningful architectures. It produces acyclic or recurrent architectures, depending on the initial graph. Both of these closure properties ensure that the cellular code produced by the GA always give interesting neural networks. The GA does not throw away any codes. This increases efficiency.

Modularity If a network can be decomposed into subnetworks, the code of network is the concatenation of the codes of subnetworks and a code that describes how to connect the subnetworks. This facilitates the formation of building blocks. The resulting regularities in the final architecture make it clear and interpretable.

Scalability A fixed size code encodes a family of neural networks. For any reasonable family of problems, there exists a code such that the ith neural network solves the ith problem. A parameterised problem of arbitrary size (e.g. 1,000,000) can be as easy to solve as the same problem of size 2 or 3 because the code of the corresponding neural net is the same, only the size parameter changes.

Power of expression The CE can be seen as a "neural network machine language". Its power of expression is greater than Turing machines and cellular automata. It is another proof of the extreme compactness of the coding.

Abstraction It is possible to compile a cellular automaton program into the code of a neural network that simulates the computation of the program. The wide search space of neural networks is transformed into a smaller search space of programs. It helps the GA to find neural networks for problems that can be solved with a short program.

These seven properties shows that cellular encoding is like a language for describing neural networks. First, a programming language must be able to describe algorithms (power of expression, abstraction). One wants to be able to program any algorithm (completeness). A program with the correct syntax can always be run (closure). There must be enough constructs so that on can describe your algorithm in a concise way (compactness). One can define procedures that can be reused many times afterwards (modularity). Loops or recursive calls allows to solve problem of parameterised size (scalability). So these seven properties shows that the genetic search of the space of cellular encoding is really a genetic programming approach where genetic representation uses the same tree data structure. Therefore we call the combination GA+CE genetic programming of neural networks.

We now compare pure genetic programming and genetic programming of neural networks. Koza has successfully used LISP-like language as a genetic language in a broad range of problems [Koza92]. However, his impressive results are mainly from problems that involve manipulation of symbols. We do not mean mathematical symbols, but entities with a high-level meaning. This success may be due to the fact that LISP is a language adapted to symbolic manipulation. On the other hand, when Koza and Rice [Koza92] attempted to search the space of neural nets using LISP, the results are not impressive.

The combination of genetic programming with neural network using Cellular Encoding is totally different from the one proposed by Koza and Rice. In cellular encoding, it is not a program that simulates a neural network, but a neural network that behaves in a program-like manner, since a program can be compiled towards a neural net. This new type of combination, genetic programming of neural networks allows to combine a symbolic search led by the GA, with a connectionist search obtained through learning.

From the GPII book of John Koza [Koza94] it seems that the most interesting point in GP is to provide a method for automatic problem decomposition. Koza uses a list of LISP S-expression were each S-expression encodes a function and functions call each other in a hierarchical way. With cellular encoding, a subnetwork is encoded in a subtree and included using the JUMP program symbol. Subnetworks includes themselves in a hierarchical manner. Koza writes:

> For a variety of different problem, from a variety of different fields, automatic function definition enables genetic programming to automatically and dynamically decompose a problem into subproblems, to automatically and dynamically discover a solution to the subproblems, to automatically and dynamically discover a way to assemble the solution to the subproblems into a solution to the overall problem.

We write : For a variety of different boolean problem, cellular encoding enables genetic programming to automatically and dynamically decompose a problem into subproblems, to automatically and dynamically discover sub-neural networks that computes the subproblems, to automatically and dynamically discover a way to assemble the sub-neural networks into neural networks that compute the overall problem. In this paper we have reported the symmetry, but we have also solve the decoder and the parity.

In order to use automatic function definition, Koza must define it before the run of the GA: the number of functions that will be used, the number of arguments of each of these functions, the allele set for each functions. Cellular encoding does not need any of these. In GP, the parameters are explicitly passed to a function so the number of arguments must be known in advance. Whereas with CE, the parameters are passed implicitly, by the neurons connecting the cell that will start to develop the sub-neural network to be included. We do not need to define a number of arguments for a subnetwork. We are not obliged either to define a particular set of alleles for the cellular code of each subnetworks, although we can do it to help the GA.

The general conclusion is that LISP is a better genetic language to do symbol manipulations, and cellular encoding is more adapted for neural network synthesis. We believe that neural network synthesis is a more fundamental problem than symbol manipulation. There is a gap between symbols and the real world which is noisy, fuzzy and very large. If the genetic search is made at the level of symbol, it cannot attempt to solve the so-called “symbol grounding problem” [Harn90]. Genetic search with cellular encoding searches at a lower-level of complexities and eventually to define structures like retina or cortical columns. Such structures are more in touch with the real world and not directly connected with symbols. A concrete example is the neural network found by Beer and Gallagher that solves a six leg locomotion problem [Beer93]. They encoded a large network with a few bits, using symmetries. Here the structure is represented by the symmetries, and symmetries can be captured by a cellular encoding, using a modular code. If our final aim is to breed complex cognitive structures rather than symbolic problem solving, then we believe it is better to use low-level languages such as cellular encoding rather than high-level languages like LISP.

References

[Beer93] Beer, R. and Gallagher, J., "Evolving dynamical neural networks for adaptive behaviour", *Adaptive Behavior* 1:92-122, 1993.

[Bele90] Belew, R.K., McInerney, J., and Schraudolf, N., "Evolving networks, using the genetic algorithm with connectionist learning", Technical Report CSE-CS-90-174, UCSD, 1990.

[Fahl90] Fahlmann, S. and Lebiere, C., "The cascade-correlation learning architecture." CMU-cs-90-100, CMU, 1990.

[Gold89] Goldberg, D.E., "Genetic algorithms and walsh functions: Part i, a gentle introduction." *Complex Systems*, 3:129-152, 1989.

[Grua93a]Gruau, F., "Genetic micro-programming of neural networks." In Kim Kinnear, (ed.), *Advances in Genetic Programming*. MIT press, 1993.

[Grua93b]Gruau, F., "A learning and pruning algorithm for genetic neural networks." In *European Symposium on Artificial Neural Network*, 1993.

[Grua93c]Gruau, F., "The mixed parallel genetic algorithm.", In Denis Tristram, (ed.) *ParCo 93 (Parallel Computing)*, 1993.

[Grua93d]Gruau, F., and Whitley, D. "Adding learning to the cellular developmental process: a comparative study." *Evolutionary Computation*, 1993. to appear vol 1, no. 3.

[Harn90] Harnad, S., "The Symbol Grounding Problem", *Physica D* 1990, Volume 42, Number 1-3, pp 335-346.

[Harp89] Harp, S., Samad, T. and Guha, A., "Toward the genetic synthesis of neural networks." In D.J. Schaffer, editor, *3rd Int. Conf. on Genetic Algorithms*, pp. 360-369, 1989.

[Holl75] Holland, John, *Adaptation in natural and artificial systems*, University of Michigan Press, 1975.

[Kita90] Kitano, H., "Designing neural network using genetic algorithm with graph generation system." *Complex Systems*, 4:461-476, 1990.

[Koza92] Koza, J.R. *Genetic programming: On the programming of computers by mean of natural selection*, MIT press, 1992.

[Koza94] Koza, J.R., *Genetic programming II: Automatic Discovery of Reusable Subprograms*, MIT press, 1994. to appear.

[Mill89] Miller, G., Todd, P. and Hedge, S., "Designing neural networks using genetic algorithm." In D.J. Schaffer, (ed.), *3rd Int. Conf. on Genetic Algorithms*, pp. 379-384, 1989.

[Mueh91] Muehlenbein, H., Schomish, M. and Born, J. "The parallel genetic algorithm as function optimiser." *Parallel Computing*, 17:619-632, 1991.

[Muhl89] Muhlenbein, H. and Kinderman, J., "The dynamics of evolution and learning—toward genetic neural networks.", In R. Pfeifer, Z. Schreter, F. Fogelman-Soulie, and L. Steels, (eds.), *Connectionism in Perspective*. North-Holland, 1989.

[Ray91] Ray, T. "Is it alive or is it GA ?", In Richard Belew, (ed.), *4th Int. Conf. on Genetic Algorithms*, pages 527-534, 1991.

[Ullm79] Hopcroft, J.E. and Ullman, J.D., *Introduction to automata theory, languages and computation.* Addison-Wesley, 1979.

[Whit90] Whitley, D., Stakweather, T., and Bogart, C., "Genetic algorithms and neural networks, optimising connection and connectivity." *Parallel Computing*, 14:347-361, 1990.

Part Four

Developing Hybrid Systems

14

Tools and Environments for Hybrid Systems

Sukhdev Khebbal and Danny Shamhong

14.1 Introduction

With the emerging interest in intelligent hybrid systems from both academic researchers and commercial developers, there is a growing need for techniques that can "bind" together many diverse processing components.

For intelligent hybrid systems to be fully accepted in the commercial environment they must be able integrate and communicate with conventional computing systems such as databases and spreadsheets. Within an organisation, intelligent hybrid systems must be able to interact with a variety of different company information and existing applications and systems. For these reasons it is vital that developers have a range of software tools, information exchange techniques and development environments for integrating intelligent systems with conventional computing systems.

This chapter reviews current information exchange technologies, programming tools and development environments for constructing intelligent hybrid systems. The main aim of this survey is to complement the theoretical issues and examples emerging from this new field, by offering a pragmatic review of commercial products so that developers can start to build their hybrid systems.

The chapter also examines the benefits of adopting an object-oriented approach to representing and building hybrid systems. This approach is particularly suited to representing the individual processing techniques, the information exchange technologies and the programming environments that "glue" these different components together.

14.2 Object-oriented Integration

Object-oriented programming is a software engineering methodology which forms a natural model for building intelligent hybrid systems. Object-oriented systems directly model entities from the problem domain as a set of distinct software units called objects, that interact by passing well-defined messages. For example, in a object-oriented banking application there may be Customer objects that send debit and credit messages to Account objects. These Account objects may co-operate to maintain Cash-on-Hand and Accounts-Payable objects. This representation scheme is particularly suitable for hybrid systems because objects can represent different information processing techniques, which communicate via a standard information exchange protocol.

The concept of object-oriented programming is a relatively new method for designing and implementing software systems. Its major commercial goals are to improve programmer productivity by increasing software extendability and reusability, and therefore reduce the complexity and cost of software maintenance. Object technology arose from a desire to change the way we design, develop and manage the complexity of software systems. A properly designed object-oriented application should contain objects that directly model the application domain, therefore making the application easily understood by the people involved in the development and maintenance process. Within a large application, objects abstract and distribute the dependencies that exist between data and functions into many smaller and simpler pieces. Breaking down an application into entities and relationships that are meaningful to end users is a common conventional programming-analysis technique. Unlike conventional programming, object-oriented programming preserves this same decomposition through the design and implementation phases. Figure 14.1 shows the abstraction of code and data into small separate objects that have a well-defined interface for manipulating the data.

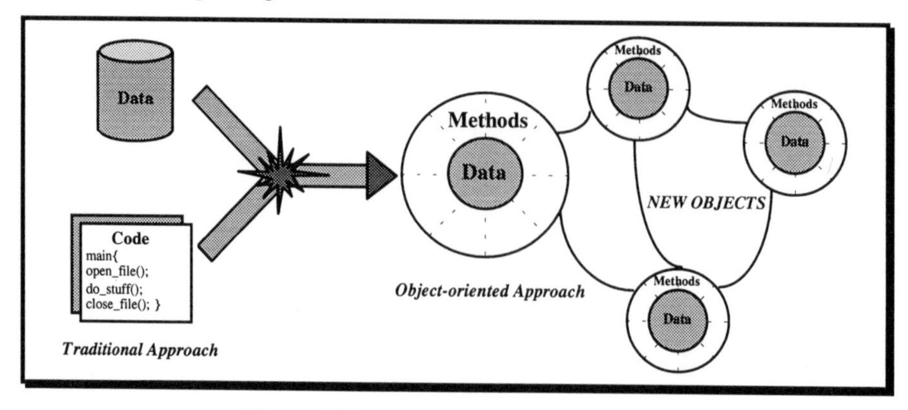

Figure 14.1 — Object-oriented approach.

Traditionally in programming systems, data and procedures are separate entities, where the programmer is responsible for applying active procedures to passive data structures. Whereas, an object-oriented programming system doesn't view an object as just passive data but as the combination of the object's current private state and the methods that manipulate that state. In a typical object-oriented language, the definition of a type is often called a *class*. A class definition defines both the instance variables (the state or data) and the methods (operations) for objects of that class. These methods form a well-defined interface to the functionality of objects of this class and forms the basis for allowing the user-access to the data and functions within an object. This well-defined modularity means that objects can be reused in similar applications in the same area. This also means that the user needs to only understand the behaviour of the objects as specified in the interface, without concern for the implementation. The user can treat the object as a "black box" that has its implementation hidden from view. This functional approach coupled with the ability to have many object instances of the same class, enables objects to be reused in the design and implementation of an application. Reusing, rather than reinventing software speeds the development and maintenance of large applications.

Maintainability is enhanced with this approach because changes in the implementation of a data structure or an operation (i.e. code within the class implementation) can be localised to the region of code that implements that particular functionality. This localisation makes detection and correction of problems much easier and faster.

14.2.1 Encapsulation, Inheritance and Polymorphism

In object-oriented programming an *object* is an instance of a specific class. Actions may be performed on such an object by invoking one or more methods defined in the class definition. The process of invoking a method is called sending a *message* to the object. Such a message typically contains the name and parameters of the method to be invoked. The invocation of a method typically accesses or modifies the data in the particular object.

Encapsulation is the backbone of object-oriented programming and is the process by which individual software objects are defined. Encapsulation involves the definition of :

1. A clear boundary that encompasses the scope of all the object's internal software.
2. An interface that describes how the object interacts with the other objects.
3. A protected internal implementation that gives the details of the functionality provided by the software object. These implementation details are not accessible outside the scope of the class that defines the object.

The unit of encapsulation is the object, which has properties described by its class description. These properties are shared with other objects of the same class.

The object-oriented approach to problem-solving uses objects for encapsulation and defines objects to be instances of classes. These classes can be structured into hierarchies where some classes are subordinate to others and are called *subclasses* or *derived classes*. Subclasses are considered to be special cases of the class under which they are grouped in the hierarchy. The lower levels in a class hierarchy usually represent an increased specialisation, and higher levels usually represent generalisation.

A subclass definition characterises the behaviour of a set of objects that *inherit* some of the characteristics of the parent class. *Inheritance* is utilised in these class hierarchies so that new objects (children) can be derived from the older objects (parents). The new objects can have their own data or operations and can modify the existing ones inherited from their parents. Inheritance serves as the basis of code reuse and sharing in an object hierarchy.

The cost and complexity of software development can be reduced by the creation of subclasses because inheritance can lead to incremental problem solving where instead of modifying existing software components, or rewriting them, new subclasses are created from a set of baseline classes.

In most object-oriented languages, if class ***P*** is a parent of subclass ***S***, then an object ***s***, of subclass ***S***, can be used wherever an object ***p***, of parent class ***P***, can be used. This implies that a common set of messages (i.e. operations) can be sent to objects of class ***P*** and class ***S***. When the same message can be sent to objects of a parent class and objects of its subclasses, this is defined as *polymorphism*. Because of polymorphism, it is very easy to plug new objects into a system if they respond to the same set of messages as existing objects. For example, if a Window object expects all graphical objects to be able to draw themselves. Then it is very easy to add a Rectangle object to this system that previously had only Lines and Circles. This is possible because polymorphism would allow the new object to respond to a draw message in its own manner.

14.2.2 Advantages of Object-oriented Integration

Due to the many modular characteristics (i.e. encapsulation, inheritance, polymorphism) and the well-defined message interface, object-oriented programming naturally accommodates many possibilities for representing intelligent hybrid systems. These properties enable different information processing techniques to be represented as objects and integrated with other processing components via a well-defined message protocol. These properties also allow the easy exchange or addition of new processing techniques.

The use of object-oriented message passing as a well-defined communications protocol allows links between different components to be established at runtime. Message passing provides the basis of the most flexible form of dynamic linking in a object-oriented hybrid system. At the lowest level, a message is simply data passed from one computational unit to the next. Combined with queues and the ability to send messages asynchronously, each object in a hybrid application could process a stream of data relatively independently of other units.

Employing object-oriented techniques allows the ability to manage concurrency because message passing systems are adaptable to many kinds of hardware environments, including parallel multi-processor machines and distributed systems. In the case of distributed systems a hybrid application may have components running on a variety of different machines, making use of the specialised architecture's of these machines. Object-oriented programming can be exploited as a means of creating a system for parallel architectures without having to explicitly deal with processes and inter-process communication. The use of objects provides an abstraction that hides processes and inter-process communications within the object and the message protocol used by the processing components.

There are other added advantages with respect to code reusability when adopting an object-oriented approach to building hybrid systems. Code reusability allows for multiple copies of the same processing technique to be active at the same time. Therefore these objects can be linked together for chained processing. For example, several expert systems or neural networks could be developed for sub-tasks of a problem and chained together to form the final solution.

With the many advantages that object-oriented integration offers, there are clearly many possibilities for implementing hybrid systems using object-oriented techniques. The three types of hybrid systems (function-replacing, intercommunicating and polymorphic) [Goon92], can use object-oriented programming at various levels, ranging from the actual programming of the processing technique using an object-oriented language such as C++ [Stro91], to the use of an object-oriented programming environment and an information exchange protocol.

For example, function-replacing hybrids (i.e. hybrids where a principle function in one technique is replaced by another processing technique) can be represented at the lowest level. Both of these processing techniques could be programmed as a collection of objects, where internally they represent their principle functions as component objects. This object-oriented representation means that a principle function object such as the pattern matcher in an expert system, could be easily replaced by a another object that encodes a processing technique that is more efficient at pattern matching, such as a neural network. This replacement would therefore enhance the abilities of the host technique.

An intercommunicating hybrid (independent techniques that communicate by exchanging information) is naturally suited to object-oriented programming because techniques can be represented as objects and communication can be conducted via object-oriented message passing or via an industry standard linking and communication protocol, such as Object Linking and Embedding (OLE) from Microsoft [Micr91]. Components within intercommunicating hybrids can be easily distributed across a network of computers where they communicate using one of the many emerging platform independent communication protocols, such as the Object Request Broker (ORB) from the Object Management Group [Obje92].

Lastly, polymorphic hybrids (systems that use a single processing architecture to achieve the functionality of different processing techniques) can be programmed at the lowest level using object-oriented methods. From the definition, the lowest level is the single architecture of a technique, for example the neurons and connections of a neural network. Object-oriented programming could be used to build the low-level structures of the architecture, in the neural network example, neurons could be easily modelled as objects and connections stored in a list or matrix of pointers to these neuron objects. This object representation allows symbolic processing mechanisms such as variable binding or logical inferencing, to be built into the low-level neural representation by simply extending the functionality of the neuron objects.

The use of object orientation in integrating diverse processing components within hybrid systems is also characteristic of recent developments in application integration software, such as Hewlett-Packard's NewWave, Object Linking and Embedding (OLE) in Microsoft Windows, and in many planned Distributed Computing Environments (DCE). The requirements of building intelligent hybrid systems and distributed computing environments are extremely similar in that they require a representation scheme and a well-defined communication protocol. The goals of using object technology in all of these systems is to provide the "glue" which allows applications to share information with one another in an easy and consistent manner.

The following sections review the key state-of-the-art integration technologies and programming environments that use object-oriented techniques as the foundation for integration, communication and co-ordination of components.

14.3 Information Exchange Technologies

The concept of an object as a self-contained package of data and functions, lends itself to distributed computing and fits well with recent trends towards client/server systems. A client/server system is one that is made up of components called *servers* that perform specific tasks and respond only to requests from other components called *clients*. The self-contained independence of these objects means that they can be distributed across a network of computers and communicate by sending messages to each other. A central philosophy of object-oriented and client/server computing, is that applications can be built from independent components that can be modified and assembled quickly and easily. In the perfect distributed system, the end user will not be aware whether the application is powered by the desktop PC or by a remote server.

The mechanisms used to pass messages between objects are extremely important and have been the focus of attention for many organisations wishing to create a standard protocol for object communication and manipulation. Obviously, it is crucial to the success of object-oriented computing that there is agreement between manufacturers and software developers over the way objects communicate in distributed systems.

Software vendors and systems suppliers have recognised the importance and virtues of intercommunicating systems and this has sparked a "software battle" to establish dominance of the distributed systems market. Currently the two main contenders in the battle are Microsoft with their Object Linking and Embedding (OLE) protocol and the rest of the software industry with their proprietary implementations of the Common Object Request Broker Architecture (CORBA) protocol from the Object Management Group (OMG). The other main technologies to join the battle are from individual or consortiums of systems vendors, such as the OpenDoc consortium, which includes IBM, Apple and Hewlett Packard. Some of these major technologies are detailed in this section.

Recently the focus of this conflict has shifted in dominance to the emerging client/server operating system market where OS/2 and UNIX have been joined by Microsoft's Windows NT. Added to this is the recently announced fact that Microsoft is joining the OMG — which means that it also backs the Common Object Request Broker Architecture (CORBA) specification — and has joined Digital to form an initiative to bridge the gap between the OLE and ORB.

In the future, it seems likely that OLE and CORBA will gradually converge and become no more than different flavours of the same mechanism. This competition to establish a standard will be of benefit because software suppliers will gain from having a large heterogeneous market for small application objects or "applets" and customers will have the luxury of a wide choice of applets to build hybrid systems, but with the confidence that the mechanisms are in place to integrate objects from many different sources. The move towards object-oriented client/server operating systems that are designed to allow different applications to interact and exchange information, forms a natural environment for building intelligent hybrid systems.

14.3.1. Object Linking and Embedding (Microsoft)

Object Linking and Embedding (OLE) is a major feature of Microsoft's Windows environment (from 3.1 to more recent versions), and allows real hybrid integration in an easy to use manner. OLE is an enhanced method for combining information from a number of different applications into a single area called a "compound document". This "document oriented interface" (DOI) allows the construction of applications by "mixing and matching" small specialised tools. OLE coupled with its DOI orchestrates transparent access to, and co-operation among, these specialised tools. OLE uses objects within compound documents, to store and manipulate exported data, independent of format, from other tools and applications. Figure 14.2 shows an example of a compound document, that enables you to view and manipulate various forms of information (for example, spreadsheets, text, and graphics).

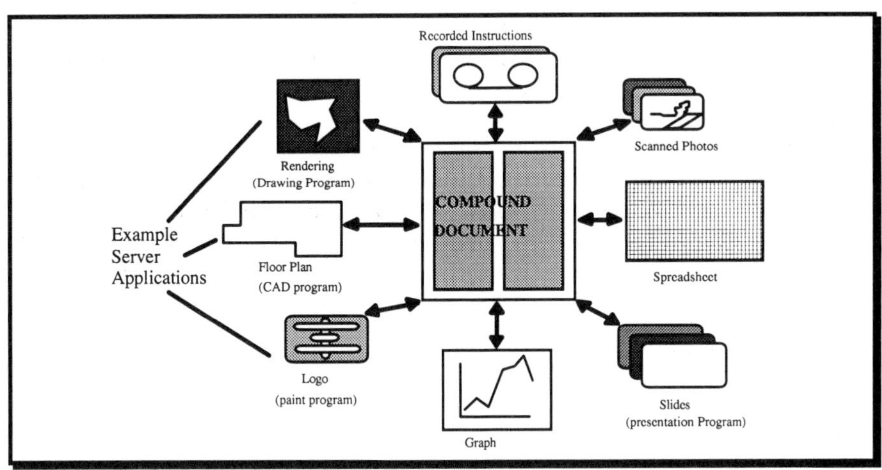

Figure 14.2 — A compound document.

Rather than having to manually switch between different applications, OLE allows the automatic editing and updating of objects, this is referred to as "edit-in-place". If a graph is generated based on the contents of a spreadsheet, and the graph is placed into a word processor, then the contents of the word processor document will be automatically updated when numbers in the spreadsheet are changed. Information can be altered by simply selecting and double-clicking the object of interest. This will launch the application that created the information and load the information back into the application, so that it is ready for modification. The process of launching applications is automated and is invisible to the user.

OLE works via a system of servers (e.g. Paintbrush a drawing program in Windows 3.1) and clients (e.g. Microsoft Word, a word processor), that manipulate OLE objects. A server is an application that provides editing and rendering services for a linked or embedded object within a client application. When an object is activated (double-clicked) in a client document, the server application is invoked. A client is an application or document that receives, stores, and displays objects created by servers. If Microsoft Word is used to create a report and import your spreadsheet chart from Microsoft Excel, then Word would be the client and Excel would be the server in the exchange.

An object is any data that can be displayed in a compound document and manipulated by a user. An object can be anything from a single cell in a spreadsheet to an entire document or illustration. When an object is incorporated into a document, it maintains an association with the application that produced it. OLE applications are "opaque"; that is, the client application never looks inside the object and therefore doesn't need to understand its contents. The client simply calls functions in the OLE libraries when it needs to display an object or invoke the server application for editing purposes.

An embedded object is an object that is completely contained within the client document. For example, when a range of spreadsheet cells from Microsoft Excel is embedded in a Microsoft Write file, all the data associated with the cells, including any formulas, is saved as part of the Write file. The name of the server for the spreadsheet (Microsoft Excel) is saved along with the data. While working with the Write file, double-click on this embedded object (the range of cells), will start Microsoft Excel automatically so that you can edit those cells.

A linked object is a reference, or link, to data shared by a server application and a client. If the range of spreadsheet cells from our example is a linked object instead of an embedded object, the data is not stored in a Microsoft Excel file but only the link to the data is saved with the Write file. The server then "presents" the data, in the proper format, to the client document. Double-clicking on the object will still start up the server application (in this case, Microsoft Excel) and edit the information.

14.3.1.1 Benefits of OLE

With these abilities for application integration, there are many benefits of using OLE, the following being the most important :

- An application that implements OLE allows you to move away from an application-centred view of computing, in which the tool used to complete a task is a single application, and move toward a document-centred view, in which you employ as many tools as necessary to complete the job.
- OLE allows applications to specialise in performing one job very well. A word processing application that implements OLE, for example, does not necessarily need any illustration tools; users can put illustrations into their documents and edit those illustrations using any illustration application that supports OLE. An example of such a relationship is the one between Microsoft Word for Windows and the illustration application Microsoft Draw, which is included with Word, but any vendors illustration application could be used if it supports OLE.
- Applications are automatically extendable for future data formats because the content of an object does not matter to the containing document.
- You can concentrate on the tasks you are performing, instead of on the applications required to complete the tasks.
- Files can be more compact because linking to objects allows a file to use an object without having to store that object's data.
- A document can be printed or transmitted without using the application that originally produced the document.
- Linked objects in a file can be updated dynamically.

14.3.1.2 The Evolution to OLE

The need to integrate information from different applications is not new. In the past, users have taken advantage of the Windows Clipboard, and more recently, dynamic data exchange (DDE) to accomplish this.

The Clipboard (an area of memory in the Windows environment) provides a basic means for moving data between applications. You select the desired information in the source document, transfer the information to the Clipboard by choosing Copy from the Edit menu, then place the information in the target document using the Paste command on the target Edit menu.

Data transfers through the Clipboard are static. This means that if the source information changes, you must manually copy the revisions to the target document if you want the target document to stay up to date. Additionally, the target application must understand the format of the data being transferred to it through the Clipboard. A word processor that accepts an ASCII text string, for example, will be unlikely to understand a set of cells from a spreadsheet.

DDE complements the Clipboard's static data exchange capability by creating a link between the target document and the source document. The target document uses this link to dynamically retrieve the associated data whenever it is updated in the source document. Microsoft created OLE because there are scenarios that are not served by either Clipboard or DDE linking — in particular:

- situations that involve compound documents.

- when using DDE, the data is retrieved from the application into your compound document, but not necessarily the formatting of the data from the application. For this reason, the receiving application has to able to understand the data it gets via DDE.
- when using DDE it is only good for bringing data into the document, it does not give an easy way to access the application itself, so to make quick changes to the original data the second application still must be opened, the source data file loaded and adjustments made.

Figure 14.3 shows the evolution of application integration from clipboard to OLE. Note that OLE does not replace either DDE or the Clipboard, but rather complements them by providing additional functionality.

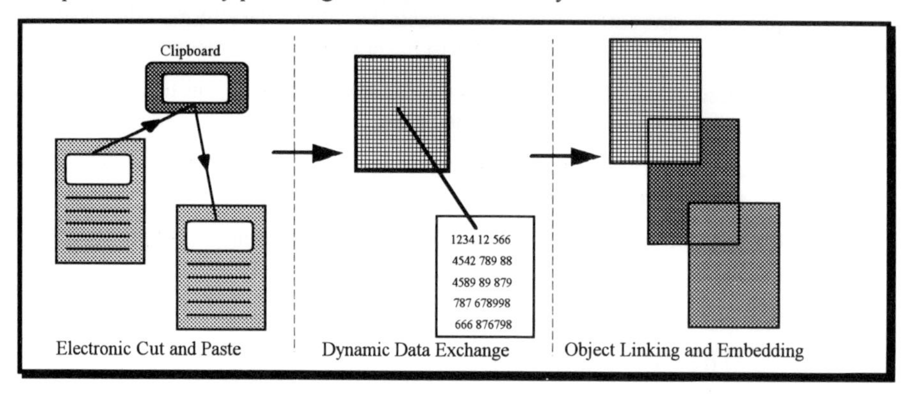

Figure 14.3 — The Evolution of application integration.

OLE is the next step in the evolution of application integration. By using OLE to embed information, you can combine text, charts, spreadsheet sections, even voice instructions into a single document, regardless of data format. With OLE, the data you are combining can actually be stored in the file, so that you don't have to find the original data file before making changes.

Microsoft is trying to get around the problem of linking applications across networks by developing a distributed management system capable of handling a mix of data and communication mechanisms. Microsoft's code name for this project is *Cairo*.

14.3.2 Object Request Broker (Object Management Group)

The Object Management Group (OMG) is an international organisation of 150 systems vendors, software developers and users, founded to promote the theory and practice of "object management" technology in the development of software. Its goal is to provide a common framework for object-oriented applications based on industry guidelines. The adoption of this framework will make it possible to develop hybrid applications across all major hardware and operating systems. The Common Object Request Broker Architecture (CORBA) currently exists only as a specification of a standard for Object Request Brokers, which provide a mechanism for objects in a distributed system to talk to each other.

The OMG defines object management as software development that models the real world through the representation of objects. The objectives of the OMG are to foster the technology's growth and to influence its direction in the following five areas:

- **Overall Architecture, Reference Model and Terms:** Industry-wide consensus on an *Object Management Architecture* (OMA) which includes the definition of terms and a common model for objects and their attributes, relationships and methods.
- **Application Programming Interfaces (API) for Objects and Applications:** Object management facilities for a common API across DOS, OS/2 and UNIX including distribution, class libraries, document content architectures and methodology.
- **Distributed Object Management:** Applications and APIs for the distribution of objects across heterogeneous networks, remote procedure calls (RPCs) and operating systems.
- **Interfaces to Object-Oriented Databases:** Pragmatic interfaces and abstractions to existing database management systems and object-oriented databases.
- **Common Services:** Specifications for interfaces and common services such as security authentication and system management.

14.3.2.1 Distributed Object Management System (DOMS)

An organisation may use many diverse object-oriented products such as databases and programming languages. To allow these products to co-operate and communicate requires a management mechanism. A distributed object management system (DOMS) addresses these issues by providing:

- a single interface to manage the complexities of a heterogeneous environment
- a uniform framework, based on standards and extendability, to build, integrate, and deploy open distributed-computing applications,
- a method for creating location independence for client applications.

DOMS lets you build applications using a standardised interface while reusing the system's existing objects. With the advent of DOMS the OMG has sponsored the Object Request Broker (ORB). The ORB has attracted much attention because it promises to deliver the benefits of object-oriented development as well as those of client/server computing. The combination of network based systems and object technology will let developers create hybrid applications that are truly independent of the system and the network architecture. The ORB is the kernel of a standardised DOMS and provides the interoperability and reuse of a system's existing objects. The ORB enables client applications to access services and other objects that exist anywhere in the distributed system.

Figure 14.4 shows how the ORB component within the OMG's evolving Object Management Architecture (OMA) will enable objects to transparently make, receive and respond to requests within an object-oriented environment. As the heart of the standard, the ORB will provide the "communication highway", enabling objects to interact over a network of different systems.

The ORB is the central message handling mechanism allowing "objects" to communicate independent of operating systems, hardware architecture's or network protocols. The OMG has also started work on an Object Model and standard class libraries which will be language independent.

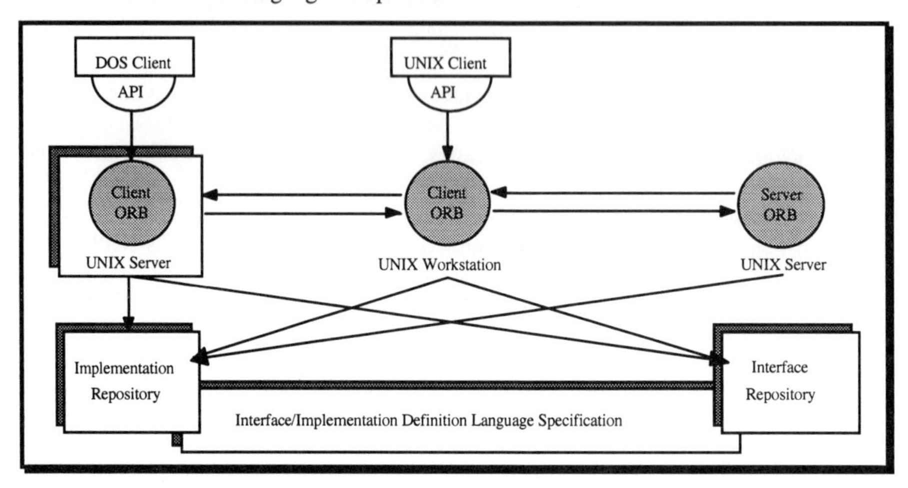

Figure 14.4 — The Object Request Broker in a distributed system.

The main task of the ORB is to register the location of objects, receive, accept, and route requests for services. In addition the ORB ensures that the request is processed properly. When a request for a service is made, an ORB locates the object, starts the specified operation (method), and passes it the parameters found in the request message. Since objects can exist anywhere on a network, you can locate them via a name service or a unique identifier - *a handle.*

Any object-oriented system needs to support subclassing, inheritance and polymorphism. These object-oriented capabilities are supported within the ORB by including primitive base classes as part of the basic ORB. Built into these classes are methods that provide the functionality of subclassing, inheritance and polymorphism. By adding more of these basic operations to the base classes, you can continue to enhance the systems capabilities. These basic classes are defined and loaded by a bootstrapping process that uses a mechanism called the *interface and implementation definition language compiler*. To maintain compatibility among different implementations of an ORB, the OMG is defining a standard for the basic functions and classes. The base classes and their methods are known as the *type repository*.

Advanced interactive client/server applications need the ability to operate across a diverse mix of hardware and software platforms. Figure 14.5 shows a client application running on a particular CPU and operating system (e.g. a Sun Sparcstation with SunOS and Motif) with some other object services (e.g. a CAD object) running on a different machine (maybe a Cray). This application requires calculations from the Cray to be displayed on the Sun. Using a DOMS, it is easy to separate the objects operations into those that are display independent and those that are display dependent. This eliminates the need for environment dependent code. In the example, the client application discovers the object handle of object FOO. The implementation of FOO resides on machine B. The client application issues a request to FOO for the operation *calc*. The ORB on machine A routes the request to the ORB on machine B, which selects the correct method and returns the results to the client application. Next, the client application issues a display request for FOO, which is automatically routed to machine B, where the local ORB determines the display method required for display of FOO on machine A and returns the appropriate user-interface-description file. The object FOO can have multiple display methods, for instance, one for each display environment (e.g. Motif, Open Look and Microsoft Windows).

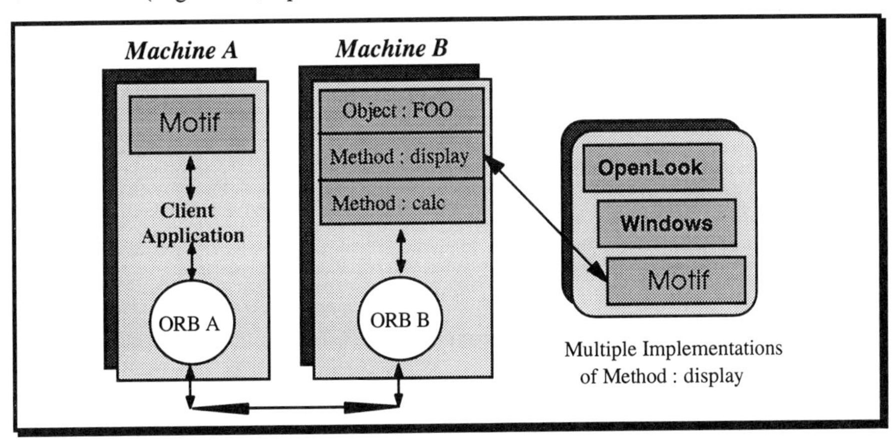

Figure 14.5 — An example client/server application.

There are many commercial advantages to adopting this complete DOMS, ORB approach to hybrid systems, these include :

- An organisation can retain its existing hardware and software assets while integrating new applications and solutions easily.
- An organisation can take advantage of extendability and reusability of objects to cut down development time and deliver distributed applications more easily.
- An organisation can integrate existing applications economically. Even applications written in non object-oriented languages (e.g. COBOL or FORTRAN) can be encapsulated with a single object often referred to as an *object wrapper* [Diet89], therefore preserving past software investment.

A complete ORB approach will bridge the operating system and communications void that exists today and would enable developers to bring applications and systems together into a truly cross-platform hybrid system.

The ORB architecture is fully specified and has been used by many companies to produce their own ORB compliant applications and frameworks for cross-platform hybrid systems.

14.3.3 Common Object Model (Microsoft)

Microsoft developed its Common Object Model (COM) as the underlying object model at the foundation of OLE 2.0. It defines protocols that objects use to communicate with one another. COM provides a single programming model for distributed and local objects but is strongly biased towards the object-oriented C++ language. COM does not explicitly support inheritance but instead uses a mechanism called *aggregation*, that requires objects to explicitly have pointers to objects higher in the hierarchy. The developers of OLE claim that this is a safer mechanism than inheritance and allows the programmer to control the inheritance process [Wayn94].

Although the COM is not CORBA compliant at present, a recent agreement between Microsoft and Digital Equipment Company (DEC) to jointly develop an open architecture for distributed computing will ensure that COM and therefore OLE, will become a major player in the open systems market. This agreement will link their two leading object systems, ObjectBroker from DEC and OLE from Microsoft. ObjectBroker [Digi94] is a platform-independent object request broker, that has OLE and DDE interfaces to allow applications built around these protocols to communicate across a network of different computers.

This approach will provide a new level of interoperability between Windows and other popular platforms, including Windows NT, Macintosh, Open VMS, DEC OSF/1, AIX, HP-UX, SunOs and ULTRIX operating systems.

14.3.4 ObjectBroker (Digital Equipment Company)

DEC's integration products such as ObjectBroker, and integration tools, such as FBE (Framework-Based Environment), provide the means to integrate different application software across a network of different computers. Integrating via a "broker" allows companies to re-engineer without re-writing, for measurable savings in time and money.

With ObjectBroker, applications on different platforms can work together seamlessly — interoperating and exchanging data. Employing object-oriented technology, ObjectBroker shields the developer from the cumbersome intricacies of platform specifics, networking protocols and operating systems. ObjectBroker provides:

- **Registration** — records capabilities and locations of applications comprising the environment in an object-oriented class repository.
- **Invocation** — client applications locate and start up server pre-processes while considering user preferences stored in context objects.
- **Interoperability** — it controls and translates messages from invoked processes.

The ObjectBroker is available on almost all operating systems. With its ability to allow OLE and DDE to operate across a network of machines, the ObjectBroker will help to make Microsoft's communications protocols even more popular than they are already. Using ObjectBroker also allows existing "legacy" applications to be integrated into networked client/server systems, without source code changes. This means that existing applications can be reused without the need to recompile or re-link, and once registered an application is universally available to all clients on all platforms.

ObjectBroker is available as a Development Option allowing a developer to compile and link an application based on ObjectBroker and to integrate an application in the runtime system. There is also a runtime option to allow users to run an already developed ObjectBroker-based application on a target machine. An ObjectBroker Advanced Development Kit (ADK) is available on Windows and UNIX systems to give maximum flexibility.

FBE: Client/Server Application Integration

The Framework-Based Environment (FBE) is DEC's advanced client/server facility for integrating business applications [Digi94]. This environment gives companies the flexibility to choose the commercial applications they want, to intermix operating systems and hardware, even to quickly alter the work-flow between applications.

The FBE is an integration environment, within which traditional applications are component subsystems. This provides the rapid development, accessibility, and information management capabilities required by commercial hybrid systems. FBE uses the CORBA IDL, the interface definition language from the Object Management Group, and ObjectBroker, DEC's implementation of CORBA.

Both the ObjectBroker and the FBE deliver improvements in productivity while providing a solid architectural infrastructure for future software engineering in a client/server environment.

14.3.5 ToolTalk (SunSoft)

The ToolTalk service [SunS91] is SunSoft's inter-application communication mechanism. Independent utilities use ToolTalk to communicate with each other, via an agreed-upon set of messages. ToolTalk compliant hybrid systems allows the "plug and playing" of tools, that is, any ToolTalk-enabled tool can be added to or replaced (plugged) by any tool that supports the same ToolTalk message protocol so as to perform (play) the tasks specified by the messages. Several messaging styles are supported, including messaging to a particular process, to an object or to a type of object. ToolTalk is a migration path to the DOE (Distributed Objects Everywhere) project, Sunsoft's CORBA compliant environment. ToolTalk is currently available on and interoperates between all the major UNIX platforms including Sun, HP, IBM, DEC, Cray, NCR and SGI. Utilities can use ToolTalk in two ways. They can employ it directly, calling functions from the ToolTalk API library to create, send and receive messages; or they can use a "service" built on top of the ToolTalk service, thereby using the ToolTalk service as a communication backbone.

SunSoft along with leading software suppliers have defined some standard message sets to ensure that utilities developed will automatically integrate with each other. These standard message sets include:

- **The Desktop Services Message Set** — supporting the management of files, windowing and processing,
- **The Document and Media Exchange Message Set** — supporting the incorporation and manipulation of text, graphics, audio and video,
- **The Data Access and Manipulation Message Set** — supporting transparent seamless data integration and
- **The CASE Interoperability Message Sets** — supporting the exchange of control information for computer-aided software engineering and electronic design automation.

There are two types of messages that can be sent using the ToolTalk service, *notices* and *requests* [Juli93]. A notice is a way for a process to announce an event. Processes that receive a notice need not return any results. A request is a call for an action. A reply to a request has the results of any action recorded in the message.

Once the interaction between tools in a hybrid application have been determined, the specific messages can be defined. Each tool must register a message pattern (information stating explicitly how messages are laid out) with the ToolTalk service, describing all the types of messages that it can process. A Tool also declares whether it can process the message (therefore becoming a *handler*) or only wants to observe, but not process the message (therefore becoming an *observer*). In the case where there are several potential handlers, ToolTalk ensures that the message is handled by only one handler. Figure 14.6 shows the routing of notices and Figure 14.7, shows the routing of requests through the ToolTalk service.

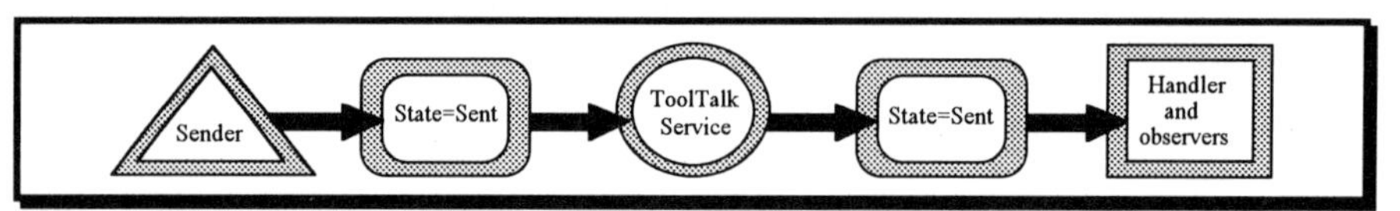

Figure 14.6 — Notice routing in tooltalk.

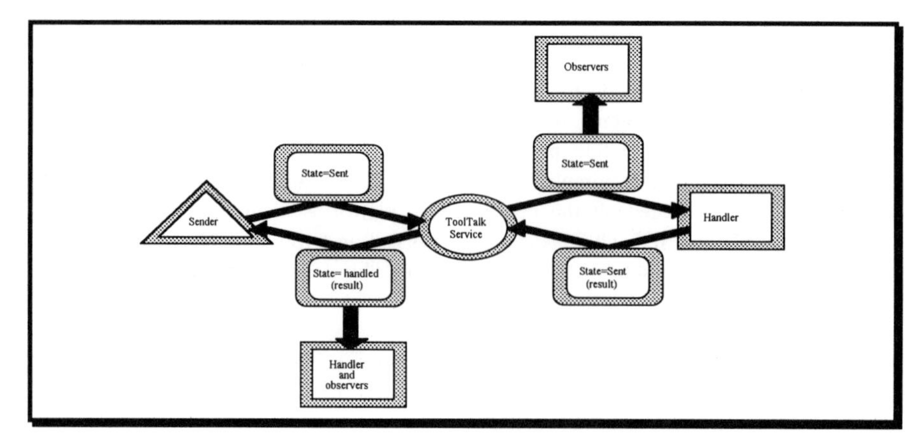

Figure 14.7 — Request routing through the tooltalk service.

ToolTalk handles all the low-level machine connections, network transport and message routing internally and therefore makes all communications network transparent to both developers and users. The ToolTalk capabilities are provided to an application through the C programming lanaguage application programming interface which is consistent across all the ToolTalk supported platforms.

The goals of both ToolTalk and other popular protocols such as OLE are to provide the infastructure for easy and consistent exchange of information. OLE from Microsoft is currently the established approach to providing application interoperability for the PC marketplace. OLE holds its own in a single user PC environment, the sheer number of Windows 3.1 installations and OLE-enabled applications available today, makes OLE a major force in the marketplace. Unfortunately, OLE does not inherently provide support for distributed networked applications, a matter being rectified with Microsoft's collaboration with DEC who is providing the bridge between OLE and CORBA.

Unlike OLE, ToolTalk was designed with a true distributed architecture in mind. A major design goal of ToolTalk was to enable universal "plug-and-play" application integration for large numbers of users doing a wide variety of tasks across the network. SunSoft has announced that ToolTalk will be the vehicle for its CORBA compliance. Since the ToolTalk API will be supported in DOE, utilities developed using ToolTalk will automatically inherit DOE capabilities, the attractive element being the compliance to the CORBA standard.

14.3.6 HD-DOMS (HyperDesk)

HyperDesk HD-DOMS is a distributed object management system, which is CORBA compliant and currently available on Sun, Data General, Hewlett-Packard, and IBM UNIX platforms. On Microsoft Windows, currently only client utilities are supported. HD-DOMS consists of an authentication service, location service and at least three databases [Mow93]. The authentication service authenticates client applications when they log into the environment. The location service keeps track of all ORBs that are running. The databases consist of an interface repository containing class definitions, an implementation repository containing method definitions and one or more databases with objects that users have created. Figure 14.8 shows that a client application on a UNIX host accesses objects through its own ORB on the same host. A client application on a Windows platform accesses objects through its own private ORB on a UNIX host. After logging in, a client application receives a list of object references available to the user such as the location service, the user's desktop information and any objects that others users have made available.

With Novell bundling HD-DOMS with Netware, HyperDesks's distributed object management system provides an object-based interface to services and resources within Novell Netware. The combination of Hyperdesk HD-DOMS on Novell Netware creates a powerful standardised framework for distributed object management across networks. The bundled HD-DOMS means that NetWare services and resources can be treated as objects. These can then be accessed through a common interface which will replace the series of different interfaces currently in use.

The software development kits (SDKs) for Netware, enables Netware developers to build any component of a distributed HD-DOM enabled application.

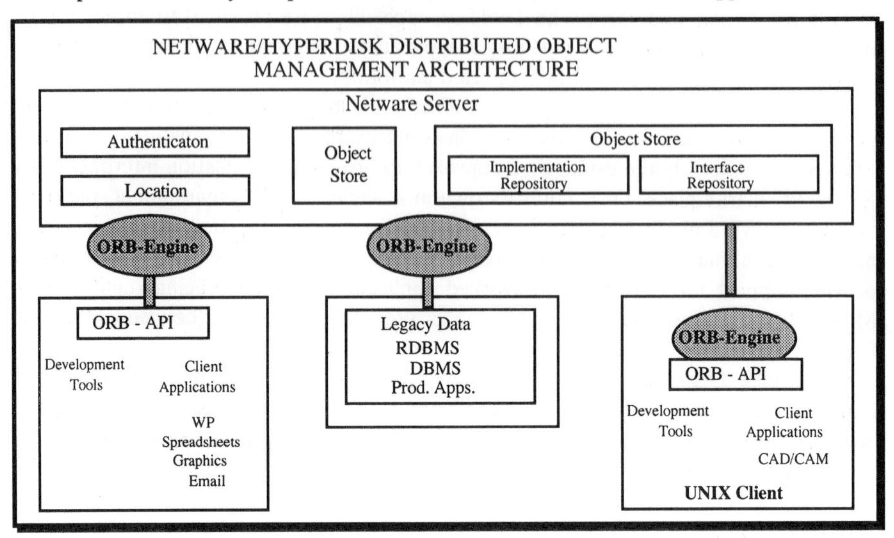

Figure 14.8 — HD-DOMS distributed architecture.

14.3.7 DSOM (IBM)

In 1992 IBM released OS/2 2.0, a operating system with an object-oriented *WorkPlace Shell* and an underlying foundation called System Object Model (SOM). SOM has received considerable recognition and has been updated by the Distributed System Model (DSOM), which extends objects across networks and now forms the core of IBM's distributed-object strategy [Udel93].

DSOM is a language-neutral mechanism for developing distributed object-oriented class libraries. DSOM is CORBA compliant and is available for OS/2, AIX/6000 and Microsoft Windows. DSOM consists of two components – the Workstation Runtime engine and the Workgroup Runtime engine. The Workstation Runtime engine enables objects to communicate across process boundaries on a single machine. The Workgroup Runtime engine enables those same objects to communicate across system boundaries over various networks. In addition to the runtime systems the SOMobjects Developer Toolkit includes among other utilities, the SOM compiler, documentation, and three sets of C++ classes that IBM calls the frameworks. These frameworks are called the emitter framework, the persistent framework and the replication framework.

The emitter framework provides interface definition language (IDL) parsing support to enable binding SOM to programming languages other than C or C++. This has been used by MicroFocus to bind SOM to its COBOL implementation to produce a object-oriented COBOL [Oden93]. The persistence framework reads and writes objects to disk and the replication framework permits safe sharing of common data between objects derived from this framework. DSOM is made available to an application by deriving from these base SOMobjects provided within these three C++ sets of classes. Figure 14.9 shows the DSOM environment where applications can take advantage of remote objects by creating a local proxy that regulates access to remote methods.

Recently, IBM and Hewlett-Packard have been involved in a technology exchange, so that DSOM and HP's Distributed Object Management Facility will be synchronised. This exchange will give HP access to DSOM, a finer grain technology than their own, and IBM will benefit from HP's Distributed Computing Environment. The ability to run DSOM over a DCE will be crucial to DSOM's long term success.

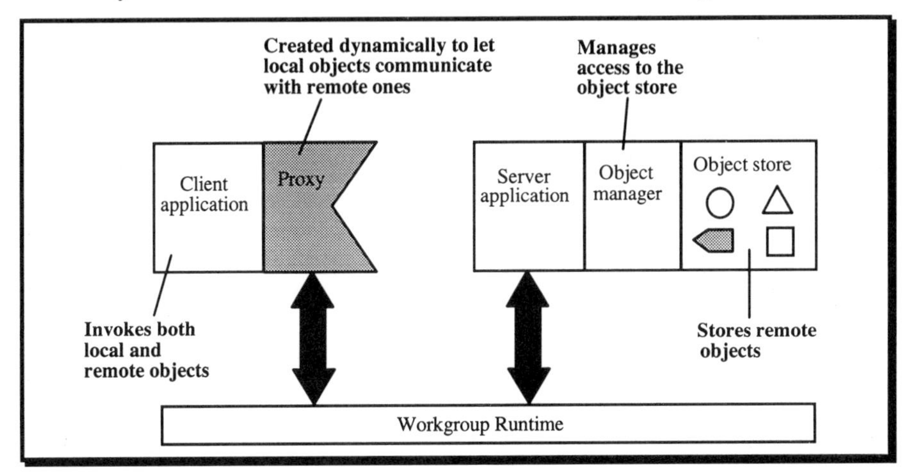

Figure 14.9 — The DSOM environment.

14.3.8 OpenDoc (Component Integration Labs)

Apple, in collaboration with WordPerfect, Novell, Sun, Xerox, Oracle, IBM and Taligent is developing a cross-platform object-oriented compound document architecture called OpenDoc. The finished product will be controlled by this consortium under the name of the Component Integration Lab (CIL), therefore making OpenDoc vendor neutral. As a developing object-oriented technology it lags behind OLE considerably and was only in alpha release in early 1994.

The core technologies in OpenDoc are the *Bento* storage mechanism [Pier94]; a scripting technology based on Apple's AppleScript; and IBM's SOM (System Object Model) [Udel93]. The Bento file format is a compound document storage format, developed by Apple, that was designed to support multimedia and work on many platforms. By using Bento, OpenDoc makes no assumptions about whether the data being stored is sequential or made up of many components. It allows both random and sequential access for inserting and deleting data. This makes the manipulation of files very efficient since updates can be made without copying massive amounts of information, as is the case in sequential based storage systems. The Bento file system, through its incremental storage capability, enables OpenDoc to support revision control and also save considerably on storage space, by saving only the differences between different versions of a file. This is a notable feature that is missing from OLE, although Microsoft say that it will appear in their planned object-oriented operating system called Cairo.

OpenDoc Scripting is designed with two aims in mind, firstly, to ensure that it presents an easy-to-use interface (not a programming language) and secondly, as a mechanism to allow *work-flow* between applications i.e. to allow several kinds of data and commands to be used by different applications to accomplish a task. OpenDoc chose to extend the Open Scripting Architecture (OSA) developed by Apple. OSA is a series of libraries that provides a standard calling convention (Apple Events) allowing utilities to call one another over a network or on a single machine. The scripting mechanism implements a set of standard verbs that are intended to be as general as possible. Fourteen core verbs will be applicable to all applications supporting OpenDoc. For example a verb might specify "move to next item", in a word processor the item would refer to a word in the text and if in a spreadsheet to a cell within a sheet.

The main benefits of OpenDoc is its language neutrality, whereas other models such as OLE are tied to an object-oriented language (e.g. C++). Through the adoption of SOM as its object dispatcher mechanism, developers are not tied to a particular programming language. And since OpenDoc is being developed as a joint venture between several major software publishers, this makes it vendor neutral too. At present its claim for cross-platform operation is not absolutely clear. Although some of its components such as Bento and SOM are cross-platform, Apple-Events and Apple Scripts at present are not. Future plans for OpenDoc encompass compatibility with Microsoft's OLE. If these plans succeed then OpenDoc will be able to wrap OLE objects with a layer of message-translation software and therefore be able to interact with OLE objects. This translation layer is being developed by WordPerfect, with help from Borland and others.

14.3.9 Taligent Application Environment (TAE)

TAE is a multi-platform object-oriented application framework which will initially be available on IBM's AIX and OS/2, Apple's Power-Open and Hewlett-Packard's HP/UX in 1995. TAE comes in the form of over 2000 C++ classes with over 30,000 member functions, making it one of the richest and most complete object systems available to software developers [Udel94]. TAE can be extended by overriding the necessary classes. For example, the data-access classes can be overridden to map them onto Open Database Connectivity (ODBC); the messaging classes onto standards such as X.400, MAPI, or document frameworks such as OLE and OpenDoc, and the file system classes onto different types of file-systems. Taligent plans to write some of these adapters but will rely on investors and third parties for many others.

Users of the TAE's document, user interface, graphics, international text, data access, and other framework classes will find them extremely flexible. For example the document classes allow multi-level object embedding (i.e. objects containing embedded objects can themselves be embedded into other objects) thus allowing the mixing of technologies (e.g. a document can contain both OLE and OpenDoc objects) and since documents inherit command objects, they are very scriptable. Currently no scripting language has been agreed upon, although as Taligent is a member of CIL (Component Integration Laboratories) there are hopes that OpenDoc's open-scripting architecture will supply the hooks for the IBM's Rexx, Apple's Applescript and others. These features along with the ability of documents to be shared across a network, makes the TAE framework a powerful tool for developing intelligent hybrid systems.

To summarise, Figure 14.10 shows the five leading manufactures and the document, object and communication models that they support. A document model, such as OpenDoc and OLE, defines how a component fits into the GUI application environment. An object model, such as COM, DSOM, DOE, defines at a high-level how components talk to other components that may be local or remote. A communication model supports conversation between components across a network. The object models shown connecting to a CORBA "cloud" are CORBA compliant.

Figure 14.10 (a) shows Microsoft's current status with OLE and COM support for Windows and Macintosh. Figure 14.10 (b) shows how IBM integrates OpenDoc with it's DSOM technology. Figure 14.10 (c) shows DEC's role in bridging OLE/COM and CORBA. DEC's ObjectBroker acts as a gateway enabling OLE components to communicate with CORBA components running on a variety of platforms. Figure 14.10 (d) shows Sun's plan to support both OpenDoc and OLE in its DOE (Distributed Object Everywhere) environment. Figure 14.10 (e) shows Apple using IBM's DSOM with it's OpenDoc architecture.

14.4 Programming Tools and Environments

Although many low-level document, object and communication methods are beginning to emerge for building intelligent hybrid systems, there are very few tools or environments for the high-level configuration and execution of these systems. Hybrid programming environments should allow the developer to configure (ideally through a graphical interface) the processing components required for the hybrid problem. The environment should be "network aware" (i.e. be aware of where processing components reside across a network), but allow the developer to configure components without concern about how and where the components reside or execute. To allow such hybrid programming environments to represent, configure and execute the processing components, requires the low-level use of the integration and communication technologies detailed previously. The next section details some of the current tools and hybrid programming environments that are helping to develop the new user-oriented, high-level integration methods required to make intelligent hybrid systems easier to build and execute.

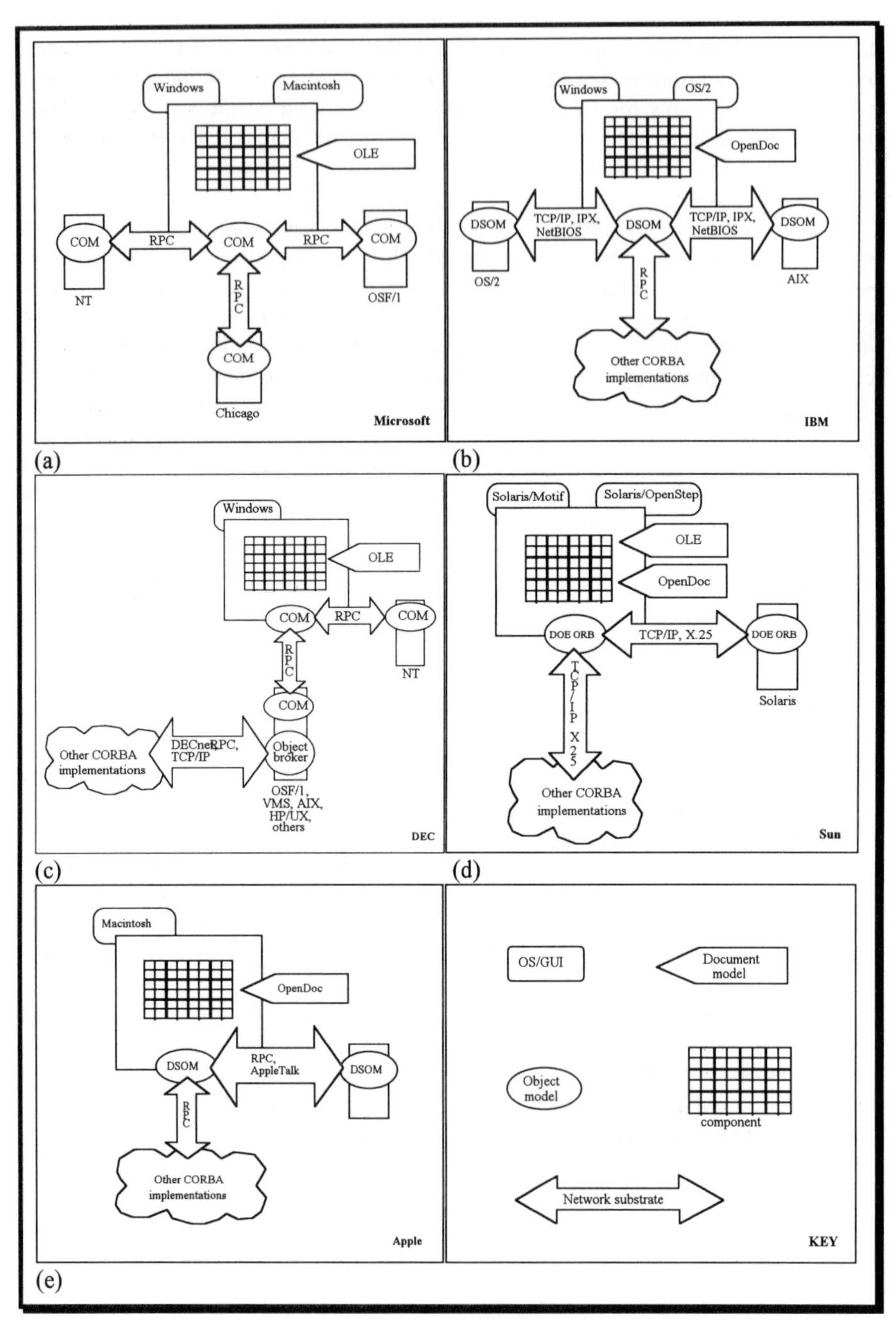

Figure 14.10 — Leading document, object and communication models.

14.4.1 HANSA

The European Esprit funded HANSA (Heterogeneous Application geNerator Standard Architecture) project is developing a framework in which a collection of independent tools can co-operate to solve complex real-world problems. The project is co-ordinated by Thorn EMI CRL, with University College London (UCL) as associate partner. The European partners include: Brainware and ARTS Consultants from Germany; MIMETICS and PROMIND from France; and O.Group Srl. from Italy.

The HANSA project uses object-oriented techniques to develop domain-specific application generators, a toolkit of application specific and industry standard tools, and a generic framework for their combination (as shown in Figure 14.11). The application generators are being constructed by the European partners in the business areas of direct marketing, banking, insurance and executive information systems.

The HANSA Framework, developed by the Computer Science Department of UCL, provides the support for integrating industry standard utilities (spreadsheets, databases, etc.) and novel artificial intelligence techniques (neural networks genetic algorithms, expert systems etc.). The framework offers the facilities to generate immediate hybrid systems by providing a rapid application generator and a communications architecture which is compliant to industrial standards such as OLE and CORBA.

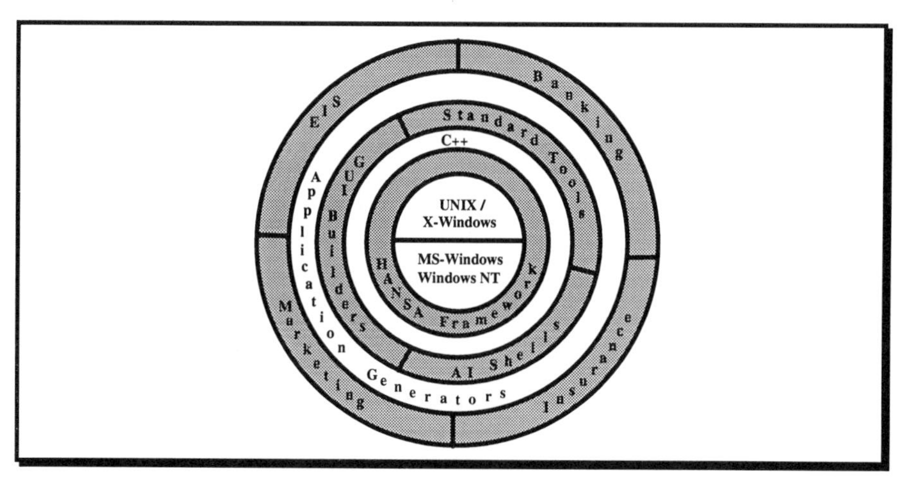

Figure 14.11 — The HANSA framework.

The HANSA framework is an object-oriented cross-platform communication architecture that allows processing components within hybrid applications to exchange information. The framework hides platform-related implementation details, promoting software reuse and rapid prototyping. On each of the target platforms (PCs running MS-Windows 3.1/NT and UNIX/X-Windows workstations) the framework implements a services interface inspired by, and compatible with, Microsoft's OLE protocol. For UNIX/X-Windows systems, HANSA is developing an OLE-like interface following closely the philosophy of the CORBA from the Object Management Group [Obje92], building the framework on top of the services supplied by the Object Request Broker. The platform-independent application programming interface allows easy porting of hybrid applications from one platform to another.

A HANSA application is composed of: one HANSA Manager (an application generator), responsible for controlling the configuration and execution of an application; and one or more HANSA Tools, the basic utilities that can be selected from a collection of HANSA compliant tools, found in the HANSA Toolkit. The HANSA Manager stores it's controlling and configuration information in an entity know as the *Blueprint*. As the name suggests this is a description of the hybrid application in terms of the tools used, their configurations, how they are linked together and the ordering of data exchange between tools.

For each application domain there is at least one HANSA Manager in charge of controlling the configuration and execution of co-operating HANSA Tools. The HANSA toolkit includes: neural network development environments such as MIMENICE from Mimetics [Mime92], the Genetic Algorithm programming environment GAME from University College London [Ribe94]; as well as any industry-standard utility (such as Excel and Access from Microsoft) that complies with the HANSA or the OLE communication protocols on PCs or UNIX workstations.

For allowing the integration of such a diverse range of Tools, HANSA constitutes a natural programming environment for the implementation of *hybrid* applications. Moreover, the possibility of using *off-the-shelf* tools, allows the user to configure and tune the application by utilising the most appropriate utilities for each task in the problem.

14.4.1.1 HANSA Communication Architecture

The core of the HANSA framework is the HANSA communication architecture (HCA). Figure 14.12, shows the HANSA Communication Architecture which comprises three communication protocols structured in two layers: application, and framework.

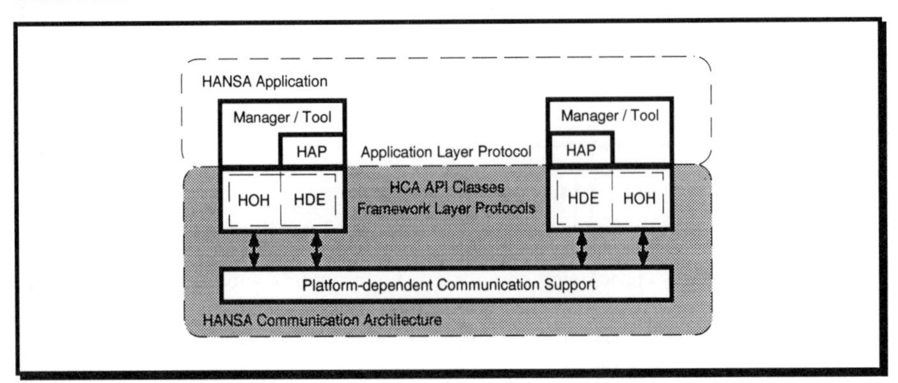

Figure 14.12 — HANSA communication architecture.

The communication protocols are the *object handling* (HOH) and *data exchange* (HDE) protocols, in the framework layer, and the *Application Protocol* (HAP), in the application layer. All protocols are implemented on top of a platform-dependent communication support hidden from the developers by a set of application programming interface classes. These constitute the platform-independent API that allows code to be portable across the HANSA platforms.

The communication architecture is based on industry standards, reflecting one of the requirements of the HANSA framework architecture. The most important of these standards is the *Object Linking and Embedding* protocol (OLE) developed by Microsoft. The Object Handling protocol is inspired by and compliant to OLE on the platforms targeted by HANSA (PCs and UNIX workstations). An important contribution of HANSA is the implementation of an OLE-style protocol for UNIX platforms.

The choice of OLE for one of the basic supporting technologies is a direct result of the desire to achieve *rapid prototyping* of applications, and to allow the use and integration of industry standard packages. As Figure 14.13 shows, this also implies the choice of a client/server architecture. In this context, HANSA Managers are full clients, while Tools can be simple servers, or client and servers simultaneously.

The implementation of the HCA application programming interface (API) can be divided into two parts: one for the implementation of the HCA protocols and one for the manipulation of the registration database (the holder of all information concerning HANSA Tools). The API is presented to the developer as a single set of C++ classes that is common to all platforms targeted by the HANSA Project. This set of classes hides all implementation and platform-specific details.

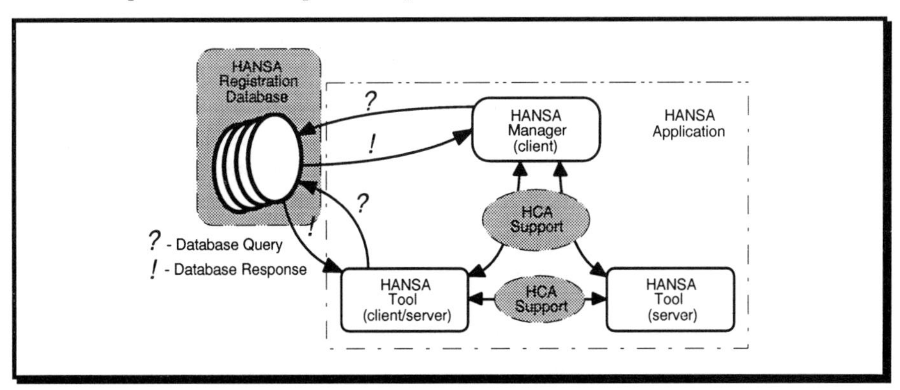

Figure 14.13 — HANSA framework operation.

HANSA Object Handling Protocol (HOH)

The Object Handling protocol is aimed at generic data exchange, configuration, and utility execution control. It is an extendable protocol that permits one utility to use the services of another. Through the Object Handling protocol a HANSA Manager can use HANSA Tools for specific tasks. Some HANSA Tools will also be able to benefit from the facilities offered by the protocol, by also using the services of other Tools.

The HOH protocol is the supporting mechanism for: handling the links between Manager and Tools, and between individual Tools; manipulating the permanent registration of services held by the registration database; and for dealing with point-to-point execution requests and user defined messages.

The utilities developed with the HOH application programming interface (API) are platform independent and source code portable across platforms.

HANSA Data Exchange Protocol (HDE)

The HDE protocol is the lowest level communication protocol of the HANSA communication architecture.

As happens in the HOH protocol, a HDE server is the provider of information to a HDE client. Server and clients have *conversations* about *topics*, and exchange data related to *items*. A conversation is an exchange of information related to a specific item of a topic. For example, a Tool that retrieves quotations from stock markets can support multiple stock markets, and multiple stocks within that market. The topic names that the Tool uses can be *NYSE*, *FT100*, etc. The item names will identify stocks such as ICL, Siemens, etc.

HANSA Application Control Protocol (HAP)

The HANSA Application Control Protocol (HAP) is a collection of HANSA-specific messages to be used in conjunction with the HDE or HOH protocols. These message definitions can be used for the control of HANSA Tools as well as OLE/DDE enabled Tools.

Upon registration with the HANSA registration database, Tools provide a message map, known as the HAP Map, which translates the HAP messages to the Tool's native commands. The main purpose of the HAP is to standardise on a set of control instructions that are common among HANSA registered Tools, thus establishing a common control language. The HAP is similar to the scripting language in OpenDoc, in that, there is a standard set of messages that all tools must be able to support.

14.4.2 Synergy (Prodea)

ProdeaSynergy is a visual programming tool that is aimed at aiding inter-utility integration. Conceptually, it is equivalent to a sophisticated general purpose HANSA manager. The product is made up of four major elements: the ProdeaSynergy Package, the ProdeaSynergy Library, the Variable Server and the Application Services Database.

A ProdeaSynergy Package is equivalent to a HANSA Blueprint. Packages are created by placing icons (which represent utilities and system objects that retrieve, process and store information), in the workspace. The movement of data between these objects are then represented by placing arrows from source to destination icons. Conditional statements may be attached to each data flow arrow allowing a different flow direction for each condition. Once defined the Package can be saved for use on any machine having the ProdeaSynergy environment.

The Library is a Package manager. It provides a password protected hierarchy tree structure to allow the storage of similar packages together and also provide facilities to change a package's representation attributes as well as a facility to launch packages.

The Variable Server provides 260 predefined global variables that can be set and referenced by each of the utilities running in a package. These variables can be manipulated via DDE, OLE and utility menu add-ins. These variables when used within utility scripts and conditional arrows allow runtime configuration.

The Application Services Database contains essential characteristics such as utility name, location, type or class, messaging protocols supported, file formats supported, and key sets of functions and commands that ProdeaSynergy needs for DOS and MS-Windows utilities. This again is similar to the HANSA registration database.

When a package is executed on a different machine, by using the Application Services Database, ProdeaSynergy is able to substitute one utility by another should the expected utility not be present on the system. This coupled with the ability for the system to act as mediator between utilities, selecting the best common format and optimum interchange method, makes the product an attractive choice for application integrators on MS-Windows. However, since ProdeaSynergy is not a programming language, it can't add extra functionality nor assume control of a utility, it can only help to link together utilities and start predefined scripts within the utilities. [Gill94].

14.4.3 Galaxy (VISIX)

Galaxy [Gala93] by VISIX enables application developers to build distributed, large-scale hybrid applications which take full advantages of the features of each platform. Galaxy is a comprehensive virtual operating environment, written in and using the C/C++ programming languages. It is available for all the main platforms including MS-Windows, Macintosh, VMS, OS/2 and UNIX. The environment consists of a set of libraries, services and tools. These collectively provide a uniform multi-platform API, portable network transparent operation, many development aids to help tool construction, "look and feel" independence and an extendable architecture for future operating environments.

The Galaxy libraries are available as a rich set of C functions and C++ classes implemented within a class hierarchy. The libraries have multiple abstraction levels for the underlying systems, namely the window system abstractions and operating system abstractions.

The window system abstractions provides API calls for the drawing model which supports rendering in PostScript; font management allowing font selection, scaling and rotation; colour and image management allowing the selection of colour and the conversion between different graphics format (e.g. GIF, BMP, MacPaint); and the look-and-feel which allows a Galaxy application to select its GUI look and feel at runtime. This means that an application can assume the MS-Windows look and feel whilst running on a Sun Sparcstation or an Apple Macintosh.

The operating system abstractions provide API calls for I/O, file and directory services (such as creating, deleting and searching of files) and communications. using the communications API, Galaxy will determine the best method for establishing and maintaining a connection-oriented network connection, shielding developers from the intricacies of comm·nication development. The Galaxy Backplane protocol accessed through the API allows seamless integration between Galaxy tools giving the developer "plug and play" functionality. There is also a Distributed Application Services (DAS) mechanism which is an inter-utility communication method. Galaxy provides an implementation of an asynchronous, symmetric, peer-to-peer protocol allowing utilities on different platforms to communicate with each other. Distributed Services such as a Hypertext help server are available to all Galaxy-based utilities running on the network through the use of DAS.

Since the libraries are compiler independent they enable native features such as shared libraries to be used where available. An example of this is the use of DLLs within MS-Windows and OS/2.

With the availability of the Galaxy Backplane protocol, VISIX has created an entry-point for third-party tools to join the initial suite of tools for developing applications. Currently there is a visual resource builder which helps speed up developing the graphical user interfaces, or front ends, for applications, a colour image editor for creating or modifying bit mapped images, and a help compiler for converting plain text into Galaxy resource files. Future tools will include a programmable text editor, help writer and project browser.

The combination of platform independence, network transparent communications and a host of tools to aid application construction, makes Galaxy a powerful development and execution environment for hybrid applications.

14.4.4 Orbix (IONA Technologies)

Orbix [Iona93] is IONA's implementation of a CORBA compliant Object Request Broker. It is one of the first commercially available ORBs, with its implementation available for several operating systems including SunOS, MS-Windows 3.1/NT, HP UX, IMB AIX, OSF/1, etc.

Orbix was designed to be a lightweight implementation of an ORB and has a layered architecture, consisting of a pair of C++ libraries (client and server), an activation daemon, an IDL compiler, interface repository and various utilities used to simplify the management of the implement repository where activation information for object handlers are kept.

The Orbix communication layer provides the essential transport facilities to allow utilities to communicate. The default implementation uses TCP/IP as its network communication protocol and XDR, a platform-independent data representation scheme, although implementations of other transport services are also possible.

The Orbix runtime layer implements the CORBA client and server APIs and the mechanisms needed to support both the static and dynamic invocation interfaces present in the CORBA specification.

Filtering is a mechanism that allows facilities not provided by Orbix to be added by the developer. These facilities are added by deriving from the filter class and implementing the extra functionality.

Orbix is one of the first commercially available ORBs and its implementation on different platforms make it very attractive. Since CORBA does not specify the message structure or transport mechanism to be used, gateways have to be built between different vendors' ORBs. IONA's link with SunSoft (who has a small stake in the company) and the fact that a gateway between Orbix and DOE (Sun's CORBA compliant environment) has already been implemented, makes Orbix an establish technology which will gain IONA blue chip status. Although Orbix does not come with many features such as persistence, authentication, encryption and threads, hooks are made available so that developers can add these features if and when they are needed. Unlike SOM/DSOM which are language neutral, Orbix is a C++ product. This ties the developer to the C++ language. IONA is very keen to have Orbix established as a standard for ORB implementation and has recently released the news that they will be integrating Microsoft's OLE within Orbix. This will enable Windows applications to transparently communicate across networks and use CORBA objects without worrying about their origin.

14.4.5 IntelliSphere (InferOne)

IntelliSphere from InferOne is one of the few, if not the only, commercial polymorphic hybrid development environment. It is a full graphical development environment for integrating symbolic and neural processing, within a neural architecture. The technical details of this approach are explained in Chapter 12.

The IntelliSphere approach is based on *Macro-connectionism*, a neuro-symbolic approach based on basic functional entities, called *neurOagents.* A neurOagent is an assembly of neurons that allows the integration of both symbolic and neural functionalities for analysing and developing knowledge-based systems. Each neurOagent is a knowledge agent which represents a piece of global knowledge. These neurOagents form a multi-agent approach to problem solving. The construction of knowledge bases with neurOagents is characterised by:

- a modular and incremental knowledge acquisition using InferOne's Local Knowledge Acquisition Methodology (LKAM),
- a top-down and/or bottom-up acquisition strategy,
- a capacity for learning from examples.

IntelliSphere offers various symbolic reasoning mechanisms coupled with neural properties, all within the neurOagent architecture. For example deduction and/or reasoning by hypothesis is achieved by using backward, forward and mixed propagation of values between neurOagents. Properties such as constraint satisfaction and processing of non-perfect information is achieved through propagation of constraints, information or numeric information between neurOagents.

IntelliSphere allows the combination of explicit modelling of knowledge and the ability to learn from examples, to be available for knowledge acquisition. This is important because it allows the modelling of expertise which can be difficult to formalise and allows historical examples within corporate databases to be utilised as knowledge.

IntelliSphere is also designed to integrate with conventional systems (such as databases, multimedia systems, document management systems, etc.) via neurOagents.

14.4.6 Blackboard Systems

A blackboard system is a group of knowledge sources collaborating with each other by way of a shared database called the *blackboard*, in order to reach a solution to a problem. This blackboard architecture is analogous to a group of human experts seated next to a large blackboard, trying co-operatively to solve a problem. The problem solving starts when the initial data is written on the blackboard. The specialists look for an opportunity to apply their expertise to the developing solution and when sufficient information is available on the blackboard, the experts, contribution is recorded on the blackboard. This process of adding contributions continues until the problem is solved.
As shown by Figure 14.14, the basic components of a blackboard system are :

- The Blackboard, a global database containing data, partial solutions, and other data in various problem-solving states. The blackboard serves as a communication medium and buffer, and a trigger mechanism for knowledge sources.

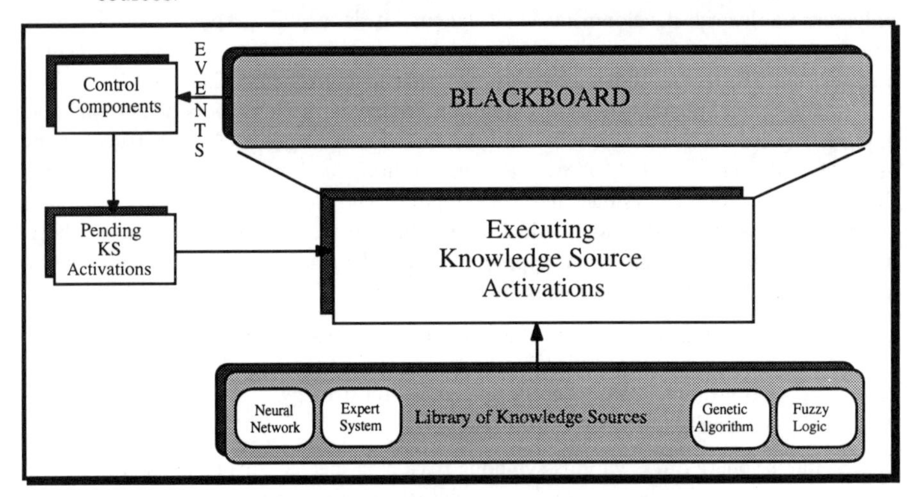

Figure 14.14 — A basic blackboard system.

- The knowledge sources (KS) are independent modules that contain the knowledge needed to solve the problem. KSs can be diverse in representation and inference techniques. A KS needs no knowledge of the expertise, or even the existence of the other KSs; however, it needs knowledge of the state of the problem-solving process and the representation of relevant information on the blackboard. It declares, to the control mechanism, the conditions under which it can contribute to the solution and when executed, attempts to contribute toward solving the problem.
- A control mechanism, is separate from the individual KSs and directs the problem-solving process by deciding which KS is most appropriate for

execution at each step in the solution process. Based on the state of the solution, the control mechanism can apply any type of KS, whether it be goal driven, data driven, or model driven. In some blackboard systems, the control component itself is implemented using a blackboard approach involving control KSs and blackboard areas devoted to control.

Knowledge sources are much larger grained than the individual rules used by expert systems. While expert systems work by firing a rule in response to a stimuli, a blackboard system works by firing an entire knowledge module, or KS. These knowledge sources can be expert systems, neural networks, genetic algorithms or any other information processing technique.

The knowledge sources are not actually active agents, instead KS activations (or KS instances) are the active entities that are competing for the execution resources. KS activations are the combination of KS knowledge and a specific triggering context. The distinction between KSs and KS activation is important for applications in which numerous events trigger the same KS knowledge (focusing on the appropriate data context), rather than among different KSs (focusing on the appropriate knowledge to apply). KSs are static repositories of knowledge; KS activation's are active processes.

The control mechanism directs the problem-solving process by letting KSs opportunistically respond to blackboard database changes. A blackboard system uses an incremental reasoning style, that is, the problem's solution is built one step at a time. At each step, the system can execute any triggered KS and choose a different focus of attention on the basis of the solution's state. Typically a control approach, would allow the executing KS activation to generate events as it makes contributions to the blackboard. These events are maintained and possibly ranked until the executing KS activation is completed. At this point the control component uses the event to trigger and activate the highest ranked or most appropriate KS activation. This control process continues until the problem is solved. Blackboard systems support a variety of control mechanisms and algorithms, so that a choice of opportunistic control techniques are available to the application developer. Blackboard problem-solving architecture is suitable when:

- Many diverse, specialised knowledge representations are needed. KSs can be developed in the most appropriate representation for the data they handle.
- Application development involves many developers. The modularity and independence provided by large-grained KSs lets each KS be designed, implemented, tested and maintained separately.
- Uncertain knowledge or limited data inhibits absolute determination of a solution. The blackboard systems incremental approach will still allow progress in such cases.
- Multilevel reasoning or flexible, dynamic control of problem-solving activities is required in an application.

Various systems based around a blackboard architecture [Cork91] have been adopted to achieve design, communication and execution autonomy. These systems have been applied in many areas including process control, planning and scheduling, computer vision and other areas. The most successful of these have been systems such as Hearsay-II [Erma80] for speech understanding, HASP [Nii82] for sonar signal interpretation, CRYSALIS for determining spatial constituents of proteins and OPM for planning multiple-task sequences in areas of conflicting goals and constraints.

Many blackboard systems utilise object-oriented technologies to implement the independent knowledge sources and it is expected that a growth of specialised blackboard application development environments will begin avialable to developers of hybrid systems.

14.5 Future

As the field of intelligent hybrid systems matures, developers of these systems will require representation, communication and execution technologies for integration of different information processing components. The recent interest in object-oriented programming methods and in building distributed computer systems, has provided some of this integration infastructure. Currently these integration technologies are being developed and accepted into mainstream computing.

With recent trends in computing, such as client/server architectures, application software constructed from components and a move towards modular object-oriented design, it will not be long before the practical requirements of intelligent hybrid systems will be fully met by the various integration and communication techniques that are becoming computing standards. For example the current race to develop and have accepted, a standardised object-oriented communication protocol, such as CORBA or OLE, will ensure that different processing components will be able to communicate across heterogeneous networks.

As the need for hybrid systems becomes apparent, there will be a need for tools and environments that make the development of hybrid systems easier for the developer or the end user. These environments should allow the user to visually configure and execute intelligent hybrid systems, across a heterogeneous network of computers. The environment should shelter the developer from any need to know exactly where components reside on the network. The various integration and communication protocols that are being developed could be used to internally implement the communication requirements of the environment.

The development of tools and programming environments coupled with the standardisation of object-oriented integration technologies will finally converge to offer easy to use, "network aware" intelligent hybrid systems.

References

[Cork91] Corkill, Daniel, "Blackboard Systems", *AI Expert*, pp. 41-47, Sept 1991

[Diet89] Dietrich, W. C, Nackman, L. R, and Gracer, F, "Saving a legacy with objects", ed. N Meyrowtz, Addison-Wesley, Reading, MA, *Proceedings of OOPSLA89*, 1989

[Digi94] Digital Equipment Company, *An Object-oriented Framework for Flexible Integration*, 1994

[Erma80] Erman, L. D., , F. Hayes-Roth, Lesser, V. R., and Reddy, D. R., "The HEARSAY-II Speech-Understanding System: Integrating Knowledge To Resolve Uncertainty", *Computer Survey*, vol. 12, pp. 213-253, 1980.

[Gala93] Visix Software Inc. *Galaxy Application Environment.* Visix Software Inc., 11440 Commerce Park Drive, Reston VA 22091, U.S.A. 1993.

[Gill94] Gillmor Steve. "A New Synergy for Windows." *Byte*. 1994.

[Goon92] Goonatilake, Suran and Khebbal, Sukhdev, "Intelligent Hybrid Systems ", *Proceedings of First International Conf. on Intelligent Systems, Singapore*, pp. 207-212, Sept 1992.

[Iona93] IONA Technologies Ltd. *The Orbix Architecture.* IONA Technologies Ltd., O'Reilly Institute, Westland Row, Dublin 2, Ireland. 1993.

[Juli93] Julienne, A., Russell, L., "Software Development : ToolTalk", *SunExpert Magazine*, pp 51-58, March 1993.

[Nii82] Nii, H. P., Feigenbaum, E. A., Anton, J. J., and Rockmore, A. J., "Signal-to-Symbol Transformation: HASP/SAIP Case Study", *AI Magazine*, vol. 3, pp. 23-35, 1982.

[Micr91] Microsoft Corporation, "Object Linking and Embedding", 10 June 1991.

[Mime92] Mimetics S.A. *Mimenice 2 – User Manual.* Mimetics S.A., Centrale Park, Avenue Sully Prud'Homme, 92298, Chatenay-Malabry, France. 1992.

[Mow93] Mowbray, T., Brando, T., "Interoperability and CORBA-based Open Systems", *Object Magazine*, pp. 50-54, October 1993.

[Obje92] Object Management Group, *The Common Object Request Broker: Architecture and Specification*, Object Management Group, 1992.

[Oden93] Odin, R., *Object-Orientation: An introduction for COBOL Programmers*, Micro Focus Publishing, 1993.

[Pier94] Piersol, Kurt, "A Close-Up of OpenDoc", *Byte*, pp.183-188, January 1994.

[Ribe94] Ribeiro-Filho, J.L., Treleaven, P, "GAME: A Framework for Programming Genetic Algorithms Applications", *Proceedings of the First IEEE Conference on Evolutionary Computing, IEEE Congress on Computational Intelligence*, 26th June-2nd July 1994, Orlando, USA, vol. 2, pp. 840-845, 1994.

[Stro91] Stroustrup, Bjarne, *The C++ Programming Language,* 2nd Ed, Addison-Wesley Publishing Company, 1991.

[SunS91] SunSoft. *The ToolTalk Service.* SunSoft, 2250 Garcia Avenue, Mountain View, CA 94043, USA. 1991.

[Udel94] Udell Jon. "A Taligent Update." *Byte*. 1994.

[Udel93] Udell, Jon. "IBM's Assault on Distributed Objects." *Byte*. 1994.

[Wayn94] Wayner, Peter, "Objects on the March", *Byte*, pp.139-150, January 1994.

Technology Suppliers

Microsoft Corporation (OLE and COM)
One Microsoft Way
Redmond, Washington 98052-6399
U.S.A.
Tel: +1 (206) 882 8080

Object Management Group, Inc. (CORBA)
492 Old Connecticut Path
Framingham, MA 01701
U.S.A
Tel: + 1 (508) 820 4300

Digital Equipment Co. Ltd (ObjectBrocker, DCE)
P.O. Box 110, Reading
Berkshire RG2 0TR
U.K.
Tel: +44 (0) 734 868 711

Sun Microsystems Inc.(ToolTalk)
2550 Garcia Avenue
Mountain View
California 94043-1100
U.S.A.
Tel: +1 (415) 336 0678

Hyperdesk Corp. (HD-DOMS)
2000 West Park Drive, Suite 300
Westborough
Massachusetts 01581
U.S.A.
Tel: +1 508 366 5050

IBM UK Ltd. (SOM/DSOM)
National Enquiry Centre
414 Chiswick High Road
London W4 5TF
U.K.
Tel: +44 (0)81 747 0747

Component Integration Laboratories (OpenDoc)
688 Fourth Avenue
San Fransisco, California 94118
U.S.A.
Tel: +1 (415) 750 8352

Taligent Inc. (TAE)
10201 North De Anza Boulevard
Cupertino CA 95014-2233
U.S.A.
Tel: +1-800-288-5545

University College London (HANSA)
Computer Science Department
GowerStreet
London WCE1 6BT
U.K.
Tel: +44 (0)71 378 7050 x 3695

Prodea Software Corp.(Synergy)
11095 Viking Dr., Suite 225
Minneapolis, MN 55344
U.S.A.
Tel: + 1 (612) 942 1000

Visix Software Inc. (Galaxy)
11440 Commerce Park Drive
Reston VA 22091
U.S.A.
Tel: +1 (703) 758 8230

IONA Technologies Ltd. (Orbix)
O'Reilly Institute
Westland Row, Dublin 2
IRELAND
Tel: + 353 1 679 0677

Inferon (IntelliSphere)
Immeuble Les Marades
1, Bd de l'Oise
95030 Cergy Pontoise Cedex
FRANCE
Tel: + 33 (1) 30.75.00.42

Blackboard Technology Group Inc (Blackboard Systems)
401 Main Street
Amherst, Massachusetts 01002
U.S.A.
Tel: +1 (413) 256 8990

Glossary

Neural Networks

Activation Level: The level of activity of a network node.

ANN: An acronym for Artificial Neural Network.

Backpropagation Algorithm (Backprop): A popular supervised neural network learning algorithm. The weights of the network are adjusted by backward propagation of the error signal from outputs to inputs.

Connectionist Networks: The general class of processing networks in which tasks are performed by several processes which interact by sending information to each other through point-to-point connections.

Distributed Representation: A representation scheme in which concepts are encoded by the collective values of nodes in a network.

Error: The difference of comparing a network's actual activity pattern and the desired pattern.

Feedforward network: A network ordered into layers with no feedback paths.

Flow-Of-Activation: The point-to-point transmission of activation levels between network nodes.

Generalisation: The ability to perform approximate matches between an input pattern and a set of known patterns.

Gradient Descent: Algorithm for minimising some error measure by making small incremental weight adjustments proportional to the gradient of the error.

Hidden Units: In a multilayered neural network, the nodes which are not inputs or outputs of the network.

Learning: The term used to describe the neural network adaptation process.

Link: A directed communication channel connecting two nodes.

Neurocomputer: A computer specialised for the implementation and execution of neural nets.

Neuron: A biological nerve cell.

Node: The fundamental processing element of a neural network.

Pattern: An ordered set of inputs (vectors) that are presented to a neural network (input patterns).

Perceptron: Simple early neural network that consists of a input layer connected to a single node.

Self-Organisation: The capability of an unsupervised neural network to modify its internal organisation or parameters through a learning mechanism.

Semantic Network: A knowledge representation scheme in which concepts are represented by a network of nodes and relationships between the nodes.

Squashing Function: Function whose values is always between finite limits, even when the input is unbounded.

Synapse: Contact point between neurons in which dendrites of several neurons are attached to an axon.

Firing-Threshold: The node input value which much be exceeded before the node will produce an output value.

Training: The process of exposing an adaptive system to a selected set of training inputs for which specific system outputs are desired.

Weight: A measure of strength of a connection linking two nodes.

Weight Matrix: An ordered collection of all of the weights present in a neural network.

Genetic Algorithms

ARTIFICIAL LIFE: The study of the study of simple computer generated hypothetical life forms.

CHROMOSOME: An individual or member of the population. This is a datastructure which holds a set of task parameters. This may be stored as binary bit-strings, arrays of integers or lists.

CLASSIFIER SYSTEM: An extension of genetic algorithms in which the population consists of a co-operating set of rules which are used to learn to solve a problem by the presentation of training cases. Between each generation the population as a whole is evaluated and fitness is assigned to each rule typically by using the "bucket-brigade" algorithm. This algorithm aims to reward or punish rules by adjusting the individual rule's fitness according to how much the rules contribute to solving the problem at hand.

CONVERGED: Generally a genetic algorithm is said to have converged when a large proportion of the chromosomes in the population are the same.

COEVOLUTION: Two or more populations are evolved at the same time. Usually the separate populations compete against each other.

CROSSOVER: A reproduction operator which forms a new chromosome by combining parts of two "parent" chromosomes. The simplest form is single-point crossover in which an arbitrary point in the chromosome is picked.

DECEPTION: The condition where the combination of good building blocks leads to reduced fitness, rather than increased fitness.

EPISTASIS: A biological term used to denote that the fitness of an individual depends upon the interaction of a number of their genes. In genetic algorithms this would be indicated by the fitness containing a non-linear combination of components of a chromosome.

EVOLUTIONARY STRATEGIES: An approach similar to genetic algorithms but does not use crossover operators.

EXPLOITATION: The process of using information gathered from previously visited points in the search space to determine which places visit next.

FITNESS: A value assigned to an chromosome (individual) which reflects how well it solves the task in hand. A "fitness function" is used to map a chromosome to a fitness value.

GENE: A subsection of a chromosome which encodes the value of a single parameter.

GENERATION: An iteration of the measurement of fitness and the creation of a new population by means of reproduction operators.

GENETIC OPERATOR: A search operator acting on a coding structure that is analogous to a genotype of an organism (e.g. a chromosome).

GENETIC PROGRAMMING: The creation of computer programs through the use of genetic algorithm operators. The members of the population are parse trees of computer programs. Genetic programming (GP) is most easily implemented where the computer language is tree structured and therefore LISP is often used for GP implementations.

GENOTYPE: The genetic composition of an organism.

GENOME: The entire collection of genes possessed by an organism.

INDIVIDUAL: A single member of a population which represents a possible solution to the problem being solved.

MIGRATION: The transfer of an individual from one sub-population to another.

MUTATION: A reproduction operator which forms a new chromosome by making small alterations to the values contained in a chromosome.

OFFSPRING: An individual generated by a process of reproduction.

PARENT: An individual which takes part in reproduction to generate one or more other individuals, known as offspring, or children.

PHENOTYPE: The expressed traits of an individual.

POPULATION: The group of individuals or chromosomes that interact together by mating to produce new offspring.

PREMATURE CONVERGENCE: This is when a genetic algorithm population converges to a solution which is not the one that is desired.

REPRODUCTION: The creation of a new individual from two parents (sexual reproduction). Asexual reproduction is the creation of a new individual from a single parent.

REPRODUCTION OPERATOR: The mechanism by which genetic information is passed on from parent(s) to offspring during reproduction. Reproduction operators fall into two main categories: crossover and mutation operators.

SEARCH SPACE: If the solution to a task can be represented by a set of N real-valued parameters, then the task of finding this solution can be thought of as a search in an N-dimensional space. This space of possible solutions is referred to as the search space.

SELECTION: The process by which some individuals in a population are chosen for reproduction, typically on the basis of favouring individuals with higher fitness.

Fuzzy Logic

Crisp Logic: Another name for traditional logic to differentiate it from fuzzy logic. In crisp logic, the logic operations AND, OR, and NOT return a 1 or 0 (true or false).

Crisp Set: The traditional definition of a set in classical or symbolic logic. Traditional or crisp sets have strict membership criteria in which an object is either completely included or excluded from the set.

Defuzzification: A process in which fuzzy output is converted into crisp, numerical results. There are several popular methods for defuzzification, the "centre of area" and "mean of maximum" being two of the most widely used.

Fuzzification: The process of converting observed values into fuzzy values through the use of fuzzy membership functions.

Fuzziness: A term that expresses the ambiguity that can be found in the definition of a concept.

Fuzzy Logic: The process of solving problems using imprecise linguistic concepts such as "low", "medium", "young" etc. In fuzzy logic, the three logic operations AND, OR and NOT return a degree of membership that is usually a number between 0 to 1.

Fuzzy Control System: A control system based on using fuzzy IF-THEN rules and fuzzy inference mechanisms.

Fuzzy Inference: The process of deriving conclusions from a given set of fuzzy rules acting on fuzzified data. In fuzzy inferencing all rules whose antecedents are true or partially true will contribute to the final results.

Fuzzy Rules: These specify the relationships among the fuzzy variables. The rules are specified in the familiar IF-THEN format. These rules are usually specified by a domain expert.

Fuzzy Set: A non-traditional type of set that allows an element to have gradual or partial degrees of membership. In fuzzy logic, traditional Boolean values of true and false (1 and 0) are replaced by continuous-set membership values ranging from 0 and 1.

Fuzzy System: A system whose variables range over states that are fuzzy sets.

Membership: The degree of inclusion in a set. Fuzzy sets have values between 0 and 1 that indicate the degree to which an element has membership.

Membership Function: The mapping that associates each new value with its grade of membership.

Expert Systems

Bi-directional Search: An inference method that combines forward and backward chaining.

Breadth-first Search: A search strategy in which the rules or objects on the same level of the hierarchy are examined before any rules or objects on the next-lower level are checked.

Certainty Factor: A number that measures the certainty or confidence in a fact or rule.

Declarative Knowledge: Facts about objects, events, or situations.

Depth-first Search: A search strategy in which one rule or object on the highest level is examined, and then rules or objects immediately below that one are examined, in contrast to a breadth-first search.

Domain Expert: A person with expertise in a particular domain.

Expert System: A computer program containing knowledge and reasoning capability that imitates human experts in problem-solving in a particular domain.

Explanation Facility: The component of an expert system that explains how solutions were reached and justifies the steps used to reach them.

Forward Chaining: An inference technique in which the IF portion of rules are matched against facts to establish new facts.

Frame: A knowledge representation method associates an object with a collection of features (e.g., facts, rules, defaults, and active values). Each feature is stored in a slot.

Heuristic: A rule-of-thumb approach that suggests a procedure to solve a problem; a heuristic approach does not guarantee a solution to a specific problem.

Inference: The process by which new facts are derived from known facts.

Inference Engine: The component of an expert system that controls its operation by selecting the rules to use, executing those rules, and determining when an acceptable solution has been reached.

Knowledge Acquisition: The process of identifying, interviewing, collecting and refining knowledge.

Knowledge Base: The component of an expert system that contains the system's knowledge.

Knowledge Engineer: A programmer responsible for capturing and encoding human expert knowledge.

Knowledge Representation: The technique used to encode and store facts and relationships in a knowledge base.

Monotonic Reasoning: A reasoning method based on the assumption that once a fact is determined it cannot be altered during the course of the consultation process.

Nonmonotonic Reasoning: A reasoning method that allows multiple approaches to reach the same conclusion and the retraction of facts or conclusions given new information.

Production Rule: A rule in the form of an IF-THEN or CONDITION-ACTION statement.

Search: The process of examining the set of possible solutions to a problem to find an acceptable solution.

Semantic Network: A method for representing associations between objects and events.

Working Memory: The component of an expert system that is composed of all attribute-value relationships temporarily established while the inference process is in progress.

Rule Induction

Chunking: Grouping lower-level descriptions (patterns, operators, goals) into higher-level descriptions.

Concept Description: A symbolic data structure defining a concept describing the class of all known instances of the concepts.

Concept Formation: A form of learning in which the learner generates concepts useful in characterising a given collection of objects or facts or subsets of them.

Constructive Induction: An inductive learning process that generates new descriptors not provided in the description of initial facts or observations.

Decision Tree: A discrimination network in the form of a tree structure.

Generalisation: Extending the scope of a concept description to include more instances: the opposite of specialisation.

ID3: A popular decision tree induction algorithm.

Negative Example: In learning from examples, a counter example of a concept that may bound the scope of generalisation.

Positive Example: An example or instance of a concept to be learned in learning from examples.

Schema: A symbolic structure that can be filled in by specific information (instantiated) to denote an instance of the generic concept represented by the structure.

Specialisation: Narrowing the scope of a concept description, thus reducing the sets of instances: the opposite of generalisation.

Version Space: The set of alternative candidate concept descriptions that are consistent with the training data, knowledge, and assumptions of the concept learner.

Intelligent Systems Resource Guide

This resource guide aims to give the reader main references to Intelligent Systems and pointers to further information. While there are still few dedicated sources for Intelligent Hybrid Systems information, relevant material is often found in sources that describe the individual intelligent techniques. The intelligent techniques covered in this guide are neural networks, genetic algorithms, fuzzy logic, expert systems and rule induction.

For each intelligent technique we include the best books, journals, introductory articles and conferences. As internet access is increasing rapidly, we also include listings of relevant bulletin boards, email lists and Mosaic home pages on the World Wide Web.

The World Wide Web (WWW) is a hypermedia document that spans the Internet. NCSA Mosaic[1] provides a convenient way to access the WWW. Mosaic is available by anonymous ftp from ftp.ncsa.uiuc.edu:/Mosaic/ as source code and binaries for Sun, SGI, IBM RS/6000, DEC Alpha OSF/1, DEC Ultrix, and HP-UX. Queries should be directed to mosaic-x@ncsa.uiuc.edu (X-Windows version), mosaic-mac@ncsa.uiuc.edu (Macintosh), and mosaic-win@ncsa.uiuc.edu (Microsoft Windows).

[1]NCSA Mosaic is a World Wide Web browser developed at the National Center for Supercomputing Applications (NCSA).

Neural Networks

Books

Aleksander, I. and Morton, H., An Introduction to Neural Computing. Chapman and Hall, 1990. ISBN 0-412-37780-2.

R. Beale and T. Jackson, Neural Computing: An Introduction. Adam Hilger, Bristol, UK, 1990.

McClelland, J. L. and Rumelhart, D. E., Explorations in Parallel Distributed Processing: Computational Models of Cognition and Perception. The MIT Press, 1988.

Wasserman, P. D., Neural Computing: Theory & Practice. Van Nostrand Reinhold: New York, 1989. ISBN 0-442-20743-3.

Caudill, M. and Butler, C., Naturally Intelligent Systems. MIT Press: Cambridge, MA, 1990. ISBN 0-262-03156-6.

Hertz, J., Krogh, A., and Palmer, R., Introduction to the Theory of Neural Computation. Addison-Wesley: Redwood City, CA, 1991. ISBN 0-201-50395-6 (hardbound) and 0-201-51560-1 (paperbound)

Articles

Geoffrey E. Hinton, "How neural networks learn from experience", Scientific American 267(3):144-151, 1992.

Scott Fahlman and Geoffrey Hinton, "Connectionist Architectures for Artificial Intelligence", IEEE Computer 20(1):100-109, January 1987.

Journals

Neural Networks, Pergamon Press, Pergamon Journals Inc., Fairview Park, Elmsford, New York 10523, USA and Pergamon Journals Ltd. Headington Hill Hall, Oxford OX3 0BW, UK. ISSN #: 0893-6080 - Official Journal of International Neural Network Society (INNS).

Neural Computation, MIT Press, MIT Press Journals, 55 Hayward Street Cambridge, MA.

IEEE Transaction on Neural Networks, Institute of Electrical and Electronics Engineers (IEEE), Service Center, 445 Hoes Lane, P.O. Box 1331, Piscataway, NJ. 08855-1331, USA.

Connection Science: Journal of Neural Computing, Artificial Intelligence and Cognitive Research, Carfax Publishing Company, P. O. Box 25, Arbingdon, Oxfordshire OK14 3UE, UK.

Conferences

IEEE International Conference on Neural Networks. Proceedings available from: IEEE Service Center, 445 Hoes Lane, P.O. Box 1331, Piscataway, NJ 08855-1331, USA.

Neural Information Processing Systems (NIPS) Proceedings available from: Morgan Kaufman Publishers, 2929 Campus Drive, Suite 260, San Mateo, CA 94403, USA.

Free Software Packages

***GENESIS* 1.4.1:**

GEneral NEural SImulation System is a general purpose simulation platform which was developed to support the simulation of neural systems ranging from complex models of single neurons to simulations of large networks made up of more abstract neuronal components. ftp from genesis.cns.caltech.edu [131.215.137.64].

Mactivation:

Anonymous ftp from bruno.cs.colorado.edu [128.138.243.151]. Directory: /pub/cs/misc. File: Mactivation-3.3.sea.hqx.

***Aspirin/MIGRAINES* 6.0:**

Consists of a code generator that builds neural network simulations by reading a network description (written in a language called "Aspirin"). ftp sites: CMU's simulator collection on pt.cs.cmu.edu [128.2.254.155]. Directory: /afs/cs/project/connect/code. File: am6.tar.Z.

***Nevada Backpropagation* (NevProp):**

NevProp is a user-friendly backpropagation program written in C for UNIX, Macintosh, and DOS. ftp from the University of Nevada, Reno: ftp from unssun.scs.unr.edu [134.197.10.128]. Directory: pub/goodman/nevpropdir. File: nevprop1.16.shar.

Commercial Software Packages

NeuralWorks Professional II Plus:

NeuralWare Inc. Pittsburgh, PA 15276-9910. Phone (US) 412 787-8222, Fax: (US) 412 787-8220.

BrainMaker:

BrainMaker, BrainMaker Pro, Company: California Scientific Software, 10024 Newtown rd, Nevada City, CA 95959. Phone (US) 916 478 9040, Fax (US) 916 478 9041, Email: calsci!mittmann@gvgpsa.gvg.tek.com.

SAS Software:

Neural Net add-on, SAS Institute, Inc. SAS Campus Drive, Cary, NC 27513, USA. Fax (US) 919 677-8000, Email: saswss@unx.sas.com.

MATLAB Neural Network Toolbox:

For use with Matlab 4.x from The MathWorks, Inc., 24 Prime Park Way, Natick, MA 01760. Phone (US) 508-653-1415, Fax (US) 508-653-2997, Email: info@mathworks.com.

Bulletin Boards, Email lists & WWW Home Pages

Neuron Digest: Internet Mailing List. Moderated by Peter Marvit. To subscribe, send email to neuron-request@cattell.psych.upenn.edu.

comp.ai.neural-nets : Usenet group.

Central Neural System Electronic Bulletin Board: Modem: 509-627-6CNS; Sysop: Wesley R. Elsberry; P.O. Box 1187, Richland, WA 99352; welsberr@sandbox.kenn.wa.us.

http://www.erg.abdn.ac.uk/projects/neuralweb/

(Neural Web WWW Page)

http://www.erg.abdn.ac.uk/projects/neuralweb/digests/ (Neuron Digest WWW Page)

http://www.neuronet.ph.kcl.ac.uk/ (NeuroNet WWW home page). NeuroNet is the European 'Network of Excellence' for Neural Networks.

A neural networks Frequently Asked Questions (FAQ) is available as

http://www.eeb.ele.tue.nl/index.html

Genetic Algorithms

Books

Davis, Lawrence (editor), "Handbook of Genetic Algorithms", Van Nostrand Reinhold, New York, 1991, ISBN 0-442-00173-8.

David E. Goldberg, "Genetic Algorithms in Search, Optimisation, and Machine Learning", Addison-Wesley, Reading, MA, 1989, 412 pages. ISBN 0-201-15767-5.

Davis, Lawrence (editor) "Genetic Algorithms and Simulated Annealing", Morgan Kaufmann, 1989.

Koza, John R., "Genetic Programming: On the programming of computers by means of natural selection", MIT Press, 1992, 819 pages. ISBN 0-262-11170-5.

Articles

Holland, J.H. (1992) "Genetic Algorithms", Scientific American, 267(1), 66-72.

Peter Wayner (1991), "Genetic Algorithms: Programming takes a valuable tip from nature", BYTE, January, 361--368.

Goldberg, D. (1994), "Genetic and Evolutionary Algorithms Come of Age", Communications of the ACM, 47(3), 113-119.

J.R. Filho, C. Alippi and P. Treleaven, (1994), "Genetic Algorithm Programming Environments" IEEE Computer, February 1994 issue.

Journals:

Evolutionary Computation, Published quarterly by: MIT Press Journals, 55 Hayward Street, Cambridge, MA 02142-1399, USA.

Artificial Life, Published quarterly by: MIT Press MIT Press Journals, 55 Hayward Street, Cambridge, MA 02142-1399, USA.

Conferences

ICGA: International Conference on Genetic Algorithms. Proceedings available from Lawrence Erlbaum Associates, Hillsdale, New Jersey, USA.

IEEE Conference on Evolutionary Computation. Proceedings available from IEEE Service Center, 445 Hoes Lane, P.O. Box 1331, Piscataway, NJ 08855-1331, USA.

PPSN: Parallel Problem Solving from Nature, Proceedings published by North-Holland, Elsevier Science Publishers, Sara Burgerhartstraat 25, P.O. Box 211, 1000 AE Amsterdam The Netherlands.

FOGA: Foundations of Genetic Algorithms, Proceedings available from Morgan Kaufmann Publishers, Inc., P.O. Box 50490, Palo Alto, CA 94303-9953, USA.

Free Software Packages

Genesis:

Genesis is a GA system written in C by John Grefenstette. Genesis is available by ftp from ftp.aic.nrl.navy.mil:/pub/galist/src/ga/genesis.tar.Z.

GAucsd:

GAucsd is a Genesis-based GA package incorporating user interface improvements with facilities to distribute experiments over networks of machines. GAucsd is written in C for UNIX systems, and is available by ftp from cs.ucsd.edu:/pub/GAucsd/GAucsd14.sh.Z. [132.239.51.3].

GAME:

GAME (GA Manipulation Environment) has been developed by University College London as part of the PAPAGENA project of the European Community's Esprit III initiative. GAME allows the simulation of sequential as well as Parallel Genetic Algorithms and is available from the ftp server bells.cs.ucl.ac.uk:/pub/papagena/. For further information, email: <zeluiz@cs.ucl.ac.uk>.

Evolution Machine:

The Evolution Machine (EM) can be run on PC's with MS-DOS and Turbo C, Turbo C++. The EM software is available by ftp from ftp-bionik.fb10.tu-berlin.de:/pub/software/Evolution-Machine/ [130.149.192.50].

GANNET:

GANNET (Genetic Algorithm / Neural NETwork) is a software package for evolving neural networks. GANNET is available by ftp from fame.gmu.edu:/gannet/source/ [129.174.1.140]. For further information email, <dduane@ fame.gmu.edu>.

LibGA:

LibGA is a library of routines written in C for developing Genetic Algorithms. LibGA runs on UNIX, DOS, NeXT, and Amiga. LibGA Version 1.00 is available by ftp from ftp.aic.nrl.navy.mil:/pub/galist/src/ga/libga100.tar.Z. For further information email, <corcoran@ penguin.mcs.utulsa.edu>.

Commercial Software Packages

Evolver:

Evolver is a spreadsheet add-in which incorporates a genetic algorithm. Evolver can be customised through a macro language, and is available for the Excel, WingZ and Resolve spreadsheets on the Mac and PC computers. For further information, contact: Axcelis, Inc., 4668 Eastern Avenue North, Seattle, WA 98103-6932, USA. Phone (US) 206 632-0885.

EvoFrame:

EvoFrame is an object oriented programming tool for evolution strategies. EvoFrame is available as Version 2.0 in Borland-Pascal 7.0 and Turbo-Vision for PC's and as Version 1.0 in C++ for Apple Macintosh using MPW and MacApp.

For further information, contact: Wolfram Stebel, Optimum Software, Braunfelser Str. 26, 35578 Wetzlar, Germany. Email: <optimum@ applelink.apple.com>

MicroGA:

MicroGA allows programmers to integrate GAs into their software. It is an object-oriented C++ framework that comes with source code and documentation. MicroGA is available for Macintosh

II or higher with MPW and a C++ compiler, and also in a Microsoft Windows version for PC compatibles. For further information contact: Steve Wilson, Emergent Behaviour, 635 Wellsbury Way, Palo Alto, CA 94306, USA. Email: emergent@aol.com.

Bulletin Boards, Email lists & WWW Home Pages

Genetic Algorithm Digest: Send administrative requests to <ga-list-REQUEST@aic.nrl.navy.mil>. The anonymous ftp archive: ftp.aic.nrl.navy.mil:/pub/galist/ contains back issues, GA-code, and conference announcements.

Genetic Programming Mailing List: Admin requests: <genetic-programming-REQUEST@ cs.stanford.edu>. The anonymous ftp archive: ftp.cc.utexas.edu:/pub/genetic-programming/ includes a FAQ.

Evolutionary Programming Digest: To subscribe to the digest, send mail to <ep-list-REQUEST@magenta.me.fau.edu> and include the line "subscribe ep-list" in the body of the text.

comp.ai.genetic : Genetic Algorithm Usenet Group.

Fuzzy Logic

Books:

Cox, Earl, "The Handbook of Fuzzy Logic and Fuzzy System Modelling", Academic Press, 1994

Kosko, B., "Neural Networks and Fuzzy Systems", Prentice Hall, Englewood Cliffs, NJ, 1992.

Klir, George J. and Folger, Tina A., "Fuzzy Sets, Uncertainty, and Information", Prentice Hall, Englewood Cliffs, NJ, 1988, 355 pages. ISBN 0-13-345984-5

R. Yager and L. Zadeh (editors) "An Introduction to Fuzzy Logic: Applications in Intelligent Systems" Kluwer, 1992.

Didier Dubois, Henri Prade, and Ronald R. Yager, (editors) "Readings in Fuzzy Sets for Intelligent Systems", Morgan Kaufmann Publishers, 1993, 916 pages. ISBN 1-55860-257-7, $49.95.

Articles:

Cox, Earl, "Fuzzy Logic Fundamentals", IEEE Spectrum, October 1992

Zadeh, Lotfi, "Outline of a New Approach to the Analysis of Complex Systems", IEEE Trans. on Sys., Man and Cyb. 3, 1973.

Bezdek, James C, "Fuzzy Models --- What Are They, and Why?", IEEE Transactions on Fuzzy Systems, 1:1, pp. 1-6, 1993.

Journals

International Journal Of Approximate Reasoning (IJAR). For subscription information contact: Elsevier Science Publishing Company, Inc., 655 Avenue of the Americas, New York, NY 10010, USA.

International Journal Of Fuzzy Sets And Systems (IJFSS). ISSN: 0165-0114

IEEE Transactions On Fuzzy Systems, ISSN 1063-6706

International Journal of uncertainty, fuzziness and knowledge-based systems (IJUFKS). For subscription information contact: World Scientific Publishing Co Pte Ltd, Farer Road, PO Box 128, SINGAPORE 9128

Conferences

IEEE International Conference on Fuzzy Systems - FUZZ-IEEE. Proceedings available from IEEE Service Center, 445 Hoes Lane, P.O. Box 1331, Piscataway, NJ 08855-1331, USA.

Bulletin Boards, Email lists & WWW Home Pages

comp.ai.fuzzy : Usenet group.

NAFIPS Fuzzy Logic Mailing List: This is a mailing list for the discussion of fuzzy logic. Email: listserv@gsuvm1.gsu.edu or listserv@ gsuvm1.bitnet and the name of the mailing list itself is nafips-l@gsuvm1.gsu.edu or nafips-l@gsuvm1.bitnet

Use the "gsuvm1.gsu.edu" addresses if you're on the Internet, and the "gsuvm1.bitnet" addresses if you're on BITNET.

The fuzzy logic Frequently Asked Questions (FAQ) is available on the WWW at http://www.cs.cmu.edu:8001/Web/Groups/AI/html/faqs/ai/fuzzy/part1/faq.html

Free Software Packages

FuzzyCLIPS 6.02:

FuzzyCLIPS is a fuzzy logic version of the CLIPS rule-based expert system shell. The system uses two main inexact concepts, fuzziness and uncertainty. Versions are available for UNIX systems, Macintosh systems and PC systems. FuzzyCLIPS is available via WWW (World Wide Web). It can be accessed indirectly through the Knowledge Systems Lab Server using the URL

http://ai.iit.nrc.ca/home_page.html or more directly by using the URL

http://ai.iit.nrc.ca/fuzzy/fuzzy.html or by anonymous ftp from ai.iit.nrc.ca:/pub/fzclips/. For more information about FuzzyCLIPS send mail to fzclips@ai.iit.nrc.ca.

Commercial Software Packages

NeuralWare:

The Fuzzy Systems Professional (FuzzySP). A comprehensive fuzzy systems modelling environment. NeuralWare Inc. Pittsburgh, PA 15276-9910, USA. Phone (US) 412 787-8222, Fax: (US) 412 787-8220.

ByteCraft, Ltd.:

Fuzz-C: "A C preprocessor for fuzzy logic" according to the cover of its manual. Translates an extended C language to C source code. Byte Craft Limited, 421 King Street North, Waterloo, Ontario. Canada N2J 4E4, Phone (US) 519-888-6911, (US) Fax: 519-746-6751

FuziWare:

FuziCalc: An MS-Windows-based fuzzy development system based on a spreadsheet. FuziWare, Inc. 316 Nancy Kynn Lane, Suite 10 Knoxville, TN 37919, USA. Phone (US) 615-588-4144, Fax (US) 615-588-9487.

HyperLogic, Inc.:

CubiCalc: An MS-Windows-based fuzzy development environment. HyperLogic Corp, 1855 East Valley Parkway, Suite 210, P. O. Box 3751, Escondido, CA 92027, USA. Phone (US) 619-746-2765 Fax (US) 619-746-4089.

Inform:

fuzzyTECH 3.0: A graphical fuzzy development environment. INFORM GmbH, Geschaeftsbereich Fuzzy-Technologien, Pascalstraese 23, W-5100 Aachen, Germany. Phone (Ger) 02408 6091, Fax (Ger) 02408 6090.

Fril Systems Ltd:

FRIL (Fuzzy Relational Inference Language) is a logic-programming language which incorporates a consistent method for handling uncertainty. Fril has a list-based syntax, similar to the early micro-Prolog from LPA. Prolog is a special case of Fril, in which programs involve no uncertainty. Fril runs on UNIX, Macintosh, MS-DOS, and Windows 3.1 platforms.

Dr B.W. Pilsworth, Fril Systems Ltd, Bristol Business Centre, Maggs House, 78 Queens Rd, Bristol BS8 1QX, UK.

Togai InfraLogic, Inc.:

Togai InfraLogic supplies software development tools, board, chip and core-level fuzzy hardware, and engineering services. Togai InfraLogic, Inc, 5 Vanderbilt, Irvine, CA 92718, USA. Phone (US) 714-975-8522. Fax: (US) 714-975-8524, Email: info@til.com.

Expert Systems

Books

P. Jackson, Introduction to Expert Systems, Addison Wesley, 1986.

Samuel J. Biondo, "Fundamentals of Expert Systems Technology: Principles and Concepts", Ablex, Norwood, NJ, 1990.

Joseph Giarratano and Gary Riley, "Expert Systems Principles and Practice", PWS Publishing, 1989.

Firebaugh, Morris W., "Artificial Intelligence: A Knowledge-Based Approach", PWS-Kent, Massachusetts, 1989. ISBN 0-87835-325-9

Articles

Frederick Hayes-Roth, "The knowledge based expert system: A tutorial", IEEE Computer 17(9):11-28, 1984.

Journals

Expert Systems: The International Journal Of Knowledge Engineering: Publisher Learned Information Ltd., Woodside, Hinksey Hill, Oxford OX1 5AU, UK. Tel: +44 (0)865-730275, Fax: +44 (0)085-736354.

International Journal Of Expert Systems : JAI Press Inc., 55 Old Post Road -- No. 2, PO Box 1678, Greenwich, CT 06836-1678, USA.

Expert Systems With Applications : Pergamon Press Inc., 660 White Plains Road, Tarrytown, NY 10591-5153, USA.

Free Software Packages

CLIPS 6.0 :

CLIPS (C Language Integrated Production System) is an OPS-like forward chaining production system written in ANSI C by NASA. The CLIPS inference engine includes truth maintenance, dynamic rule addition, and customisable conflict resolution strategies. CLIPS runs on many platforms including IBM PC compatibles (including Windows 3.1 and MS-DOS 386 versions), Macintosh, VAX 11/780, Sun 3/260, and HP9000/500. For more information, contact: University of Georgia, 382 East Broad Street, Athens, GA 30602, USA. Email: service@cossack.cosmic.uga.edu, Phone (US) 706-542-3265, Fax (US) 706-542-4807.

DYNACLIPS :

DYNACLIPS (DYNAamic CLIPS Utilities) is a set of blackboard, dynamic knowledge exchange, and agent tools for CLIPS 5.1 and 6.0. It is implemented as a set of libraries that can be linked with CLIPS 5.1 or CLIPS 6.0. Agents use the blackboard to communicate with other intelligent agents in the framework. Each intelligent agent can send and receive facts, rules, and commands. Rules and facts are inserted and deleted dynamically while the agents are running. For more information: PO Box 10412, Daytona Beach, FL 32120-0412, USA. Email to cengelog@dab.ge.com.

WindExS :

WindExS (Windows Expert System) is a fully functional Windows-based forward chaining expert system. Its modular architecture allows the user to substitute new modules as required to enhance the capabilities of the system. WindExS sports Natural Language Rule Processor, Inference Engine, File Manager, User Interface, Message Manager and Knowledge Base modules. It supports forward chaining, and graphical knowledge base representation. Write etoupin@aol.com for documentation and operational system.

GEST

GEST (Generic Expert System Tool) Version 4.0 is a flexible expert system shell applicable to a variety of problem domains. Its knowledge representation schemes include facts, frames, rules, and procedures. Object files are available by anonymous ftp to ftp.gatech.edu in the (binary) file pub/ai/gest.tar.Z. Register your copy with John Gilmore, Georgia Tech Research Institute, ITL/CSITD 0800, Atlanta, GA 30332, USA. Internet: john.gilmore@gtri.gatech.edu.

Commercial Software Packages

Aion Development System (ADS):

Runs on numerous platforms, including DOS, OS/2, SunOS, Microsoft Windows, and VMS. It includes an object oriented knowledge representation, graphics and calls to/from other languages. For more information contact: Aion Corporation, 101 University Avenue, Palo Alto, CA 94301, USA. Phone (US) 415-328-9595, Fax (US) 415-321-7728.

ART*Enterprise and CBR Express:

The ART*Enterprise (Inference Corporation) development environment for enterprise-wide applications, incorporating rules, a full object system which includes features currently not present in C++, and a large collection of object classes for development across platforms. CBR Express family of products supports case-based retrieval of information. For further information contact Inference Corporation, 550 N. Continental Blvd., El Segundo, CA 90245, USA. Phone (US) 310-322-0200, Fax (US) 310-322-3242, Email marketing@ inference.com.

ECLIPSE :

Runs on personal computers (DOS, Windows) and UNIX systems. Features include data-driven pattern matching, forward and backward chaining, truth maintenance, support for multiple goals, relational and object-oriented representations. For more information: The Haley Enterprise, Inc., 413 Orchard Street, Sewickley, PA 15143, USA. Phone: (US) (412-741-6420), Fax (US) 412-741-6457.

Exper-OPS5-Plus :

Runs on the Apple Macintosh. For more information: ExperTelligence, 5638 Hollister, Suite 302, Goleta, CA 93117. Phone: (US) 805-967-1797.

EXSYS :

EXSYS runs under MS-DOS, MS-Windows, Apple Macintosh, Sun Open Look, UNIX and Vax. For more information: Exsys, Inc., 1720 Louisiana Boulevard, NE, Suite 312, Albuquerque, NM 87110, USA. Phone (US) 1 800-676-8356), Fax (US) 505-256-8359.

GURU :

GURU is a real-time expert system that runs on personal computers. For more information: Micro Data Base Systems, Inc. (MDBS), PO Box 6089, Lafayette, IN 47903-6089, USA. Phone (US) 317-463-2581, Fax 317-448-6428.

ILOG RULES :

ILOG RULES is a high performance embeddable rule-based inference engine. It is a forward chaining tool, written in C++ (hence it is object-oriented and supports inheritance mechanisms) and is also provided as a C++ library.. For more information: ILOG, Inc., 2073 Landings Drive, Mountain View, CA 94043, USA. Tel (US) 415-390-9000, Fax (US) 415-390-0946, E-mail info@ilog.com.

KEE, ProKappa, and Kappa :

Run on personal computers, workstations, and Lisp machines. For more information, write to IntelliCorp, Inc., 1975 El Camino Real West, Mountain View, CA 94040-2216, USA. Call 415-956-5660/5500/5777 or fax 415-965-5647. In Europe call +44-344-305305. See also CACM 31(4):382-401, April, 1988.

KnowledgeWorks :

Runs on workstations (Sun3, Sparc, DEC (MIPS), HP 400, 700, IBM RS6000, Intergraph, MIPS). It includes a graphical programming environment, user-defined conflict resolution, and a SQL interface for relational databases. For more information: Harlequin Inc., One Cambridge Center, Cambridge, MA 02142, USA. Phone (US) 617-252-0052, Fax (US) 617-252-6505. In Europe: Harlequin Ltd., Barrington Hall, Barrington, Cambridge CB2 5RG, UK. Phone (UK) 0223-872522, Fax (UK) 0223-872519. E-mail: knowledgeworks-request@ harlqn.co.uk or works@harlequin.com.

NEXPERT OBJECT :

Runs on many platforms including PCs, workstations, and mainframes. Nexpert Object includes a graphical user interface, and knowledge acquisition tools. For more information: Neuron Data, 156 University Avenue, Palo Alto, CA 94301,

USA. Phone (US) 415-321-4488, Fax (US) 415-321-3728.

Bulletin Boards, Email lists & WWW Home Pages

comp.ai.shells : Expert System Shells news group.

comp.ai : Artificial Intelligence news group.

A Frequently Asked Questions (FAQ) is available by anonymous ftp from ftp.cs.cmu.edu:/user/ai/pubs/faqs/expert/expert_1.faq [128.2.206.173] using username "anonymous" and password "name@host" (substitute your email address).

The FAQ is also accessible by WWW. The URL for the Expert Systems FAQ is http://www.cs.cmu.edu:8001/Web/Groups/AI/html/faqs/ai/expert/top.html

Rule Induction

Books

Quinlan, J.R., C4.5:A Machine Learning Paradigm, Morgan Kaufmann, 1993.

S.M. Weiss and C.A. Kulikowski, Computer Systems that Learn: Classification and Prediction Methods From Statistics, Neural Nets, Machine Learning and Expert Systems, 1991, Morgan Kaufmann Publishers Inc., San Mateo, California.

R. Forsyth, "Machine Learning: Applications in expert systems and information retrieval", Ellis Horwood Ltd. 1986.

Articles

J.R. Quinlan, "Induction of Decision Trees", Machine Learning, 1,81-106, (1986).

Nitin Indurkhya and Sholom M. Weiss, Iterative Rule Induction Methods, , Applied Intelligence: The International Journal of Artificial Intelligence, Neural Networks, and Complex Problem-Solving Technologies, 1,1, 1991

R.Michalski, R.L. Chilansky, "Learning by being told and learning from examples", J. Policy Analysis & Information Systems", 4, 1980.

M. Shaw and J. Gentry, "Inductive Learning for Risk Classification", IEEE Expert February 1990.

Journals

Machine Learning: Contact Academic Publishers, PO Box 322, 3300 AH Dordrecht, The Netherlands, or Kluwer Academic Publishers, PO Box 358, Accord Station, Hingham, MA 02018-0358, USA.

Conferences

ML: International Conference on Machine Learning. Submissions to ML conferences from researchers in machine learning or related fields, such as psychology, statistics, or neuroscience.

ECML: European Conference On Machine Learning. ECML provides the major European forum for presenting the latest advances in the area of Machine Learning.

Software Packages

C4.5 Rule Induction software is available with the book *C4.5: Programs for Machine Learning,* Morgan Kaufmann Publishers, 2929 Campus Drive, Suite 260, San Mateo, CA 94403, USA. ISBN 1-55860-240-2.

Bulletin Boards, Email lists & WWW Home Pages

comp.ai : Usenet group has discussions on rule induction and machine learning.

University of California at Irvine - Machine Learning archive : http://www.ics.uci.edu/ArtificialIntelligence/ML/Machine-Learning.html

Machine Learning Archive at the German National Research Center for Computer Science (GMD): ftp://ftp.gmd.de/MachineLearning/ML-groups.html

Index